QUANTUM COHERENCE

Yakir Aharonov (left) and David Bohm at the Banquet
15 December 1989 (Photo taken by Paul Frampton)

Proceedings of the International Conference on
Fundamental Aspects of Quantum Theory
to Celebrate 30 years of the Aharonov-Bohm Effect

QUANTUM COHERENCE

University of South Carolina, Columbia, USA
14 – 16 December 1989

Editor

JEEVA S. ANANDAN
University of South Carolina
Columbia, USA

World Scientific
Singapore • New Jersey • London • Hong Kong

Published by

World Scientific Publishing Co. Pte. Ltd.
5 Toh Tuck Link, Singapore 596224
USA office: 27 Warren Street, Suite 401-402, Hackensack, NJ 07601
UK office: 57 Shelton Street, Covent Garden, London WC2H 9HE

British Library Cataloguing-in-Publication Data
A catalogue record for this book is available from the British Library.

QUANTUM COHERENCE
Proceedings of the International Conference on Fundamental Aspects of Quantum Theory — to Celebrate 30 Years of the Aharonov-Bohm-Effect

ISBN-13 978-981-02-0256-9
ISBN-10 981-02-0256-3
ISBN-13 978-981-02-0257-6 (pbk)
ISBN-10 981-02-0257-1 (pbk)

CONTENTS

5. QUANTUM COSMOLOGY AND QUANTUM GRAVITY

6. QUANTUM STATISTICS

7. QUANTUM MEASUREMENT

8. NON LOCAL ASPECTS AND BELL'S INEQUALITIES

Foreword

This is the proceedings of the Conference on Fundamental Aspects of Quantum Theory that was held at the University of South Carolina in 1989 to celebrate 30 years of the Aharonov-Bohm effect. It is also a tribute to two truly outstanding physicists, Yakir Aharonov and David Bohm, who have made great contributions to the foundations of quantum theory for more than three decades. They have studied practically every aspect of quantum theory; but their best known work is perhaps still the effect they discovered at the University of Bristol in 1959, contained in their paper entitled "Significance of Electromagnetic Potentials in the Quantum Theory", which is commemorated by this conference.

In its original form, the Aharonov-Bohm effect is the phase shift in electron interference due to the electromagnetic field, determined by the phase factor

$$\exp\left(\frac{ie}{\hbar c}\oint_{\gamma} A_{\mu}dx^{\mu}\right)$$

where the integral of the electromagnetic potential A_{μ} is along a closed curve γ passing through the beams along which the field strength $F_{\mu\nu}$ vanishes. This prediction by Aharonov and Bohm fell like a bombshell in the physics community, destroying the cherished notion that the influence of a classical electromagnetic field on a charged particle can only occur through the local action of the field strength. Perhaps as much as ninety five percent of the physics community at that time did not believe it. But it was first experimentally confirmed in 1960 by R. G. Chambers, and most recently by a very precise experiment by A. Tonomura which is described in his article in this volume.

There are three aspects of the Aharonov-Bohm effect which have made it so important. First, it is non local in the sense that it exists even when the interfering beams pass through a field free region and is associated with the entire closed curve γ. Second, it is topological in the sense that the phase shift is unaffected when γ is deformed within the field free region. Finally, it is geometrical in the sense that the above phase factor represents parallel transport (holonomy transformation) around γ with respect to the electromagnetic connection or gauge field. Now the gauge field is a geometric concept which originated from a study of the foundations of quantum mechanics with the work of Hermann Weyl in 1929, for an abelian gauge group, and was generalized to a non abelian group by Yang and Mills in 1954. It has proved to be very useful in describing three of the four fundamental interactions in nature while the remaining interaction, gravitation, is described classically also by a geometric theory. This geometric aspect is present more generally than the other two aspects even when the electromagnetic field strength $F_{\mu\nu}$ does not vanish along the beams. This suggests that the geometric aspect of the Aharonov-Bohm effect may be much more fundamental and important than the other two aspects. The connection between the Aharonov-Bohm effect and gauge fields is investigated further in the article by Wilczek who says that even "thirty years later we are still far from exhausting the full meaning of this deep concept."

In this connection, I might mention a conversation I had with Yakir Aharonov two years ago, one of several conversations which led me to the idea of holding this conference. He remarked how depressing it was when he was a graduate student because of so many physicists working on S-matrix theory. I replied that during this time he, along with David Bohm, discovered the effect which provided direct evidence of the fact that the electromagnetic field is a gauge field. So, the geometric aspect of this effect should have convinced people that they should focus on gauge theories instead of S-matrix theory. But if we disregard the geometry and focus instead on the non locality, then S-matrix theory (which can incorporate the non locality) would appear as important or even more important than gauge theories which, like general relativity, are deeply rooted in the concept of locality. Physicists seem to have, however, made up for lost time with a flurry of activity on gauge theories beginning with the unification, in 1967, of the electromagnetic and weak interactions (S. Glashow, S. Weinberg and A. Salam).

On this historic occasion, mention must be made also of W. Ehrenberg and R. E. Siday who in their pioneering work on electron optics, published in 1949, obtained the phase shift due to a magnetic field which is proportional to a cyclic integral of the vector potential **A**. But they did not obtain the phase shift due to the scalar potential A_0, unlike Aharonov and Bohm. While they recognized the non locality of the effect of **A**, their effect was not geometric because it was not expressed in terms of the above phase factor which is a space-time scalar, i. e. independently of the Lorentz frame in which it is viewed. In general, geometric effects should be expressible entirely in terms of space-time scalars or tensors.

When the Aharonov-Bohm effect is viewed geometrically, not only is it clearly superior to earlier work on this subject, but it also becomes the prototype of geometric phases discovered since then. The most famous of these is the geometric phase factor discovered in adiabatic evolutions by Michael Berry in 1983 (see the articles by Berry, Pines and Mueller, Chiao, and Christian and Shimony). This can be understood as a holonomy transformation in the projective Hilbert space, in which case the restriction of adiabaticity is not necessary, as pointed out in 1987 by Aharonov and Anandan. But the curvature in the projective Hilbert space, which plays the role of the electromagnetic field strength $F_{\mu\nu}$ in the Aharonov-Bohm effect, is non zero so that this effect is neither non local nor topological. However it is geometric like the Aharonov-Bohm effect.

Mention may also be made, in this context, of the phase shift due to the electric field in neutron interference, obtained by Anandan in 1982, which Aharonov and Casher pointed out as being "dual" to the Aharonov-Bohm effect in 1984 (see the articles by Opat et al, and Reznick). This effect has also been generalized to include the magnetic field, and expressed by a path ordered operator similar to the above phase factor (Anandan, 1989). It is then associated with a closed curve in space-time, independently of the inertial frame in which it is viewed. Since the "Yang-Mills field strength", which plays the role of $F_{\mu\nu}$ in the Aharonov-Bohm effect, does not vanish along the interfering beams, and there is a torque on the neutron which causes the phase shift, this effect cannot be regarded as non local or topological, except in the two dimensional plane perpendicular to the electrode enclosed by the interfering beams. But it is a geometric phase similar to the Aharonov-Bohm phase or Berry's phase, and is richer because of its non abelian character.

Ideas from the foundations of quantum mechanics, such as the Aharonov-Bohm effect, are now being applied to many fields such as condensed matter physics, quantum statistics, optics, and quantum cosmology (see, for example, the articles by Webb, Horne et al, Zee, and Hartle). Moreover many experiments have tested foundational concepts which were previously merely gedanken experiments (see, for example, the articles by Tesche, and Matteucci and Pozzi). There is also the still unsolved measurement problem, on which we had some talks varying from a deeply philosophical one by Albert to a practical discussion by Stodolsky. Moreover, difficulties in constructing a quantum theory of gravity have made many theorists reexamine the foundations of quantum theory (see the articles by 't Hooft, and Ne'man and Chang-Yeaong Lee). All these factors have made fundamental aspects of quantum theory even more important in recent years. We therefore hope that this conference and the proceedings will be of benefit to the physics community.

I would like to thank very much the members of the local organizing committee, John Safko, Jim Knight, Chi-Kwan Au, Frank Avignone, Lev Vaidman, Rick Creswick and Horacio Farach, for their help in organizing this conference. On behalf of the organizing committee, I gratefully acknowledge generous support for this conference from the University of South Carolina, the National Science Foundation, Hitachi Ltd., the Department of Energy, IBM Corporation, Doty Scientific and Southern Bell.

University of South Carolina, Columbia
September 1990

Jeeva S. Anandan

QUANTUM COHERENCE

Some Global Problems in Gauge Theories (Variations on a Theme of Aharonov and Bohm)

FRANK WILCZEK[†]
School of Natural Sciences,
Institute for Advanced Study,
Olden Lane,
Princeton, N.J. 08540

ABSTRACT

Several situations are discussed, in which the sort of global considerations made famous by Aharonov and Bohm in their discussion of the interaction of charged particles with magnetic flux tubes have important physical implications. It is argued that discrete gauge symmetries in the continuum make sense, and manifest themselves most clearly in Aharonov-Bohm type scattering of charged particles off string singularities. The existence of such discrete symmetries has important implications for the quantum mechanics of topologically non-trivial space-times in general and black holes in particular. It is argued that in the non-abelian case essentially new features arise, most notably that the symmetry group of the homogeneous ground state generally ceases to be globally defined in the presence of a string. When continuous rather than discrete symmetries are involved, a variety of fascinating and as yet poorly understood dynamical effects occur. Perhaps the most striking is a new form of string superconductivity, that exists for purely topological reasons, and is *not* well modeled by regarding the string as a superconducting wire.

In their famous paper, to which we pay tribute today, Aharonov and Bohm emphasized that one may have effects associated with the *global* structure of gauge fields, even in cases where these fields are trivial locally. I would like to describe for you some recent work, which shows that 30 years later we are still far from exhausting the full meaning of this deep concept.

---

† Research supported in part by DOE grant DE-FG02-90ER40542

1. Discrete Gauge Theories in the Continuum

1. Upon first hearing the concept of a discrete gauge theory in the continuum might sound rather silly. Since any path can be continuously deformed to a trivial path, the corresponding parallel transports can be continuously deformed to the identity. But for a discrete group this means we must have the identity all along: thus all parallel transport is trivial.

Also, if the discrete gauge group arises as the remnant of an initially continuous group, after all continuous symmetries have been wiped out by the Higgs mechanism, one is accustomed to think that charge is completely screened.

2. Upon further reflection, however, one realizes that each of these arguments has serious limitations.

First, there are situations in which one cannot deform all paths to the identity. This occurs, by definition, on spaces that are not simply connected.

Second, there are situations in which the parallel transport is not a continuous function of the path, or is ill-defined for some paths. An important class of examples, which will be extensively considered below, concerns flux tubes in the Higgs phase of gauge theories.

What about screening? One should appreciate that the screening occurring in the Higgs mechanism is not a mystical process, but essentially a special case of the homely phenomenon of dielectric polarization as occurs in ordinary dielectrics or plasmas. It is true that we are concerned, in the Higgs mechanism, with a particularly effective screening, produced by a condensate of massless charged particles. Such particles are capable of producing a dipole distribution, to exactly cancel any imposed electric field. The particles in the condensate are singlets under the remnant discrete symmetry, however, and it is unreasonable to expect that any distribution of them can perfectly mimic a non-singlet source.

3. As a definite example, let us consider the case of a $U(1)$ gauge theory spontaneously broken to a discrete Z_p subgroup. That is, we imagine that some charge p field ϕ condenses, but that there are additional unit charge particles, produced by a field ψ, in the theory. The case $p = 2$ is realized in ordinary BCS superconductors, where the doubly charged Cooper pair field condenses, and there are additional singly charged fields to describe the "normal electron" excitations.

Such a theory supports vortex solutions, where the ϕ field behaves asymptotically as a function of the angle θ as

$$\phi(r,\theta) \to ve^{i\theta}, \; r \to \infty \tag{1}$$

where v is the value of ϕ in the homogeneous groundstate. Along with this asymptotics for ϕ we must have for the gauge potential

$$A_\theta(r,\theta) \to \frac{1}{pe} \tag{2}$$

so that the covariant derivative $D_\theta\phi = (\partial_\theta\phi - peA_\theta)$, which appears (squared) in the energy density, vanishes at infinity.

In this set-up the field strength also vanishes asymptotically. Indeed we note that since under a gauge transformation the fields transform as

$$\begin{aligned} \phi'(x) &= \exp(iQ\Lambda(x))\phi(x) = \exp(ipe\Lambda(x))\phi(x) \\ A'_\mu(x) &= A_\mu(x) + \partial_\mu\Lambda(x) \end{aligned} \tag{3}$$

we can formally, by making the choice $\Lambda = -\theta/pe$, remove the space dependence of ϕ in (1) and make A_θ in (2) vanish altogether – we have transformed back to the homogeneous groundstate. Of course the gauge transformation function Λ is not quite kosher, since the angle θ is not a legitimate single-valued function. The correct statement is that the vortex asymptotics is trivial and can be gauged away locally, but not globally. Since we can pick a well defined branch of θ in any patch that does not surround the origin, all local gauge invariant quantities must reduce to their groundstate values – this explains, if you like, why $D\phi$ and F vanish. However the line integral of A around a closed loop surrounding the origin, which according to Stokes measures the flux inside, cannot be changed by any legitimate gauge transformation, and it is definitely *not* zero for the vortex. Indeed we find the basic flux unit is $\Phi_p = \frac{2\pi}{pe}$.

Another perspective on the global non-triviality of the vortex, is that our putative gauge transformation $\Lambda = -\theta/pe$ transforms a unit charge field ψ into something that is not single-valued: following (3) we find that $\psi'(\theta + 2\pi) = exp(-\frac{2\pi}{p})\psi'(\theta)$.

What has all this got to do with discrete gauge theories in the continuum? Well, the condensate $\phi = v$ is not invariant under a general gauge transformation, but it is invariant under the discrete subgroup generated by $\Lambda = \frac{1}{pe}$. This discrete subgroup acts trivially both on ϕ and, it would seem, on the gauge field (Λ, since it takes only discrete values, cannot change continuously at all). However, Λ and its various powers are definitely not trivial acting on ψ, which gets multiplied by powers of pth roots of unity. Thus there is a discrete but non-trivial gauge subgroup left. Moreover, ***the gauge transformation associated with winding around a vortex is precisely an element of the residual discrete gauge symmetry group*** – this is just a restatement, in our new interpretation, of the result of the previous paragraph.

Alternatively we could say that the Wilson loop for parallel transport around the vortex defines an element of the residual gauge group:

$$\exp(\oint ieA_\mu dx^\mu) \in Z_p. \tag{4}$$

We anticipated that the discrete gauge group might come into its own in the presence of singularities. Now we see how this is quite simply realized. The required "singularity" is here supplied by the vortex core. In that core, where the condensate vanishes, the discrete gauge symmetry blows up into a full scale continuous $U(1)$, and the vector potential is unfrozen. The only trace of the vortex visible outside the vortex core is the total flux – which is none other than an element of the residual discrete gauge symmetry group.

4. Now that we have convinced ourselves that the idea of discrete gauge symmetry in the continuum makes sense, the next step is to elucidate its physical implications. The first of these, has to do with the angular momentum of particles in the presence of the vortex.

The asymptotics (1) of the scalar order parameter seems not to be rotationally invariant: a scalar field should be unchanged by a rotation, whereas in (1) it acquires a phase. However the phase of ϕ is gauge dependent and we cannot infer from (1) that any physical property of the vortex violates rotation symmetry. In fact it is easy to see that if we supplement the naive rotation generator J_z with an appropriate gauge transformation:

$$K_z = J_z - \frac{1}{pe}Q \tag{5}$$

then K_z leaves both the action, and the asymptotic scalar field configuration (1) invariant.

Thus – assuming the core is invariant – a true rotation symmetry of the vortex is generated by K_z. (If the core is not invariant, since the asymptotics is invariant the solution will still have a finite moment of inertia, and upon proper quantization we will get a spectrum of rotational excitations of the vortex similar to the band spectrum of an asymmetric molecule.) By way of comparison, the vortex associated with a global gauge symmetry truly breaks rotational symmetry, and has infinite moment of inertia.

An interesting consequence of the modification of J into K is that the quantization of angular momentum, for quanta orbiting the vortex, is modified. The angular momentum of quanta with the fundamental charge e, for example, is quantized in units of $-\frac{1}{p}+$ integer. This modification of the quantization condition, and the associated existence of states with fractional angular momentum, is one basic physical consequence of the discrete gauge symmetry.

5. Since the spin and statistics of particles are deeply related, we might be led to ask whether there is peculiar statistics of the excitations with peculiar angular momentum. There is indeed. The emergence of fractional statistics for the sort of vortex-particle composite that carries fractional angular momentum may be seen as a consequence of the same kind of Aharonov-Bohm phase that is responsible for the fractional angular momentum. Indeed as one such composite particle is wound around another, it acquires phase as the charge is transported through the vector potential, according to the action $\int A \cdot j dt = q \int A \cdot dr$.

The physics of fractional statistics particles, or anyons, has grown from a curiosity into quite a substantial subject with solid applications to the fractional quantized Hall effect and many more conjectural applications. Since I have extensively reviewed this rich and rapidly developing subject elsewhere recently, I will not further expand on it here.

6. A direct physical consequence of the modification of the angular momentum quantization condition is the existence of a deeply geometric scattering process, which is in fact basically the one studied by Aharonov and Bohm in their original paper. I will merely quote the famous result:

$$\frac{d\sigma}{d\theta} = \frac{1}{2\pi k} \sin^2(\frac{q\Phi}{2}) \frac{1}{\sin^2(\frac{\theta}{2})}, \tag{6}$$

where $q\Phi$ is the product of charge times flux, and make a few comments on the different factors. The cross-section is directly proportional to the DeBroglie wavelength of the incoming particle. This is very reasonable: all those wave packets, and only those, that actually pass through the core of the vortex have their internal phase relationships reshuffled. Secondly, there is a factor which depends on the *fractional part* of the charge-flux product. This is also the fractional part of the angular momentum, or the fractional part of the statistics, according to our other interpretations. Third, there is an angular factor, that again has a purely geometric origin. Note that this factor is singular in the forward direction (it goes as the square root of the Rutherford factor in Coulomb scattering). This singularity reflects the long-range character of the interaction responsible for the scattering, and insures that *there is a characteristic part of the scattering cross section, separable in principle from the contribution of any short-range interaction, that directly reflects the discrete gauge charge, fractional angular momentum, or fractional statistics.* This criterion can be used as an *operational definition* of these properties for the fully dressed physical particles, or quasiparticles in a condensed system, at the level of the physical S-matrix. It is similar to the strict definition of electric charge as the appropriate coefficient in low-energy Thomson scattering.

The modification of the angular momentum quantization condition in the presence of a vortex, of course modifies and may in favorable cases remove altogether the centrifugal barrier. This is very important for cosmic strings, where the core may contain condensates of particles which mediate baryon number violation, for example. We might expect generally that the cross-section for reaching the core and tapping the potential symmetry-breaking fields lurking there must depend sensitively, at low energy, on the form of the centrifugal barrier. And detailed study does indeed reveal striking dependence of this sort; cross-sections for baryon-number violating processes may come close to saturating unitarity limits (similar to the Callan-Rubakov effect for magnetic monopoles), and vary by many orders of magnitude depending on the precise value of the charge-flux product.

7. The existence of observable discrete gauge quantum numbers has significant implications for fundamental physics.

One implication is a modification of, or at least a significant addendum to, the no hair theorems of black hole physics. The quantum numbers associated with discrete gauge symmetries are well-defined, conserved, and measurable at arbitrarily large distances. Therefore a necessary part of specifying the quantum state of a black hole, is to specify its discrete gauge quantum numbers. On the other hand, these quantum numbers leave no *classical* residue at large distances. The usual no hair theorems, stating that the only properties of a black hole are its mass, charge, and angular momentum, are based on an analysis of the classical field equations. (Even at this level there is an additional possibility, axion charge, as was discussed in Giddings' talk here.) It appears, when we come to consider the quantum mechanics of black holes, that there are additional states, distinguishable by their quantum-mechanical properties: specifically, they differ in their Aharonov-Bohm scattering from cosmic strings.

Another implication concerns discrete symmetries. There are several "phenomenological" reasons why one might like to have discrete symmetries in fundamental models of particle physics. Such symmetries seem to be necessary to forbid catastrophically large rates for proton decay in models of low-energy supersymmetry. We might also like to use them in understanding the regularities of the quark and lepton mass matrices. On the other hand, it has been argued that all global symmetries, whether continuous or discrete, are strongly violated by wormhole effects. If the discrete symmetries in question are gauge symmetries, however, there is no cause for alarm. Discrete gauge symmetries cannot be violated by wormholes or by anything else.

8. The results of the previous paragraphs, and in particular the very strong

statement in the previous sentence, are best understood as consequences of a simple yet profound distinction between gauge symmetries and global symmetries. Whereas global symmetries express that the laws of physics look the same from different perspectives, or when certain variables are substituted for others, gauge symmetries express a *redundancy* in the variables: physical results can only depend on gauge invariant quantities, even though the laws may be most conveniently expressed using gauge variant quantities in intermediate expressions. Thus gauge symmetries are tautologies; they can no more be violated than can $1 + 1 = 2$. Even such exotic processes as particles disappearing past event horizons, or baby universes pinching off from and merging with our own, cannot alter tautologies.

The statement that gauge symmetries cannot be broken may sound peculiar to someone who has studied the standard gauge theory of weak interactions, or even the BCS theory of superconductivity, since the usual exposition of these theories makes heavy use of the concept of spontaneous breakdown of gauge symmetries. However a more accurate statement is that in the "broken" phases the gauge symmetries, rather than being violated, merely become useless. Whereas in the unbroken phase there are superselection sectors characterized by different values of the charge, in the unbroken phase only the $Q = 0$ sector exists. Actually, as we have seen, one may have a residual discrete gauge symmetry; in this case, what has happened is that there are a finite number of different superselection sectors, corresponding to the different possible values of the discrete gauge charge. Indeed the number and structure of the superselection sectors can be used to distinguish between possible phases of (totally or partially) "broken" gauge theories.

(A penetrating discussion of these issues, including nonabelian aspects that are at this moment not fully understood, is contained in the a recent paper by Preskill and Krauss, cited below.)

2. Nonabelian Aspects of Discrete Gauge Symmetries

1. Essentially new features appear when we generalize the considerations above to the case when the unbroken symmetry group H is nonabelian. Perhaps the most striking one is that in the presence of a flux tube the symmetry group H of the homogeneous vacuum, which is also the symmetry group observed locally by observers far from the tube, generally *ceases to be well defined* globally.

To be specific, let the flux through the tube, which is parametrized by an element of H that specifies the result of parallel transport around the tube, be h. Then, as we shall see in a moment, only those elements of H which commute with h are legitimate, globally defined symmetries. The other elements of H are not so much broken, as obliterated: they are not even defined on the physical Hilbert space. This despite the fact that off the tube, and in any region that does not actually encircle the tube, the physics is rigorously the same as – is a gauge transform of – the physics of the homogeneous vacuum groundstate.

2. The physical reason for this phenomenon, is closely tied up with the Aharonov-Bohm effect. Consider charged particles transforming according to some representation R of H. As such particles wind around the flux tube, their internal quantum numbers are transformed according to the matrix $R(h)$. Let us imagine diagonalizing this matrix, by taking appropriate combinations of the charge states. In this basis, the action of h is simply to generate a phase proportional to the eigenvalue and to the number of windings of the superposition of particles. The physics is identical to that in the abelian case, where it leads to Aharonov-Bohm scattering as in (6) , with a coefficient governed by the eigenvalue.

Now according to Schur's lemma particles lying in an irreducible representation of the centralizer $C(h)$ of h will all be characterized by the same eigenvalue. This is as it must be, if $C(h)$ is to be a legitimate symmetry group. The physical properties of particles in an irreducible representation of a legitimate symmetry group, including specifically their scattering cross-sections, must be the same.

On the other hand particles which do not lie in the same irreducible representation of $C(h)$ have no particular reason to have the same eigenvalue. Indeed one can show that given any subgroup U of H properly larger than $C(h)$, then for some irreducible representation S of U, $S(h)$ will not be proportional to the identity. This will imply that the action of U definitely alters the properties of particles, specifically their cross-sections for scattering off the flux tube, and therefore that U is not a legitimate symmetry group.

3. The basis in which the scattering off a single flux tube diagonalizes, is generally different from the basis in which the scattering off a flux tube characterized by a different flux diagonalizes. Thus considering arrays of several flux tubes, widely separated, one can construct analyzers that separate different components of a beam of non-abelian charged particles. Is it possible to determine the representation of H to which a given particle belongs? I suspect that this is possible, and Alford, March-Russell and I showed it in detail for a special example; but a general proof is lacking. It should also be possible to distinguish further quantum numbers, corresponding to any complete set of commuting observables – i.e. to any maximal abelian subalgebra of the group algebra of H. These questions are of intrinsic mathematical interest; they are, of course, also relevant to the question of possible non-abelian black hole hair.

4. Another essentially non-abelian effect, is that in the nonabelian case not only may a charge scatter off a flux tube, but also one pure flux tube can scatter

off another.

There is an essential subtlety in specifying the flux associated with a non-abelian flux tube, i.e. the parallel transport around it. That is, that this flux is not gauge invariant. If we perform a gauge transformation by $h\epsilon H$, then the parallel transport g around the tube is modified to hgh^{-1}. This would seem to indicate that the flux is defined only to within a conjugacy class. However, when we have two or more flux tubes the ambiguity is still only one overall conjugation. Thus it seems appropriate, and it is certainly convenient, to fix a gauge and to speak as if each flux tube is characterized by a definite group element. Of course all physical results must then be invariant against an overall conjugation.

Now we come to the heart of the matter: why do such flux tubes scatter at all? The reason is most easily grasped visually. We refer to Figure 1, where the passage of one flux-tube–charge composite (a,ξ) over another (b,η) is depicted. From this Figure, it is apparent that the effect of this passage is to make the change:

$$\begin{aligned}(a',\xi') &= (b,\eta)\\ (b',\eta') &= (bab^{-1},R(b)\xi)\end{aligned}\tag{7}$$

where the charges transform according to the representation R. Iterating this once, we find the result for a complete winding of one composite particle around another:

$$\begin{aligned}(a^{(1)},\xi^{(1)}) &= (bab^{-1},R(b)\xi)\\ &= ((ba)a(ba)^{-1},R(ba)R(a)^{-1}\xi)\\ (b^{(1)},\eta^{(1)}) &= ((bab^{-1})^{-1}b(bab^{-1}),R(bab^{-1})\xi)\\ &= ((ba)b(ba)^{-1},R(ba)R(b)^{-1}\eta).\end{aligned}\tag{8}$$

In the second step of each of these equations we have written them in such a way that the result of further iterations can be read off easily. Noting that $b^{(1)}a^{(1)}=ba$, we readily find that after p windings

$$\begin{aligned}(a^{(p)},\xi^{(p)}) &= ((ba)^pa(ba)^{-p},R(ba)^pR(a)^{-p}\xi)\\ (b^{(p)},\eta^{(p)}) &= ((ba)^pb(ba)^{-p},R(ba)^pR(b)^{-p}\eta).\end{aligned}\tag{9}$$

For pure flux tubes, a result essentially equivalent to (8) was given by Carlip in the context of 2+1 dimensional gravity. He phrased it in terms of a geometric construction known as the Dehn twist. It may be appropriate to add a word concerning the relationship between these approaches.

To investigate the dynamics of our flux tubes, we must consider how to weight the amplitudes contributed by different possible space-time trajectories, in a Feynman path integral. Since the gauge field outside the tubes is locally trivial, the very most naive reaction is to ignore it; but of course Aharonov and Bohm taught us long ago that this is wrong. The next level of sophistication is to realize that the global residue of flux outside the tubes may be localized, by appropriate gauge transformations, onto small lines emanating from the tubes (i.e. it is only within these regions that the vector potential and parallel transport are non-trivial). Now as we consider a trajectory where flux tubes wind around one another we can, by appropriate gauge transformations, change the direction of the lines emanating from them in such away that they never cross one another. In this way, it seems that once again we have defined away the interaction. However the result of shunting the singular lines around this way, is that they get tangled up, as shown in Figure 2. A test charge circling just one or the other of our composite particles in the final state will, in this description, have to cross several singular lines. In general, therefore, it will be parallel transported through a different transformation than for the original unwound configuration, which involved just crossing a single line. Similarly, a test flux winding around one of our composite particles will be affected by its passage through the surrounding spaghetti. Thus the effective fluxes and charges in the final state are definitely different from their original values in this (Carlip's) description as well. In fact it is not difficult to convince oneself that both descriptions of the process lead to the same result, although in my opinion the first is more transparent.

From this point the analysis proceeds along channels similar to the analysis of nonabelian charge-flux tube scattering: one diagonalizes the conjugation operation, and in this basis the calculation of the amplitude reduces to the original calculation of Aharonov and Bohm. One may then, if desired, reassemble the group theoretic factors to produce the amplitudes for generally inelastic scattering (i.e. with change of quantum numbers) in the basis where the flux tubes have definite H-flux.

In my opinion nonabelian charge-flux-tube composites are the proper generalization of anyons to the nonabelian case. They arise as quasiparticles in a simple generalization of the flux phase construction, as follows. The essence of the flux phase construction is the postulation of a correlation

$$\Gamma_{ij} = \langle c_i^\dagger c_j \rangle \tag{10}$$

for hopping between sites i and j, such that the total amplitude for hopping around a closed loop contains a complex phase depending on the area and the orientation (and no other properties) of the loop. With such correlations, the system has manufactured an effective magnetic field. If there are several types of fermions –

e.g. with different spin, or coming from different components of a disconnected Fermi surface – then, in this construction, the same phase is assigned to each. Generically quasiparticles around such an ordered state will be associated with defects in the ordering – vortices – and will carry fractional statistics. Now if we generalize (10) by allowing Γ to depend on indices labeling fermion types, then plausibly the quasiparticles acquire non-abelian statistics. Though we certainly cannot claim at present that ordering of this type occurs in the ground state of any realistic 2 dimensional Hamiltonian, such a possibility does not seem absurd and deserves further investigation.

3. Alice Electrodynamics

1. Now I'd like to mention some of the additional effects that occur when a nonabelian residual gauge symmetry group H contains a continuous part. In the discrete case the global non-existence of a locally valid symmetry, although perhaps surprising, does not offend our physical sensibilities – after all, the physical consequences of discrete gauge symmetries are pretty esoteric in the first place. It is quite a different matter when we are informed, say, that because of a flux tube somewhere in the Andromeda galaxy, electromagnetic gauge symmetry does not exist. *That* takes some explaining.

2. The phenomena that may arise for the general case of continuous, non-abelian H are rich and only very partially explored. To be concrete I'd like to discuss a comparatively simple case, that is however sufficient to display several striking phenomena, which I believe are rather generic. My example is the Alice string.

This string arises in an $SO(3)$ gauge theory which is broken down to $O(2)$ in the homogeneous ground state by the condensation of a Higgs field ϕ in the 5-dimensional symmetric traceless tensor, or isospin 2, representation. Suppose that $\phi_{11} = \phi_{22} = -\frac{1}{2}\phi_{33}$ are the non-vanishing components of the vacuum expectation value of ϕ in the homogeneous ground state. Then the residual $O(2)$ is generated by rotations around the 3-axis together with a parity transformation, say reflection in the 1-axis. Note that the residual symmetry is nonabelian: conjugation by the parity operation reverses the sense of a rotation.

An experimentalist in such a world might identify the residual gauge rotation as (his or her version of) electromagnetism.

Now consider a string whose flux is the parity reflection in $H = O(2)$. A charged particle, after winding around such a string, comes back with the sign of its electric charge reversed – a sort of looking-glass-world behavior. This is the Alice string, and now you can appreciate the reason for the name.

3. As our first example of a physical phenomenon associated with the Alice string, let us consider electrostatics. How is the electric field of a charge affected by the presence of an Alice string?

To find the field, it is simplest to work in a singular gauge, such that the vector potential is zero outside of a singular line (or, in three dimensions, a singular sheet) starting at the string and pointing directly away from the charge. Then the boundary condition on this line, is that the electric field must *change sign* across it. It is not difficult to solve the electrostatic problem with this boundary condition: one finds that the electric field is that of a charge in the presence of a conducting plate – with the singular line, which of course has no direct physical significance, playing the role of the plate!

The little paradox, that a fictitious line should be apparently the seat of a physical effect, is easily resolved. The point is that the sign of the electric field has no gauge-invariant meaning, being altered by the unbroken gauge transformation of parity in the internal $O(2)$. Only its magnitude is physically significant, and the magnitude of the electric field is perfectly continuous across the singular line.

Note that the charge is *attracted* to the string, by a long-range force.

4. A second striking phenomenon concerns the scattering of particles from the string.

The photon field changes sign as one circles the string: thus the *photon* is maximally Aharonov-Bohm scattered off an Alice string!

Consider the gauge bosons of the original $SO(3)$, which acquire mass from the condensate. We have every right to call the charge eigenstates W^+ and W^-, since the model is closely similar to the standard model of weak interactions. Now after parallel transport around the Alice string, one finds that W^+ has transformed into W^- while W^- has transformed into W^+. Thus the combination $W^+ + W^-$ is left unchanged, while the combination $W^+ - W^-$ changes sign after parallel transport around the string. Thus the first combination does not scatter at all, while the second combination maximally Aharonov-Bohm scatters. The net result is that when, say a W^+ beam impinges on the string, half the flux winds up going in the forward direction, but appears as a 50-50 mixture of W^+ and W^-, while the rest appears with the familiar $\frac{1}{\sin^2(\frac{\theta}{2})}$ cross-section, and at any angle again as a 50-50 mixture of both charge eigenstates.

5. A third striking phenomenon is the possible existence of nonlocalized charge in the presence of string-antistring pairs – what we like to call *Cheshire charge*.

Cheshire charge is perhaps best introduced through a thought experiment. Let there be given an Alice string and an antistring at some finite distance in two dimensions; the discussion also applies, with trivial modifications, to a large loop of Alice string in three dimensions. In this geometry the full $O(2)$ symmetry group is well-defined; the sorts of global problems which obstructed the consistent definition of some generators for a single string cancel out. In particular electric charge can be consistently defined and is rigorously conserved. Still, if we wind a charged particle slowly around a closed path enclosing the string, but not the antistring, it will return with the opposite charge. Where is the rest of the charge? Evidently, it must be left behind in a form somehow associated with the string-antistring system.

Now the winding can be done arbitrarily slowly. If the antistring were not there, one could take a circular orbit, and then the symmetry of the process would guarantee that no work is done. We should expect therefore that the energy necessary to wind around goes smoothly to zero as the string-antistring separation increases. The energy associated with the charge left behind, must become arbitrarily small as the distance between string and antistring increases. Since some of this energy is Coulomb energy (more precisely electric field energy), clearly the charge distribution (more precisely the electric field distribution) must become more and more spread out as the strings separate. While Gauss' law is not mocked, and the integrated surface integral of the electric field far from the string holds firm, the field at any fixed point goes to zero as the strings separate. It calls to mind the smile of the Cheshire cat.

The profound point is that Gauss' law can only be used to define the charge inside surfaces containing both the string and the antistring. (Or, in three dimensions, the entire loop.) Thus it cannot be used to identify the local charge density.

6. Since the broken $SO(3)$ symmetry supports not only Alice strings but also magnetic monopoles, one can conduct similar thought experiments with magnetic monopoles. The conclusion is, that a loop of Alice string can appear, viewed from afar, as a magnetic monopole. Perhaps surprisingly, it turns out that it is much easier to understand how the loop carries magnetic charge, than how it carries electric charge. I will merely state the result, leaving the proof as an exercise for the reader. (It has a pretty solution, requiring not calculation but rather that one visualize the correct picture.) Consider a loop of loops around the loop of string: i.e. draw a little circle linking the loop of string somewhere, and imagine pulling this circle around in a closed path, always linking the string loop, until it comes all the way back to its initial location. Mark off a particular point on the circle, and a smooth path of this point back to itself following the circle's orbit. Now for each

fixed position of the circle, we may compute the parallel transport for the path from the marked point around the circle and back to itself. It gives us an element of the residual symmetry group $O(2)$. Now as we follow the circle's orbit, these elements give us a closed path within $O(2)$. Since this group itself has essentially the topology of a circle, we can define the winding number of our path in $O(2)$. The exercise is, to show that this winding number is the magnetic monopole charge of the Alice string loop.

4. Zero modes and topological superconductivity

1. The various peculiarities we found in the electrodynamics of Alice strings, are quite generic and robust for flux tubes when the residual symmetry H is non-abelian.

2. One might expect that there should be a dimensionally reduced analog of the Nambu-Goldstone mechanism associated with the reduction of symmetry by a string. That is, we have found that H is reduced to some subgroup by the presence of the strings, and one might expect there to be arbitrarily low energy modes associated with the slow modulation of the string by the H generators which are no longer symmetries. More precisely, since we are concerned with gauge theories, one might expect that there is a dimensionally reduced version of the Higgs phenomenon, with the strings acting as superconducting wires.

Something close to this is true, but there is an interesting surprise. There are indeed essentially massless modes that carry charge and current, and string loops do form superconductors of a sort. However, as our discussion of Cheshire charge brought out, there is no reason to expect that the charge and current are localized on the loop. A detailed analysis of the relevant modes shows that, in fact, they are not. We are faced with a new form of superconducting electrodynamics, for which a loop of superconducting wire is not an adequate metaphor.

3. The superconductivity associated with breaking of a symmetry by a non-abelian flux tube is very robust, since it occurs for essentially group theoretic and topological reasons. It should be taken into account, in considering the possible behavior of cosmic strings produced in the early universe.

References

1. The material discussed in this lecture is contained in the following recent papers, which contain many further references:
2. L. Krauss, F. Wilczek *Discrete Gauge Symmetry in Continuum Theories*, Phys. Rev. Lett. **62** (1989) 1221.
3. M. Alford, J. March-Russell, F. Wilczek *Discrete Quantum Hair on Black Holes and the Non-abelian Aharonov-Bohm Effect*, Harvard preprint HUTP 89/A040.
4. J. Preskill, L. Krauss *Local Discrete Symmetry and Quantum Mechanical Hair* Caltech preprint CALT-68-1601.
5. M. Alford, J. March-Russell, F. Wilczek *Enhanced Baryon Number Violation Around Cosmic Strings*, Nucl. Phys. **B328** (1989) 140.
6. M. Alford, K. Benson, S. Coleman, J. March-Russell, F. Wilczek *Interactions and Excitations of Non-abelian Vortices*, Harvard preprint HUTP 89/AO51.
7. M. Alford, K. Benson, S. Coleman, J. March-Russell, F. Wilczek Harvard preprint HUTP 89/AO52.
8. F. Wilczek, Y.-S. Wu *Space-Time Approach to Holonomy Scattering*, Institute for Advanced Study preprint IASSNS 90/02.

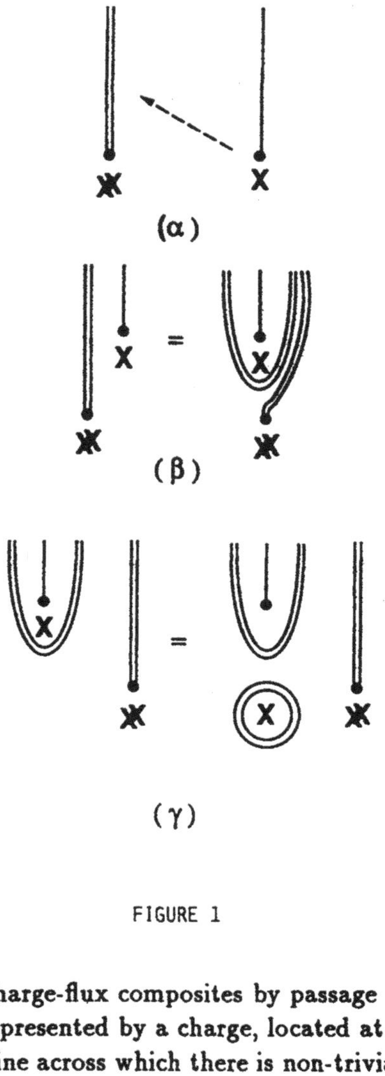

FIGURE 1

The modification of charge-flux composites by passage around one another. The composites are represented by a charge, located at the cross or double-cross, and a singular line across which there is non-trivial parallel transport. Note that these two structural elements must be slightly separated, to give an unambiguous specification of the quantum numbers. As one composite winds counterclockwise through another – operation α – it hits the singular line; to pull it through the singular line, we add a tube and antitube to the right and perform operation β. The resulting object on the left is not quite of the form of a charge split from a flux tube; to put it in that form we perform operation γ.

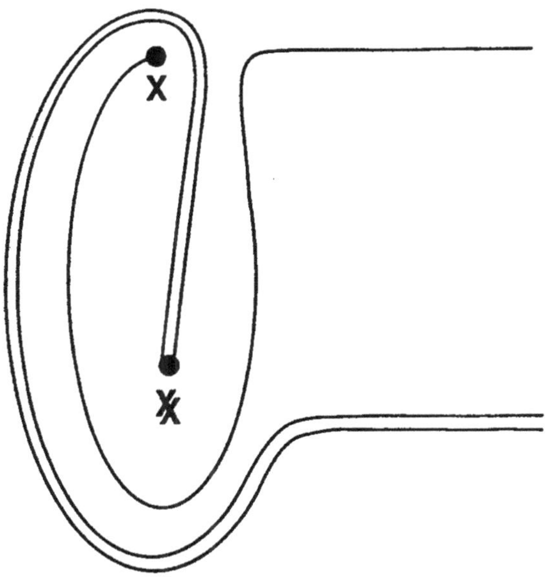

FIGURE 2

An alternative procedure to Figure 1, where the singular lines are twisted to avoid one another as one charge-flux composite winds around the other.

Experimental Confirmation of the Aharonov-Bohm Effect by Electron Holography

Akira TONOMURA

Advanced Research Laboratory, Hitachi, Ltd.

Kokubunji, Tokyo 185, Japan

Existence of the Aharonov-Bohm (AB) effect is confirmed by detecting a relative phase shift of π between two electron waves, having passed through inside and outside a tiny toroidal ferromagnet completely covered with a superconductor. The phase shifted value of exactly π comes from quantization of the existence of magnetic flux within the superconductor, which assures the complete magnetic shielding. Utilizing holographic electron interferometry based on the AB effect principle, the magnetic flux quantization process and individual magnetic flux quanta are directly observed.

1. Introduction

In 1959, Aharonov and Bohm theoretically predicted a surprising phenomenon in which an electron wave passing through a field-free region can be physically influenced by a hidden magnetic field [1]. This Aharonov-Bohm (AB) effect was soon tested by Chambers [2], Fowler et al. [3], Boersch et al. [4] and Möllenstedt and Bayh [5]. Although various assertions had been made after the prediction concerning the interpretations and implications of the AB effect [6] such as the physical significance of vector potentials, no one seemed to doubt its existence.

In 1975, Wu and Yang [7] revealed that the AB effect had a deeper implication in connection with the unified theories of fundamental physical interactions in nature. However, a dispute arose in 1978 concerning even the existence of the AB effect: Bocchieri and Loinger [8] asserted that the AB effect was imaginary and a purely mathematical concoction. Furthermore, it was claimed [9],[10] that in all previous experiments electrons must have been affected by the inevitable fringing fields from the finite whiskers or solenoids used in those experiments.

The author was convinced that a firm experimental foundation should be established for the increasingly important AB effect, since the controversy seemed fated to go on indefinitely. Additionally, a "coherent" field-emission electron microscope had been developed which opened a new way to observe matter and fields in microscopic dimensions by high-precision holographic electron interferometry, and the AB effect provided the fundamental principle behind this measurement technique. It is difficult to consistently interpret electron interference micrographs, especially of magnetic samples, without incorporating the AB effect.

A series of new experiments have been made since 1982 to confirm the AB effect without any ambiguity, and results of these experiments are reported here.

2.Experimental Design

The crucial experimental point discussed during the AB controversy concerned the effect of magnetic flux leakage from finite solenoids or whiskers on an electron wave. According to Kuper [12], an infinite solenoid is experimentally unattainable. However, an ideal geometry can be realized by using a finite toroidal solenoid. Although such a geometry had often been proposed during the controversy, an experimental execution could not be made until 1982 [13]: since sample sizes have to be smaller than the coherent region of an electron wave, development of the microlithography technique made fabrication possible.

In 1982, the present study produced its first result using tiny toroidal ferromagnets with completed magnetic circuits [13]. The resultant interferogram, i. e., the interference pattern between the electron wave transmitted through a toroidal magnet and a tilted plane wave, is shown in Fig. 1. A phase difference of six wavelengths was observed inside and outside the toroid. That report asserted that the existence of the AB effect was established by removing the ambiguity of magnetic flux leakage.

Bocchieri et al. [14], however, soon disputed this result. They asserted, on the one hand, that the detected phase shift could be interpreted as a result of the Lorentz force on electrons which partly penetrated the ferromagnet and, on the other hand, that the leakage flux should be

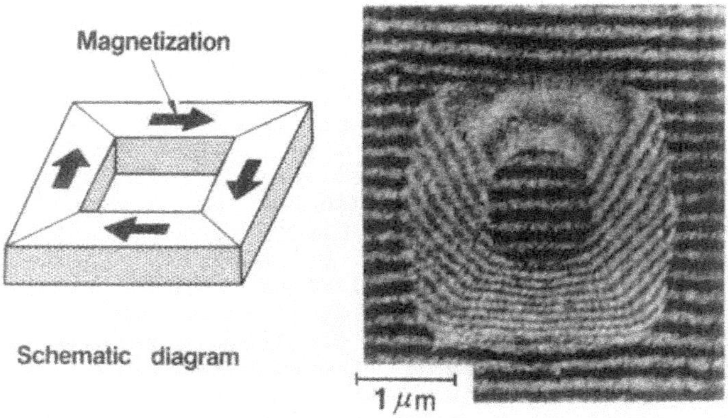

Fig. 1. Confirmation experiment of AB effect.

reduced to exactly zero for this test.

Therefore, a decisive experiment was designed to confine the magnetic field in a finite region, without allowing the slightest field leakage, and to prevent electrons from penetrating into the field region. In order to prevent electron penetration into the magnet, the magnet has to be covered with a metal layer. The resultant sample is free from electron penetration but, as a price, the absolute fringe displacement cannot be measured, but only the fractional deviation.

There is still another problem in that a small amount of flux leakage from the magnet might possibly influence the electron wave. In order to exclude this possibility, the covering metal layer is made of superconducting materials. Magnetic fields are not allowed to pass through a superconducting layer by the Meissner effect, and , therefore, no magnetic field leaks from the sample. The final sample structure is shown in Fig. 2, in which a toroidal magnet is

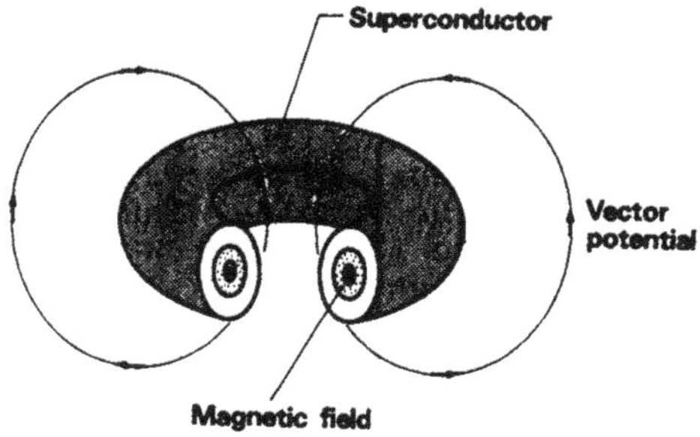

Fig. 2. Schematic diagram of toroidal sample.

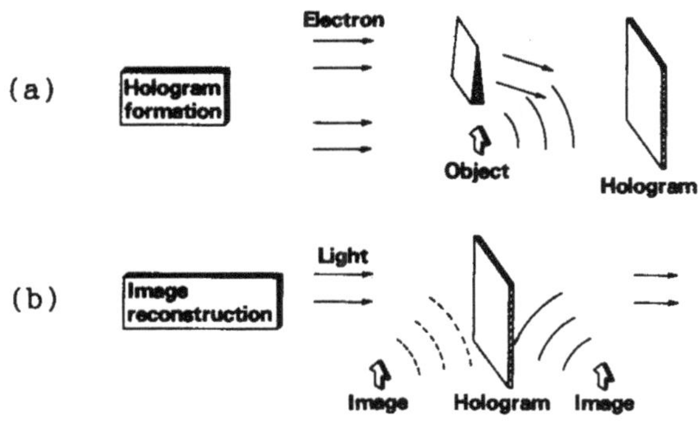

Fig. 3. Schematic diagram of electron holography:
(a) formation of electron hologram;
(b) image reconstruction.

completely covered with a superconductor. Here, there is no overlap between an electron wave and the magnetic fields. Since the coherent region of an electron wave is very narrow, the sample sizes have to be smaller than 10 μm. However, the thickness of the superconducting layer should be greater than the penetration depth to achieve magnetic shielding.

3 Experimental Procedure

- Electron holography -

Throughout the present experiments, electron holography has played an important role. Therefore, the experimental procedure is explained here.

Electron holography [15] is a two-step imaging process (see Fig. 3). An interference pattern (hologram) between an object wave and a reference wave is formed using an electron wave. The image of a sample is reconstructed by illuminating the hologram with a light wave. Once the electron wavefront is transformed into an optical one, optical devices can obtain information which is inaccessible using only electron devices.

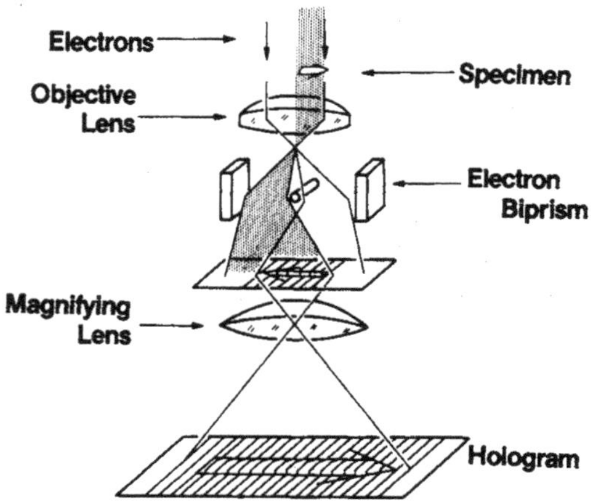

Fig. 4. Schematic diagram of electron hologram formation.

3.1 Electron hologram formation

Off-axis holography is formed by a 150 kV field-emission electron microscope [16]. This device provides an electron beam coherent enough to produce 3,000 interference fringes, in contrast to the 300 possible with a thermionic electron beam.

A schematic for hologram formation in this apparatus is shown in Fig. 4. A sample is located in only half the specimen plane. A collimated electron wave illuminates it, and the image is formed through a lens. A reference wave passing through the other half of the specimen plane reaches the image plane without disturbance from the sample. An electron biprism [17] superimposes these two waves to form an off-axis image hologram.

3.2 Optical reconstruction for interference microscopy

Optical reconstruction is performed using a He-Ne laser. The optical reconstruction system for interference microscopy is schematically shown in Fig. 5. A Mach-Zehnder interferometer splits a collimated laser wave into two waves. Each wave illuminates the electron hologram to produce two diffracted waves, from which reconstructed and conjugate images are formed. Since the hologram is an image hologram, four images are located in the hologram plane. These images are reformed through a lens onto its image plane. If the interferometer is adjusted so that only the reconstructed image of one wave and the transmitted wave of the other wave are selected to pass through an aperture, a standard interference image forms in the image plane. If the reconstructed image of one wave and the conjugate image

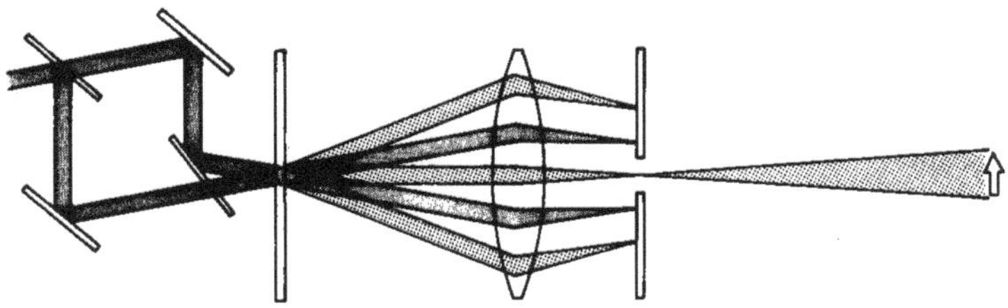

Fig. 5. Optical reconstruction system.

of the other wave are selected, an interference image with doubled phase distribution forms because the amplitudes of the two images complexly conjugate with each other. This technique can be repeated to detect a phase shift as small as $2\pi/100$.

If the object is a magnetic field, the contour lines give projected magnetic lines of force measured in h/e units [18]. This can easily be understood from the simple diagram in Fig. 6.

An incident electron is deflected by the Lorentz force during its passing through a uniform magnetic field. Let's regard the electron as a wave. Since the wavefront is perpendicular to the electron trajectory, the incident parallel electrons correspond to a plane wave, and the transmitted electrons to an inclined plane wave. In other words, the incident wavefront is rotated around a magnetic line of force as an axis.

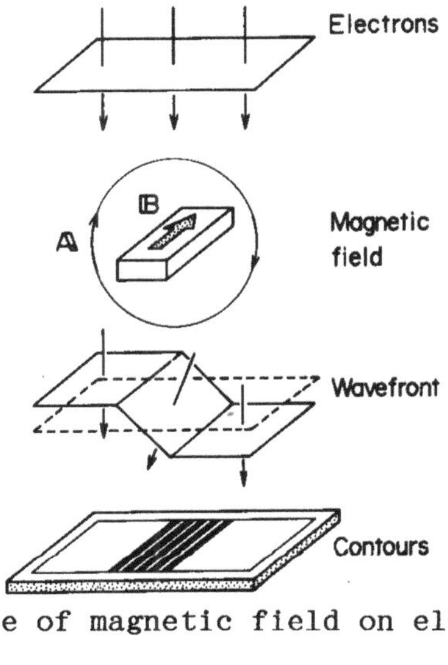

Fig. 6. Influence of magnetic field on electron wavefront.

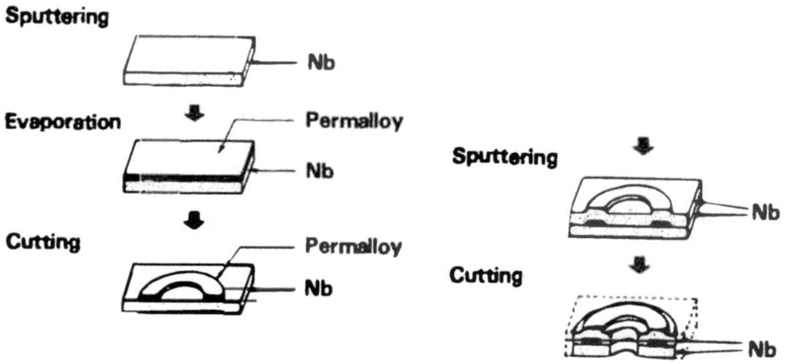

Fig. 7. Fabrication process of toroidal ferromagnets.

The contour lines of the wavefront, in fact, follow magnetic lines of force. The reason for this is very simple. The height of the wavefront is always the same along magnetic lines of force, because the magnetic line is the rotational axis. Thus, a very clear conclusion is that the contour lines of the electron wavefront are magnetic lines of force. It was also found from a simple calculation that quantitative measurement is possible. A constant magnetic flux of h/e is flowing between two adjacent contour lines.

This wavefront rotation may more easily be unstood in terms of vector potentials $\mathbb{A}$. Since magnetic fields $\mathbb{B}$ are described as $\mathbb{B}$ = rot $\mathbb{A}$, the vector potential is circulating around the magnetic line. Since the vector potential shifts the incident wavefront, the wavefront level differs on both sides of the magnetic line, but is the same along the magnetic line.

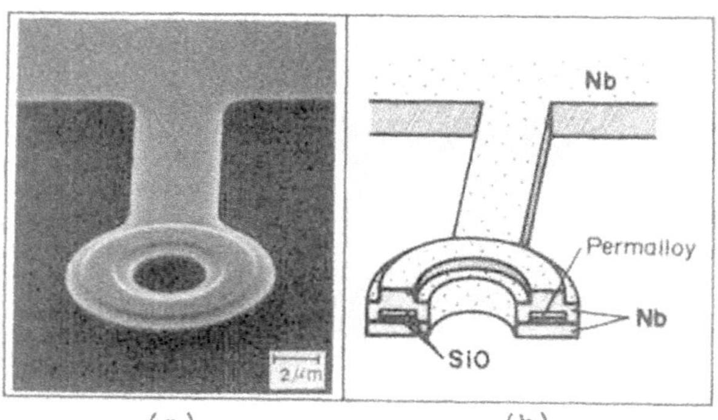

(a) (b)

Fig. 8. Toroidal permalloy covered with niobium layer:
(a) scanning electron micrograph;
(b) cross-sectional diagram.

4. Experimental results

4.1. Experimental test of the AB effect [19]

Tiny, complex samples were fabricated using state-of-the-art photolithographic techniques. The fabrication process is schematically illustrated in Fig. 7. First, niobium film, and then permalloy film were prepared on a Si wafer. The toroidal shape was cut out of only the permalloy film, and then a niobium film was again sputtered on it. Lastly, a toroidal sample was cut. A scanning electron micrograph of a fabricated sample is shown in Fig. 8. Special attention was paid to attain perfect superconducting contact between the upper and lower Nb layers (see Fig. 8(b)), in order to completely enclose the Permalloy magnet. The Nb layer thickness was set at 3000 Å to exceed the penetration depth of 1000 Å. Many toroidal samples with differing magnetic flux values ($4(h/e) \sim 6(h/e)$) were connected to a niobium plate by a tiny bridge, to achieve high thermal conductivity (see Fig. 9).

Samples were first observed at room temperature by holographic electron interferometry [11], and only those with leakage flux less than h/20e (1% of the total flux) were selected for this experiment.

The selected toroidal magnet was cooled to 5 K, and the relative phase shift was measured between two electron waves passing through inside the hole and outside the toroid

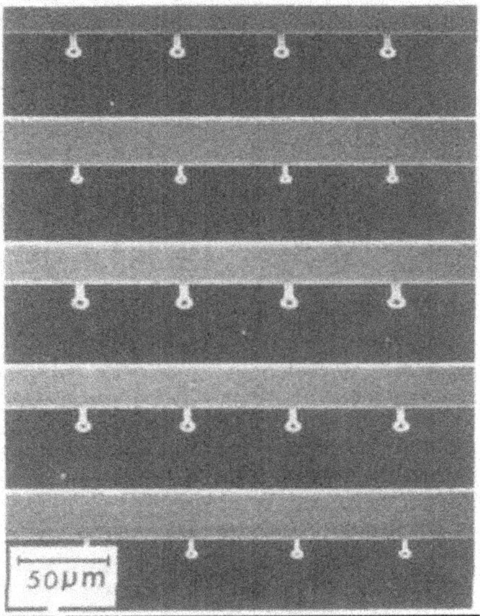

Fig. 9. Toroidal samples of various sizes connected to niobium plate.

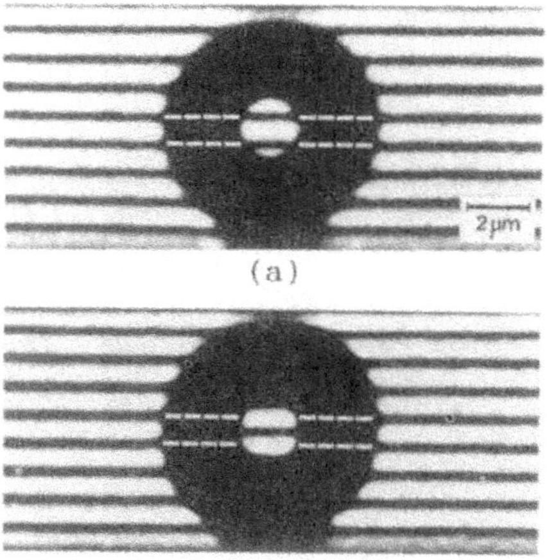

(a)

(b)

Fig. 10. Electron interferograms of toroidal samples at 4.5K (a) $\Delta S=0$; (b) $\Delta S=\pi$.

during the holography process. Although the value of the magnetic flux inside the toroid was distributed continuously, only two kinds of interferograms could be obtained, as shown in Fig. 10. The phase shift values, ΔS, was either 0 or π (mod 2π). The conclusion is obvious: a relative phase shift of π is produced even when the magnetic field is confined within the superconductor and is shielded from the electron beam. This proves that the AB effect exists.

But why does the phase shift value take either 0 or π? This fact's existence provides key evidence that the superconductor completely shields the magnetic field, as can be understood from the temperature dependence of the phase shift, an example of which is shown in Fig. 11. The phase shift at room temperature was 0.3π as shown in Fig. 11.(a). However, when sample temperature T was reduced, the phase shift gradually increased becoming 0.8π at T = 15 K (see Fig. 11(b)). When T was further reduced to below Tc (9.2 K), the phase shift abruptly changed from 0.8π to π.

This behavior can be interpreted as follows. The enclosed magnetic flux determines the phase shift above Tc. The temperature dependence of the phase shift arises from the fact that magnetization in the permalloy increases by 5% due to decreasing thermal fluctuations of the spins when T changes from 300 K to 9.2 K. When T goes below Tc, the magnetic flux jumps to the discrete values quantized in h/2e units. Therefore, the phase shift is either 0 or π depending on whether the number of fluxons is ever or odd.

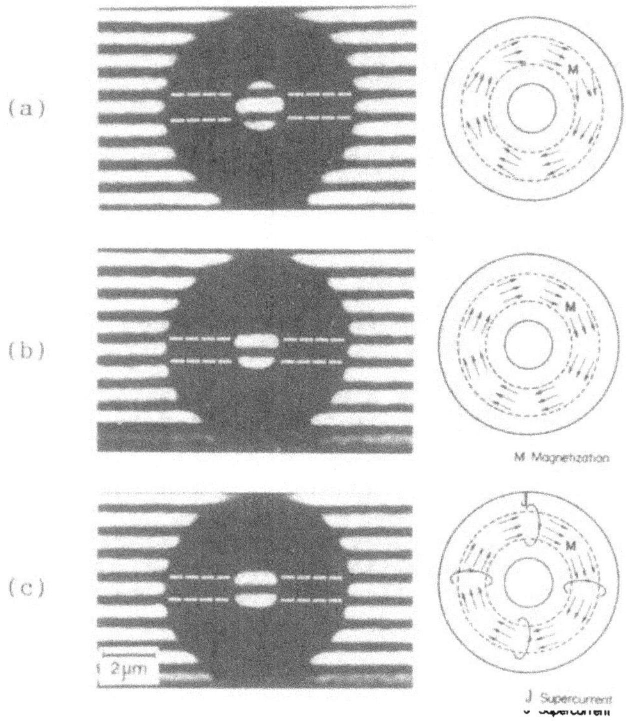

Fig. 11. Temperature dependence of electron interferogram of toroidal samples: (a) $\Delta S=0.3\pi$ at T=300K; (b) $\Delta S=0.8\pi$ at T=15K; (c) $\Delta S=\pi$ at T=4.5K

In this work, no quantization was observed in a sample having even a slightly oxidized layer between the two niobium layers. Therefore, the occurrence of flux quantization can be taken to assure that the niobium layer actually becomes superconducting, that the superconductor completely surrounds the magnetic flux, and that the Meissner effect prevents any flux from leaking outside. In Fig. 11(b), the phase shift just above Tc was 0.8π. This was closer to π than to 0, resulting in a jump to π.

Viewing this from a different standpoint, the process of magnetic flux quantization has been observed by electron interferometry. In this case, a quantized flux, a fluxon, was hidden inside the superconductor, and it could be recognized through the interaction of an electron wave with vector potentials circulating outside the sample.

4.2 Direct observation of magnetic flux quanta [20]

Fluxons, not trapped inside a superconductor but penetrating it, play an important role in fundamental

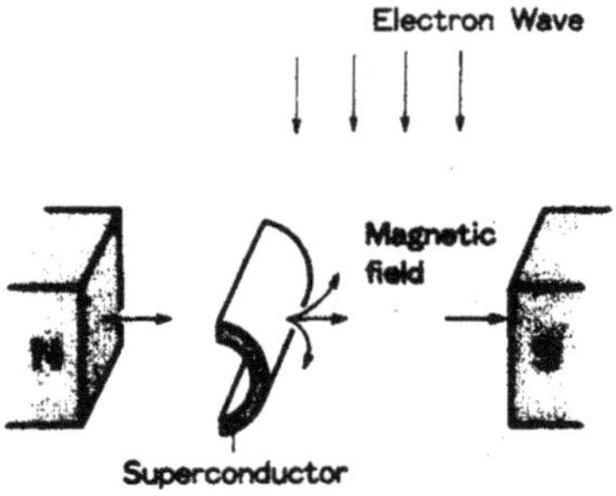

Fig. 12. Experiment of flux quantum observation.

superconductivity. To give a concrete example, the critical current of a superconductor is completely determined by the behavior of such fluxons. However, they are too small to observe directly. The only way to locate fluxons has been by indirect methods. In the Bitter technique [21], for example, fine-particle iron distribution as the fluxons' replicas can be observed by electron microscopy. More recently, new techniques using scanning electron microscopy [22] and scanning tunneling microscopy [23] have been developed.

Electron interferometry based on the AB effect has already been employed to observe a fluxon [24]. So far, however, a fluxon has been detected as a line of dislocations of parallel interference fringes by half of its spacing, the line being not sharp enough to permit a clean determination of the fluxon's image. The introduction of electron holography was expected to make it possible to directly observe magnetic lines of force of a single fluxon.

The experimental arrangement for fluxon observation is diagrammed in Fig. 12. A magnetic field is applied perpendicularly to a superconducting lead film and the film is

Fig. 13. Interference micrograph of single fluxon.

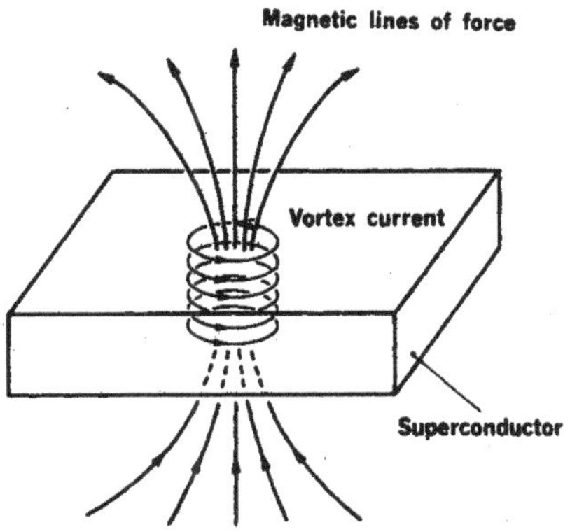

Fig. 14. Single fluxon.

cooled to become superconductive. The magnetic field, when it's weak enough, is expelled due to the Meissner effect. A stronger magnetic field can penetrate the film in thin filament form, which is the fluxon. The fluxon is illuminated by an electron wave and is observed as an interference micrograph through the electron holography process.

The resultant photograph is shown in Fig. 13. Since the photograph is phase-amplified two times by overlapping a reconstructed image not with a plane wave but with its conjugate image, the phase difference becomes two times larger. Therefore, one contour fringe corresponds to a single fluxon. A magnetic line of force of a single fluxon is produced from a very small region of the superconductor surface and fans out into free space. This magnetic line can be considered to be produced from a vortex supercurrent, which acts as a tiny solenoid as shown in Fig. 14. This was confirmed by two experiments. When the temperature of the superconductor exceeded the critical temperature, the magnetic line disappeared. It can easily be interpreted that the superconducting state was destroyed, causing the vortex current to stop flowing. This offers proof that the observed magnetic line was generated by the superconducting current.

There is further proof in that the magnetic line persisted even after the applied magnetic field was removed. This supports the fact that the observed magnetic line was not due to the applied field, but to the persistent vortex current inside the superconductor.

Details of a single fluxon can be observed using the phase amplification technique. Figure 15 shows the enlarged

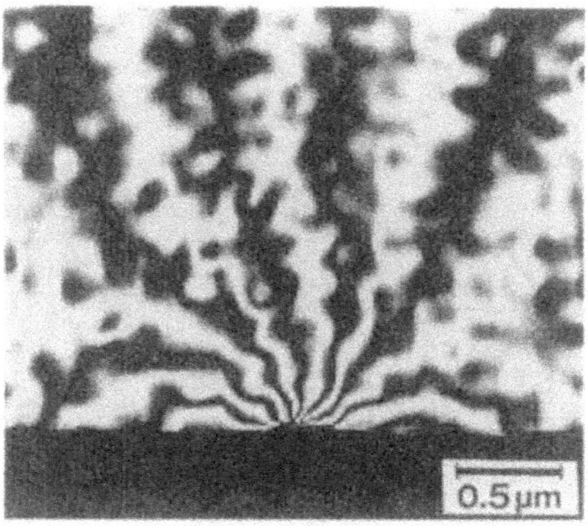

Fig. 15. Interference micrograph of single fluxon phase-amplified 16 times.

interference micrograph of a single fluxon phase-amplified 16 times. Since we can count eight fringes, the flux value is proven to agree with the theoretical value h/2e. In addition, the inner structure of a fluxon can be measured. For example, the fluxon diameter at the superconductor surface can be measured and was determined to be 0.15 μm.

In addition to this isolated fluxon, an antiparallel pair of fluxons was unexpectedly found for the first time, as shown in the left part of Fig. 16. Such an identification has not been possible before, since the polarity of a fluxon couldn't be determined by any method then available.

The reason for the pair creation is not obvious. One possible reason is as follows. Such pairs are predicted to be created and annihilated in a two-dimensional system by thermal excitation according to the Kosterlitz-Thouless theory [25]. During the cooling process, excited pairs might be frozen and pinned by some mechanism.

Fig. 16. Magnetic lines of force penetrating lead film 0.2 μm thick. One fringe corresponds to a flux of h/2e.

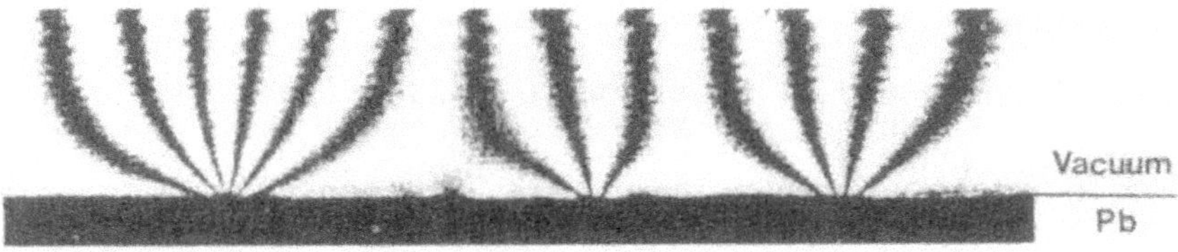

Fig. 17. Magnetic lines of force penetrating lead film 1.0 μm thick.

The film thickness was 0.2 μm in this case. However, what happens when the film becomes thicker? Figure 17 shows the result. The film thickness here is 1 μm. Quite different magnetic lines of force are observed. The magnetic flux penetrates the film, not as an isolated fluxon but as a bundle of several fluxons. No vortex pair can be found here.

This development can be interpreted as follows. Lead belongs to the type II superconductor group. Therefore, when a strong magnetic field is applied to the thick film, the lead is divided into normal and superconducting domains. Magnetic fields can pass through a normal region. The other surrounding region is superconducting and, therefore, the total magnetic flux passing through a normal region is quantized to integral multiples of h/2e. In the case of an extremely thin film, however, the lead film behaves like an entirely type II superconductor and we can look at individual fluxons as in Fig. 16.

We can conclude that individual fluxons and their behavior, inaccessible by any other method, have become observable by electron holography based on the AB effect.

5. Conclusions

The existence of the Aharonov-Bohm effect was conclusively confirmed using both technologies of electron holography and microlithography. That is, a relative phase shift of π was detected between two electron waves passing through the hole and outside a toroidal magnet, which is completely covered with a superconductor. Furthermore, the Aharonov-Bohm effect was employed to directly observe the process of magnetic flux quantization and individual flux quanta penetrating a superconductor.

References

[1] Y. Aharonov and D. Bohm: Phys. Rev. 115 (1959) 485. See also W. Ehrenberg and R. E. Siday: Proc. Phys. Soc. London B62 (1949) 8.
[2] R. G. Chambers: Phys. Rev. Lett. 5 (1960) 3.
[3] H. A. Fowler et al.: J. Appl. Phys. 32 (1961) 1153.
[4] H. Boersch et al.: Z. Physik 165 (1961) 79.
[5] G. Möllenstedt and W. Bayh: Phys. Bl. 18 (1962) 299.
[6] See S. Olariu and I. I. Popescu: Rev. Mod. Phys. 57 (1985) 339; M. Peshkin and A. Tonomura: The Aharonov-Bohm Effect, Lecture Notes in Physics Vol. 340 (Springer, Heidelberg, 1989).
[7] T. T. Wu and C. N. Yang: Phys. Rev. 12 (1975) 3845.
[8] P. Bocchieri and A. Loinger: Nuovo Cimento 47A (1978) 475.
[9] P. Bocchieri, A. Loinger and G. Siragusa: Nuovo Cimento 51A (1979) 1.
[10] S. M. Roy: Phys. Rev. Lett. 44 (1980) 11.
[11] A. Tonomura: Rev. Mod. Phys. 59 (1987) 639; A. Tonomura: "Electron Holography" in Progress in Optics Vol. 23 (North-Holland, Amsterdam, 1986) P.183
[12] C. G. Kuper: Phys. Lett. 79A (1980) 413.
[13] A. Tonomura et al.: Phys. Rev. Lett. 48 (1982) 1443; A. Tonomura et al.: Proc. Int. Symp. Foundations of Quantum Mechanics, Tokyo, 1983, ed, S. Kamefuchi et al. (Physical Society of Japan, Tokyo, 1984) P.20.
[14] P. Bocchieri, A. Loinger and G. Siragusa: Lett. Nuovo Cimento 35 (1982) 370.
[15] D. Gabor: Proc. R. Soc. London A97 (1949) 454; See also for recent applications A. Tonomura: Rev. Mod. Phys. 59 (1987) 639.
[16] A. Tonomura et al.: J. Electron Microsc. 28 (1979) 1.
[17] G. Möllenstedt and H. Dücker: Naturwiss. 42 (1955) 41.
[18] A. Tonomura et al.: Phys. Rev. Lett. 44 (1980) 1430.
[19] A. Tonomura et al.: Phys. Rev. Lett. 56 (1986) 792; N. Osakabe et al.: Phys. Rev. A34 (1986) 815; A. Tonomura et al.: Proc. 2nd Int. Symp. on Foundations of Quantum Mechanics, Tokyo, 1986, ed. M. Namiki et al. (Physical Society of Japan, Tokyo, 1987) P.97.
[20] T. Matsuda et al.: Phys. Rev. Lett. 62 (1989) 2519; A. Tonomura et al.: to be published in Proc. 3rd Int. Symp on Foundations of Quantum Mechanics, Tokyo, 1989, ed. S. Kobayashi et al. (Physical Society of Japan, Tokyo, 1990).
[21] U. Essman and H. Träuble: Phys. Lett. 24a (1967) 526; See also P. L. Gammel et al.: Phys. Rev. Lett. 59 (1987) 2592.
[22] J. Mannhart et al.: Phys. Rev. B35 (1987) 5267; W. S. Millman et al.: J. Mater. Res. 2 (1987) 543.
[23] H. F. Hess et al.: Phys. Rev. Lett. 62 (1989) 214.
[24] B. Lischke: Phys. Rev. Lett. 115 (1969) 485; H. Wahl: Optik 28 ((1968) 69) 417; H. Boersch et al.: Phys. Stat. Sol. (6) 61 (1974) 215.
[25] J. M. Kosterlitz and D. Thouless: J. Phys. C6 (1973)

1181; B. I. Halperin and D. R. Nelson: J. Low Temp. Phys. 36 (1979) 599.

The Aharonov-Bohm Effect Without Gauge Fields - A Paradox

Igor Klebanov
Department of Physics, Princeton University
Princeton, NJ 08544

Leonard Susskind †
Department of Physics, Stanford University, Stanford, CA 94305-4060

† NSF contract # PHY 86-12280

ABSTRACT

Recent claims that an effective Aharonov-Bohm Effect can occur in theories without dynamical gauge fields are examined. Certain paradoxes arise whose resolution is unclear.

I Introduction

When the Aharonov-Bohm effect was first discovered it surprised a lot of people. I remember hearing about it from a skeptical friend who said that it was like trying to find out if someone was at home in a building where all the windows are shut, no sound or light can penetrate the walls, no telephone wires come out, and so on. It seemed impossible. My friend joked that all you had to do was walk once around the building and check your phase. He felt fairly sure that it would work about as well as dowsing or voodoo and that experiments to check it would fail. I was sure the experiments would work.

In fact, a lot of serious physicists didn't believe it and tried to find ways out because they had been brought up to believe that only electric and magnetic fields are real. Since only A_μ and not $F_{\mu\nu}$ penetrate outside the solenoid it seemed that the electron would have to enter the solenoid to feel the effects of the flux. Of course, this was wrong. The point is that what is "real" is what is gauge invariant and, although A is not gauge invariant, its line integral $e \int A_\mu dx^\mu$ is. This means that a very subtle kind of non-local information does leak out.

Now, a new phenomonon has been discovered in fractional quantum Hall effect systems[1]. It is a kind of A-B effect in a theory with no gauge fields[2]. No dynamical variable like A_μ exists in the theory;

nevertheless, we can detect the presence of a localized disturbance without going anywhere near it. Again, it involves carrying a particle around the region of the disturbance and checking its phase. But this time there does not appear to be any function of the fields on the outside which registers the information. I must admit that this time I was skeptical, even though everyone else was untroubled. They would say things like "It's just like the A-B effect in gauge theories with massive gauge bosons" or "It's like the CP_N Model with a Hopf Term" and so on. Well, I think it's not like these theories. If it is correct, then it is an entirely new manifestation of quantum mechanics.

Let me begin by being clear about what the phenomenon is, and how it differs from the massive gauge or CP_N Hopf Models. Consider a 2 + 1 dimensional field theory with a Hamiltonian containing only local interactions. Suppose, furthermore, that there is an energy gap so that no massless photons, phonons, magnons or other modes propagate. Also, assume all correlations are short range. Finally, and this is the key ingredient, suppose all of this can be done without gauge fields. In particular, assume that the theory can be expressed locally without gauge fields. I don't care if somebody can find a clever way to introduce gauge fields, as long as eliminating them introduces no action at a distance.

Now, suppose particle-like excitations A and B exist. Consider transporting A around a large closed path which may or may not contain B. Since there are only short range correlations, no local information about B exists in the region where A moves. By transporting A, we mean a physical process which could be carried out by introducing potentials ("Buckets") to trap the particles and then, by hand, moving the buckets. If A is moved slowly enough, then, when it returns to its initial position, the final quantum state will be the same as the original state apart from a phase.

$$|\Psi\rangle \rightarrow e^{i\Phi_B}|\Psi\rangle \qquad 1.1$$

The phase Φ_B is called the Berry phase. If $\exp i\,\Phi_B$ depends on the presence of A, we have reproduced an A-B effect without gauge fields. It is believed to happen in Laughlin's theory of the fractional quantum Hall states. The purpose of this lecture is to explain Laughlin's theory, its A-B effect and explain why it is paradoxical.

II Berry Phase

In this section I will explain what the Berry[3] phase is. Consider a quantum system with a time dependent Hamiltonian H(t). Let us suppose that the ground state is non-degenerate, and that the time over which H changes appreciably is much longer than the inverse energy gap (H is adiabatic). Let us define the instantaneous ground state $|\Psi_I(t)\rangle$ by

$$H(t)\,|\Psi_I(t)\rangle = E_0(t)\,|\Psi_I(t)\rangle \tag{2.1}$$

where $E_0(t)$ is the lowest eigenvalue of $H(t)$. In general, $|\Psi_I(t)\rangle$ is not a solution of the time dependent Schroedinger equation. However, under the assumption of adiabatic variation of $H(t)$, if the wave function starts at $t = 0$ equalling $|\Psi_I(0)\rangle$ it will stay in the instantaneous ground state to zeroth order in the time dependence of H. To state this more accurately, we note that $|\Psi_I(t)\rangle$ is defined by 2.1 only up to a phase. Let us suppose that this phase is fixed by some arbitrary but definite convention. To do this, we can once and for all fix the phase of every wave function in Hilbert space. For example, in the case of single particle non-relativistic quantum mechanics $\psi(x)$ can be chosen real at the origin $(x = 0)$. Thus $|\Psi_I(t)\rangle$ has a definite phase. Furthermore, if at two times $H(t_1)$ and $H(t_2)$ are the same, then $|\Psi_I(t_1)\rangle$ and $|\Psi_I(t_2)\rangle$ have the same phase.

Now, the actual evolution of the solution to Schroedinger's time dependent equation is

$$|\Psi(t)\rangle = e^{i\,\Phi(t)}\,|\Psi_I(t)\rangle \tag{2.2}$$

The phase $\Phi(t)$ will have a part which is just due to the instantaneous energy of the state. Let us eliminate this by redefining $H(t)$ so that its ground state energy $E(t)$ is subtracted.

If the evolution of H eventually brings it back to its initial value after time T then $|\psi(o)\rangle$ and $|\psi(t)\rangle$ differ only by a phase

$$|\psi(t)\rangle = e^{i\,\Phi_B(t)}\,|\Psi(0)\rangle \tag{2.3}$$

The phase Φ_B is called the Berry phase. Before describing how to calculate it let me give some examples. As far as I know, the first example of the calculation of such a phase involved a single $\frac{1}{2}$ spin $\vec{\sigma}$ in a time dependent magnetic field $\vec{B}(t)$

$$H = -\mu\,\vec{B}(t)\cdot\vec{\sigma} \tag{2.4}$$

In this case the spin follows the direction of $\vec{B}$. Suppose the B field initially points along the Z axis and is then allowed to slowly rotate about the y axis by a full 2π. The initial state in the σ_Z representation is $\binom{1}{o}$ but, after rotation, it is $-\binom{1}{o}$. The Berry phase is

π.

Another example is a pair of fermions, each bound in a bucket centered at X_1 and X_2. Imagine slowly moving the buckets around in space until X_1 moves to X_2 and X_2 to X_1. That is, imagine interchanging them. The Berry phase is again π.

The third example involves transporting a charged particle in a bucket through a magnetic field. Not suprisingly, when the bucket is returned to its initial position, the phase is changed by

$$\Phi_B = e \int A_\mu \, dx^\mu \tag{2.5}$$

This phase, which is the basis for the A-B effect, is also a Berry phase.

To compute Φ_B write the Schroedinger equation

$$i \frac{\partial |\psi\rangle}{\partial t} = H(t) \; | \psi \rangle \tag{2.6}$$

Now substitute (2.2) and use the fact that $| \Psi_I (t) \rangle$ is an eigenvector of H(t) with eigenvalue zero

$$i \, [\, |\dot{\Psi}_I (t) \rangle + i \, \dot{\Phi} \, | \, \Psi_I (t) \rangle] = 0 \tag{2.7}$$

Now project this onto $\langle \Psi_I (t) \, |$.

$$i \langle \Psi_I \, | \, \dot{\Psi}_I \rangle = \dot{\Phi} \tag{2.8}$$

(I've assumed $\langle \Psi_I \, | \, \Psi_I \rangle = 1$) and we finally obtain

$$\Phi_\beta^{(t)} = \Phi \, (0) + i \int dt \, \langle \Psi_I (t) \, | \, \dot{\Psi}_I (t) \rangle \tag{2.9}$$

The Berry phase is

$$\Phi_B = \Phi \, (t) - \Phi \, (0) = i \int dt \, \langle \Psi_I \, | \, \dot{\Psi}_I \rangle \tag{2.10}$$

All you need to know to compute Φ_B is the lowest eigenvector of H at each t. You don't even have to know what H is.

There is a very interesting viewpoint about Berry phases which draws on an analogy with gauge theory. Let's suppose H contains some parameters X_i whose time dependence determines that of H

$$H(t) = H\{X_i(t)\}$$

The eigenvectors of $H(X_i)$ for given values of the X_i are

$$\Psi_I\ (X_i)$$

The evolution of H is thought of as a trajectory in X_i space. Now write the Berry phase in the form

$$\Phi_B = i \int dt < \Psi_I\ (t)\ |\ \dot{\Psi}_I\ (t) >$$

$$= i \int dx_i < \Psi_I\ (X)\ |\frac{\vec{\partial}}{\partial X_i}\ |\ \Psi_I\ (X) > \qquad 2.11$$

Thus, Φ_B is a line integral of a vector field V_i on the space of the X's.

$$\Phi_B = \int dX_i\ V_i\ (X) \qquad 2.12$$

V_i is very similar to an ordinary vector potential. The condition that the Berry phase is not generally zero is the non-vanishing of the "field strength"

$$\frac{\partial V_i}{\partial X^j} - \frac{\partial V_j}{\partial X^i}$$

Let's return to the example of the spin

$$H = -\mu\vec{B} \cdot \vec{\sigma}$$

The components of B are the parameters X_i. Let's suppose the magnitude of μB is fixed so that the "configuration" space X_i can be taken to be a unit sphere. Given the eigenvectors of

$$\vec{X} \cdot \vec{\sigma}$$

for unit vector $\vec{X}$, we can compute the vector potential V on the sphere. The result is interesting. It is just the vector potential of a unit monopole located at the center of the sphere. Thus, transporting the spin around on a sphere gives the same phase as transporting a charge in the field of a monopole.

There is a special case where the Berry phase is particularly easy to compute. Supposing that H(t) has the form

$$H(t) = e^{i\omega Lt}\, H\, e^{-i\omega Lt}$$

where ω is a frequency and L is a hermitian generator of some transformation. For example, if a particle is being transported in a uniformly moving bucket then L would be the momentum and ω the velocity of the bucket. If the bucket moved in a circle, L would be angular momentum. Similarly for the spin if the B field rotates about some axis, then L is the component of angular momentum about that axis.

In this case,

$$| \Psi_I (t) \rangle = e^{i\omega Lt} \, | \Psi_I(0) \rangle \qquad 2.13$$

$$\frac{d|\Psi_I\rangle}{dt} = i\omega L \, | \Psi_I (t) \rangle \qquad 2.14$$

and from (2.8) we get

$$\begin{aligned} \dot{\Phi} &= - \omega \langle \Psi_I (t) \,|\, \Psi_I (t) \rangle \\ &= - \omega \langle \Psi_I (0) \,|\, \Psi_I (0) \rangle \qquad 2.15 \end{aligned}$$

thus the phase grows at a uniform rate. Furthermore, if $T = 2\pi/\omega$ is the time for a full cycle, then Φ_B is

$$\Phi_B = - \langle \psi (0) \,|\, L \,|\, \psi (0) \rangle \times 2\pi \qquad 2.16$$

III Laughlin's Wave Function

The system is a bunch of non-relativistic electrons moving in 2 spatial dimensions under the influence of a uniform magnetic field perpendicular to the plane. The vector potential is

$$A_m = \frac{B}{2} \epsilon_{mn} X^n$$

and the single particle Schrodinger equation is

$$\sum_{m=1}^{2} \left(-i \frac{\partial}{\partial X^m} + e A_m \right)^2 \Psi = E \psi \qquad 3.1$$

The energy levels are discrete "Landau Levels". The lowest Landau level (LLL) is degenerate, with wave function

$$\psi = F\ (Z)\ \text{EXP} - (\frac{Z^* Z}{4}) \qquad 3.2$$

where F(Z) is any analytic function of Z (Z is a point on the complex plane). A complete set of wave functions for the L.L.L. are the angular momentum states

$$\psi_n = \frac{Z^n}{\sqrt{n!}}\ \text{EXP}\ (-\ \frac{Z^* Z}{4}) \qquad 3.3$$

which are concentrated near rings of radius

$$r_n = \sqrt{2n} \qquad 3.4$$

Now let's fill up the first N angular momentum states with electrons. The wave function depends on N complex coordinates Z_i (i = 1-N), is antisymmetric and has the form

$$\psi\ (Z_1,\ Z_2, \ldots) = A\ (Z_1^0\ Z_2^1\ Z_3^2\ Z_4^2 \ldots Z_N^{n-1})\ \text{EXP}\ \Sigma\ Z_i^*\ Z_i/4$$

where A means antisymmetrize. The result is

$$\psi\ (Z_1 \ldots Z_n) = \prod_{i<j} (Z_i - Z_j)\ e^{-Z_i^* Z_i/4} \qquad 3.5$$

Since every L.L.L. is filled out to n = N-1 and the nth L.L.L. forms a ring of radius (3.4) the result is a filled disc with radius $\sqrt{2(n-1)}$.

A very useful tool for analyzing the distribution of electrons is the coulomb gas analogue[1]. The probability for electrons at $(Z_1 \ldots Z_n)$ is given by

$$\prod_{i<j} (Z_i - Z_j)^* (Z_i - Z_j)\ \text{EXP}\ \frac{1}{2} \sum_i Z_i^* Z_i$$

$$\text{EXP} - \{\frac{1}{2}\ \Sigma\ Z^* Z - 2 \sum_{i<j} \log|Z_i - Z_j|\} \qquad 3.6$$

Now, the quantity in the exponent is the energy of N classical charges, each with charge $\sqrt{2}$,interacting through 2-dimensional coulomb potentials in a uniform background of charge density $\frac{1}{\pi}$. The uniform background should be thought of as the limit of a large ball of charge centered at Z= 0. Thus, the expression in 3.6 is the Boltzmann distribution for N charges in a neutralizing background at temperature $\kappa T = 1$. The result is a uniform ball of charges just dense enough to cancel the background. The charge density will quickly go to zero for $r > \sqrt{N}$. Such a blob of charge is degenerate since it can be centered anywhere. However, if a wall exists at $r = r_N$, then the blob fills the available volume and is nondegenerate. There is an energy gap of order $\frac{Be}{M}$. A quasiparticle (or perhaps a better word would be quasihole) can be constructed as follows. Multiply the wave function 3.6 by the factor

$$\prod_i (Z_i - Z_0) \qquad 3.7$$

where Z_o is the position of the quasihole. To see what this does, go back to the coulomb gas analogue. The new factor is equivalent to an additional non fluctuating static charge at Z_o. No real electron is present at Z_o so the extra charge is called a phantom. Now the effect of the extra phantom is to redistribute the other charges and Debye-screen the phantom. A hole is created in the distribution centered at Z_o. It looks like Fig (1). Of course, the charge that was in the hole reappears at the edge of the blob. In fact, the blob increases its radius from $\sqrt{N}$ to $\sqrt{N+1}$. To understand this better, put Z_o at the origin. Then the quasihole is generated by multiplying ψ by a factor Z_i for each electron. This can be thought of as an operation which takes each L.L.L. $Z^n e^{-Z^*Z/4}$ to the next L.L.L. $Z^{n+1}e^{-Z^*Z/4}$. Each electron gets pushed out to the next level. The level n = 0 is emptied, but a new electron appears in the (N + 1) level. The quasiparticle is just the hole at Z = 0.

The influence of the QP exponentialy decreases as we move away from Z_o. For example, the charge density relative to the uniform background is

$$\rho_{QP} = -\frac{1}{2\pi} \text{EXP} - (Z-Zo)^2/2 \qquad 3.8$$

Similarly, a current is set up in the vicinity of Z_o. The current is

$$J_m = \hat{\psi}^\dagger \left(-i \frac{\partial}{\partial X^m} + e A_m\right) \hat{\psi} + \text{he}$$

where $\hat{\psi}$ is the second quantized electron field. It is easy to prove that if all electrons are in L.L.L. then

$$J_m = \epsilon_{mn}\,\partial_n\rho \tag{3.9}$$

For a quasiparticle at $Z_o = 0$

$$J_m = \epsilon_{mn}\partial_\upsilon\rho_{QP} = \frac{1}{2\pi}\epsilon_{mn}\,X^n\,\mathrm{EXP}\,(-Z^*Z/2) \tag{3.10}$$

In other words a current circulates around the QP but falls to zero as a gaussian. Other quantities, such as density-density correlations show similar behavior. First of all the density-density correlation function $\langle \rho(Z_1)\,\rho(Z_2) \rangle$ falls gaussianly when $|Z_1 - Z_2|$ increases. Furthermore for fixed $|Z_1 - Z_2|$, the effect of a Q.P. at Z_3 decreases gaussianly as $|Z_1 - Z_3|$ inncreases. In fact all products of local operators show the same behavior. If we separate the points the correlation dies gaussianly. If we fix the separations, but increase the distance to a quasiparticle, the effect of the presence of the Q.P. dies.

Let's compute the Berry phase for carrying a Q.P. called A around another (B). Let $Z_B = 0$. Z_A will slowly be transported around a circle of radius R. A bucket can be introduced which will modify the wave function somewhat, but it can be proved that the answer is unaffected. The simplest way to do the calculation is to use Eq. (2.16). We must compute the expectation value of the total angular momentum when two QP's are located at $Z_1 = 0$ and Z_2. To do this we can assume that the angular momentum is the integral of a local density. Such local quantities have short range correlations as long as we keep away from the boundary of the blob. Let us begin with the angular momentum if there are no QP's. Call it L_o. Now put a Q.P. at $Z_1 = 0$. In the interior of the blob the L-density is unaffected except near Z_1. The vicinity of Z_1 gives a change of δ to $\langle L \rangle$. In addition, the electron which was expelled from near Z_1 is compensated for by an additional one in the N + 1 orbital which has angular momentum N + 1. Thus, the total angular momentum associated with the QP is

$$\delta + N + 1$$

Next, let us introduce a QP at Z_2. This again changes the angular momentum by a part due to the region near Z_2 and a boundary contribution. Call the contribution from near Z_2 Δ. Since all correlations in the interior are short range, Δ is just what it would be if the first QP had not been introduced. The new boundary term is N + 2. Thus, the total L is

$$(\delta + N + 1) + (\Delta + N + 1) + 1$$

Note that the increase is the sum of 3 terms. The first is the extra L due to the QP at Z_1. The second is that which would be due to the QP at Z_2 if the first was not there. The last term is an extra piece which is not additive and can be thought of as due to a kind of interaction between QP's 1 and 2. In other words, it is an A-B phase. However since it is an integer, it is unobservable.

IV. The Fractional Quantum Hall Effect Case

Laughlin[1] has given persuasive evidence that, if electrons have short range repulsions, the ground state can change over to a new form described by the wave function

$$\Psi_{LAUGH} = \prod_{i<j} (Z_i - Z_j)^m \, e^{-\Sigma Z_i^* Z_i/4} \qquad 4.1$$

The new feature is that there is a $(Z_i - Z_j)^M$ dependence on the seperation between electrons. The quasiparticle is given by

$$\Psi_{Q.P.} = \prod_i (Z_i - Z_o) \, \Psi_{LAUGH} \qquad 4.2$$

To understand the properties of these wave functions we can still use the coulomb gas analogue. The analogue charge of an electron is $\sqrt{2M}$ and the background density is $\dfrac{1}{\pi\sqrt{M}}$. Hence, for stability the electrons must have a density 1/M times smaller than before. Thus, the blob extends out to the MN^{th} orbit with a density of 1/M electrons per LLL. The phantom associated with the QP has charge $\sqrt{\frac{2}{M}}$ and is screened by making a hole with an electron deficit of 1/M electrons. Thus, the QP has fractional charge. The conclusions concerning the structure of correlations functions are the same as before.

This time the Berry phase for moving one QP about another is more interesting. Each LLL has 1/M electrons out to the MN^{th} orbit. Inserting a QP at $Z_1 = 0$ deposits 1/M electron in the (MN + 1) orbit, thereby increasing $\langle L \rangle$ by

$$\delta + \frac{NM+1}{M} \qquad 4.3$$

The second QP adds

$$\Delta + \frac{MN+2}{M} \qquad 4.4$$

giving a change

$$(\delta + \frac{MN+1}{M}) + (\Delta + \frac{MN+1}{M}) + \frac{1}{M} \qquad 4.5$$

This time, the non-additive Berry phase is a fractional multiple of 2π and is observable. We have found an observable A-B phase.

V. The Paradox

The question is whether this A-B phase proves that there is information stored far from Z = 0 (and far from boundaries) which knows about the existence of the QP at Z = 0. To be specific, consider an angular ring Γ of some width L at a large distance R from the origin. L and R are much larger than any natural scale in the problem (see Fig. 2). The moving quasiparticle and the bucket are entirely contained within Γ. We want to know if the nontrivial A-B phase requires the existence of some operator built out of the quantized electron fields in Γ to have an expectation value which depends on the existence of a QP at $Z_1 = 0$. In particular, I mean an order one effect on an operator whose eigenvalues are order one. Let me first argue that this is impossible. The expectation value of any operator which depends only on $\psi(Z)$ (Z in Γ) is determined by the density matrix describing Γ. This is defined as follows. Let the many body wave function be Ψ $(Z_1 \ldots . Z_N)$. Let coordinates in Γ be called Y and those outside be W. The density matrix $\langle Y_1 \ldots Y_\ell \,|\rho|Y_1' \,Y_2' \,Y_3' \ldots\rangle$ is defined by

$$\begin{aligned}&\langle Y_1 \cdots Y_\ell \,|\rho|Y_1' Y_2' \cdots \rangle \\ &\qquad = \int \Pi \, dW \, \Psi^*(W_1 \cdots W_{N-\ell}, \, Y_1 \cdots Y_\ell) \, \Psi \, (W_1 \cdots W_{N-\ell}, \, Y_1' \cdots Y_\ell')\end{aligned} \qquad 5.1$$

ρ is a matrix in the space of possible multi electron states within Γ. It is not difficult to prove that the analyticity properties of Laughlin's wave functions determine the entire ρ-matrix from its diagonal elements $\langle Y_1 \ldots |\rho|Y_1 \ldots\rangle$. These can be computed using the coulomb gas analog. Since the Y's are fixed and none of the other particles (W's) can go inside Γ, we can think of the region Γ as nonconducting, but with fixed immobile charges Y. The rest of space is conducting. The density matrix element is the partition function for this mixed conductor-insulator. We want to know how it changes if a QP is placed at the origin.

Putting a quasiparticle at Z = 0 creates fractional charges on the boundaries of the conductor adjacent to Γ. One might expect the charges in Γ to interact with this fractional charge thus giving rise to a dependence of ρ on the QP. However, we must remember that each W must be allowed to be in either of the two conducting regions. This is equivalent to introducing a "short circuit" between them. The result is to discharge the surface charges near the boundaries of Γ. The density matrix only feels the QP with exponentially small strength.

If the density matrix is not signficantly affected by the presence of a QP at Z = 0 then no expectation value of a functional of fields in Γ can be effected.

Now I want to prove exactly the opposite. There must be operators which feel the QP with an order 1 response. Begin with a single QP in Γ and a bucket to hold it. Let the bucket be described by a potential V $(Z, \bar{Z}, t)$.

The field theoretic interaction with the bucket is

$$H_B(t) = \int d^2Z \; V(Z, Z^*, t) \; \rho(Z, Z^*) \qquad 5.2$$

where $\rho = \psi^+\psi$. (ψ is the electron field operator) The bucket potential V can be chosen to have a support well within the boundaries of Γ. Let the initial state at t = 0 be a QP at Z, and let $V(Z, Z^*, 0) = 0$. Next, turn on V adiabatically keeping it centered at Z. This takes time T_1. Now slowly cause the central position of V to follow a circular path around the origin taking time T_2. Finally, turn off V taking time T_3. The evolution of the state is described by

$$|\Psi(t)\rangle = Te^{i \int_0^t H_B(\tau)\, d\tau} |\Psi(0)\rangle \qquad 5.3$$

where $t = T_1 + T_2 + T_3$, and the Berry phase is given by

$$\langle \psi(0) \,|\, Te^{i\int H_B\, dt} \,|\, \psi(0) \rangle \qquad 5.4$$

Now, the hard question is whether $T\, e^{i\int V\rho}$ is a functional of Schroedinger picture operators in Γ ? To answer this we would have to integrate the equations of motion which is too hard to do. What we can say for sure is that it is a functional of operators within a circle of radius Ct where C is the maximum velocity of signal propagation. This distance will generally be larger than $2\pi R$ since moving the QP with velocity faster than C violates adiabaticity. Thus, in general, $T\, e^{i\int V\rho dt}$ can depend on operators outside of Γ and, in particular, operators in the vicinity of the origin. It is no surprise that its expectation value can depend on the presence of a second QP at Z = 0.

Now, consider another experiment. Instead of a single QP in Γ, let us begin with a "telephone dial" configuration which has N ~ R quasi particles arranged on a circle within Γ (see fig (3).. It is still true that the perturbation on the density matrix due to a QP at Z = 0 is exponentially small.

Let each QP have a bucket and transport each QP by angle $2\pi/N$. It is easy to prove that the extra Berry phase due to a QP at Z = 0 is the same as for a single QP being transported by angle 2π. But now the total time required does not increase with R. Hence there is some finite width for

Γ, so that $Te^{i\int H_B}$ depends only on fields within Γ. In this case we prove the existence of a unitary functional in Γ, which has an order 1 response to the QP at Z = 0. How is this consistent with the apparent fact that the density matrix within Γ does not seem to depend on it? This is the basic paradox of the AB effect without gauge fields. If it were not for Laughlin's example I would have concluded that it (along with the related ideas[4-6] of induced Chern-Simon interactions etc.) is impossible.

Before discussing a possible resolution of the paradox, let me formulate a slightly different version of it.

VI. Cylinder Geometry

Instead of an infinite plane, let us consider a cylinder or, eqivalently, an infinite strip in the Y direction of width $2\pi R$ with periodic conditions in X. Choose the background vector potential to be

$$A_x = By \qquad Ay = 0 \tag{6.1}$$

It is easily seen that the Schroedinger equation has translation invariance in the X direction. (This is the analogue of rotational invariance in the plane).

A individual LLL is parametrized by an integer n which labels its X-momentum. One finds

$$\psi_n = e^{inx/R}\, e^{-\frac{1}{2}(y - \frac{n}{R})^2} \tag{6.2}$$

More generally, an arbitrary state in the LLL is given by

$$F\,(Z)\; e^{-1/2\, y^2} \tag{6.3}$$

where F(Z) is an analytic function of Z = X + iY, periodic in X.

$$F\,(Z + 2\pi\, R) = F\,(Z) \tag{6.4}$$

For convenience I am going to "cap off" the cylinder at Y = 0 so that no LLL with n < 0 are permitted. Now consider filling the n > 0 levels with electrons. The slater determinant for N electrons filling orbits n = 0, 1 2 .... N - 1 is

$$\prod_{i<j} [e^{i\, Z_i/R} - e^{i Z_i/R}]\; e^{-\Sigma\, Y_i^2/2} \tag{6.5}$$

The analoge of the Laughlin wave functions for fractional filling is

$$\psi_M = \prod_{i<j} [e^{i Z_i/R} - e^{i Z_j/R}]^m e^{-\Sigma Y_i^2/2} \qquad 6.6$$

A quasiparticle at the y = 0 end of the cylinder can be defined by

$$\psi_{QP} = \prod_i e^{i Z_i/R} \psi_M \qquad 6.7$$

This can be considered to be a QP distributed uniformly over the end of the cylinder. The effect of $\Pi e^{iZ/R}$ is to shift each particle to the next orbit leaving a hole at n = 0. A QP at a more general position Z_o on the surface of the cylinder is given by

$$\prod_i [e^{i Z_i/R} - e^{i Z_o/R}] \psi_M \qquad 6.8$$

The coulomb analogue for $\psi^* \psi$ is still exact with the coulomb potential given by M ln sin$[(Z_i - Z_2)/2R]$. Each electron is again a charge $\sqrt{2M}$. A QP is a charge = $\frac{\sqrt{2}}{\sqrt{M}}$. The quasiparticle at Y = 0 is represented by a charge $\frac{\sqrt{2}}{\sqrt{M}}$ uniformly distributed over the cylinder's end.

Now consider the effect of transporting a QP around a periodic path at height Y. The Berry phase depends on the existence of a second QP at Y = 0. The extra Berry phase is again $\frac{2\pi}{M}$. This is independent of the height Y. This time the bucket operator T $e^{i\int \rho V}$ can depend only on those Schroedinger picture operators within a band Γ of width ~CR (see fig 4). Since Y can be arbitrarily large, it is hard to see how $\langle T e^{i\int \rho V} \rangle$ can depend on conditions near Y = 0 unless there is some kind of long range correlation.

VII. A Possible Resolution

The only resolution of this paradox that I can see is that exponentially small influences are not ignorable. Let me return to the "telephone" dial experiment of Fig.(3). Consider the influence of a QP at Z = 0 on the density matrix within the anulus Γ. As I have argued, the screening of the phantom and the short circuiting of the inner and outer "conducting" regions insures that the influence is exponentially small in R. By this I will mean

$$\delta\rho \cong e^{-CR} \qquad 7.1$$

On the other hand, the bucket operator, $T\, e^{i\int V\rho} = B$ is unitary, which means that its matrix elements are bounded. How then can

$$\mathrm{Tr}\ \delta\rho \cdot B \tag{7.2}$$

remain ~ 1 as $R \to \infty$. The only possible answer is that the number of important terms in 7.2 increases rapidly enough with R to compensate the supression factor (7.1). To see that this is possible let us consider the size of the Hilbert space describing the region Γ. The number of LLL in a given region is proportional to its area, which in turn is proportional to the radius of Γ. Furthermore, each LLL may either contain an electron or not contain one. Thus, the dimensionality of the Hilbert space describing Γ is $\sim e^{\upsilon R}$. Evidently the number of potentially significant terms in $\mathrm{Tr}\ \delta\rho \cdot B$ is $\sim e^{2\upsilon R}$.

Now the main part of ρ is concentrated in particularly important elements near the diagonal and describe distributions of electrons with the average density. Configurations far from the mean, and significantly off diagonal terms in ρ, are much smaller.

On the other hand, $\delta\rho$ is generally small. I see no reason for its off diagonal and non average elements to be even smaller. In other words, apart from the overall exponential suppression, $\delta\rho$ could be a matrix with order 1 elements everywhere. If the elements of B are also similarly spread out, the large number of contributions to 7.2 could keep it from going to zero as $R \to \infty$. However, this explanation is not sufficient to explain the cylindrical case shown in Fig. 4. In this case, the band Γ does not increase in radius as $Y \to \infty$. Therefore, an order $e^{-\mu Y}$ suppression of a QP at $Y = 0$ cannot be similiarly compensated. Fortunately, another effect can rescue us. To see this, consider the behavior of the charge density for $R << 1$. In this case, the individual orbits are separated by distance $1/R$. The charge density for the nth individual orbit is

$$\rho_n = e^{-(y - \frac{n}{R})^2}.$$

If we fill every orbit with an electron, the charge density is

$$\sum_n e^{-(y - n/R)^2}$$

which consists of a series of well separated gaussian peaks as in Fig (5). How do we understand this from the viewpoint of the Coulomb gas? The answer is that for small R the coulomb gas becomes 1 dimensional and in one dimension the phenomenon of Wigner crystalization takes place. In other words, the charges just seek their classical positions, which arranges them in a periodic crystal. The number of particles per unit Y is just what is needed to cancel the uniform background on the average.

In the fractionally filled case the Wigner crystal still forms but this time the spacing is one particle per every M orbits, as in Fig (6). Now, adding a QP at $Y = 0$ shifts everybody up one unit leaving $n = 0$ vacant (see Fig (7)). The influence of the QP is felt arbitrarily far up the Y axis by the locations of the crests of the charge density waves.

Now consider increasing R. The charge density waves never entirely disappear. Their strength, however, diminishes like e^{-R}. Evidently then the influence of a QP at $Y = 0$ is felt with undiminished strength at arbitrarily distant Y, but exponentially ($\sim e^{-R}$) decreases as R is increased.

Transporting a QP around the cylinder takes a time $\sim$ R and the band Γ must have area R^2. The number of states within the band $\sim 2^{R^2}$ and can offset the exponential decrease in the influence of the QP at $Y = 0$.

Another way to understand this effect is through spontaneous symmetry breaking. The symmetry is translations along the Y axis. Ordinarily this symmetry could not be broken on a cylinder because it is a continous symmetry and the cylinder is essenially a one dimensional system. The Coleman theorem insures that no such symmetry can be broken. However, in our case translation symmetry is discrete. The background vector potential

$$A_x = y$$

explicitly breaks the usual invariance. If we combine Y-translation with a gauge transformation,

$$Y \to Y + Y_o$$

$$A_x \to A_x - \frac{\partial \Lambda}{\partial X}$$

$$\Lambda = Y_o X,$$

then the symmetry is restored. The periodicity of the electron wave function then requires Y_o to be quantized. In other words, translations in Y are discrete and correspond to translations among the LLL.

The fully filled case is invariant under this discrete group since each LLL is filled. For small R, the Wigner crystalization breaks the symmetry for the fractional case. A quasiparticle at a point Z is a dislocation or soliton which is felt at arbitrary Y. (see Fig 8)

The Berry phase experiment provides a kind of order parameter for this spontaneous breaking. Moving a QP around the cylinder can tell us whether the filling began at $n = 0$, $n = 1$, ... More generally it tells us how many dislocations there are between the place where the experiment is carried out and $Y = 0$.

What is surprising is that as $R \to \infty$ this order parameter does not disappear along with the density waves. Of course, the order parameter exists in a Hilbert space of exponentially growing size.

A number of comments are in order. First of all, I have assumed that both the influence of a quasiparticle and the behavior of charge density waves on a cylinder fall exponentially as $e^{-\mu R}$. This is in contrast to the explicitly known gaussian behavior in the fully filled m = 1 case. For example, a quasiparticle at Z = 0 on the infinite plane is just a missing electron in the n = 0 L.L.L. Since the wave function for this LLL is $e^{-Z^*Z/4}$, the missing charge density is gaussian. If this were the case for the fractionally filled case, it would not be possible to rely on exponential tails to solve the paradox. For example, in the telephone dial experiment, the region Γ has area ~R and therefore 2^{LR} states if the perturbation $\delta\rho$ due to a quasiparticle is gaussian, there is no way that the number of terms in Tr $\delta\rho \cdot B$ can overwhelm e^{-R^2}. It is interesting that for m > 1 Coulomb gas simulations suggeste a nongaussian tail[7]. In fact the density profile for m > 1 appears to oscillate with an envelope which could be exponential.

The oscillations are interesting because they are similar to the Wigner crystalization that takes place on the cylinder. There we found oscillations with a strength $\sim e^{-\mu R}$. The presence of a QP seems to induce similar oscillations with a similar strength. In other words, annular regions of radius R tend to behave very much like cylinders of radius R.

The oscillations in the planar case for m > 1 are in a certain sense due to Wigner crystalization. It is known that for large enough m genuine Wigner crystalization occurs in the coulomb gas. For slightly smaller m one should expect a good deal of short range crystal ine order with the fluctuations destroying the long range order exponentially. Putting a quasiparticle at Z = 0 pins the local lattice structure so that nearby oscillations should occur with exponentially decreasing strength. I suggest that these oscillations continue to infinite radius and that they are the signal that a QP is present at Z = 0. That their exponentially small magnitude can translate into an order 1 Berry phase seems possible as I have explained. The precise mechanism for this remains mysterious.

On general grounds it seems that an intimate connection must exist between the A-B effect without gauge fields and the spontaneous breaking of discrete symmetries on the cylinder. It is clear that if a quasiparticle at y ~ 0 can be detected at indefinitely large y, then the ground state for $y \to \infty$ must be different with and without the QP. Furthermore if the QP energy is to be finite then the energy density in the two states must be the same for $y \to \infty$. That the broken symmetry is discrete follows from the Coleman theorem.

The symmetry breaking phenomenon has to be unusual in that it should disappear exponentially as $R \to \infty$. Normally spontaneous symmetry breaking works the other way. As you decrease the dimensionality it usually disappears.

Although I have indicated a way out of the paradox of the A-B effect without gauge fields, I feel somewhat uneasy. I would like to see an explicit construction of the bucket operator and $\delta\rho$. Futhermore, in the system I have considered, the presence of the background magnetic field would make it increasingly difficult to measure a Berry phase as R increases. This is because the quasiparticle picks up a phase proportional to the area of its orbit due to the background field. In order to detect the extra effect of an additional QP, the area of the orbit must be defined to within a fraction of the area of a LLL. The probability of not making a mistake of area ~ 1, in a trajectory of circumference R, decreases exponentially with R. Perhaps all we have learned is that an experiment to detect exponential tails must be done with exponentially increasing care. It would be of great interest to exhibit A-B phases in a system in which the QP's moved without an effective background magnetic field.

If I have not succeeded in answering the questions I have raised, I hope I have at least convinced you that they are interesting questions and that the phenomenon of the A-B effect without gauge fields is an exceedingly subtle and new one.

Acknowledgements

The authors are very grateful to Bob Laughlin, Marvin Weinstein and Helen Quinn for many useful conversations and insights.

References

1. R. B. Laughlin, Phys. Rev. Lett. 50, (1983) 1395.
2. D. Arovas, J. R. Schrieffer, and F. Wilczek, Phys. Rev. Lett. 53, (1984) 722.
3. M. V. Berry, Proc. Roy. Soc. London, Sev. A392, (1984) 45-57.
4. V. Kalmeyer and R. B. Laughlin, Phys. Rev. Lett. 59 (1987) 2095.
5. F. Wilczek, X. G. Wen, and A. Zee, ITP preprint 88-179.
6. I. Dzyaloshinski, A. M. Polyakov and P. Wiegmann, Phys. Lett. 127A (1988) 112.
7. J. M. Caillol, D. Levesque, J. J. Weis, and J. P. Hansen, J. Stat. Phys. 28, (1982) 325.

FIGURE 1

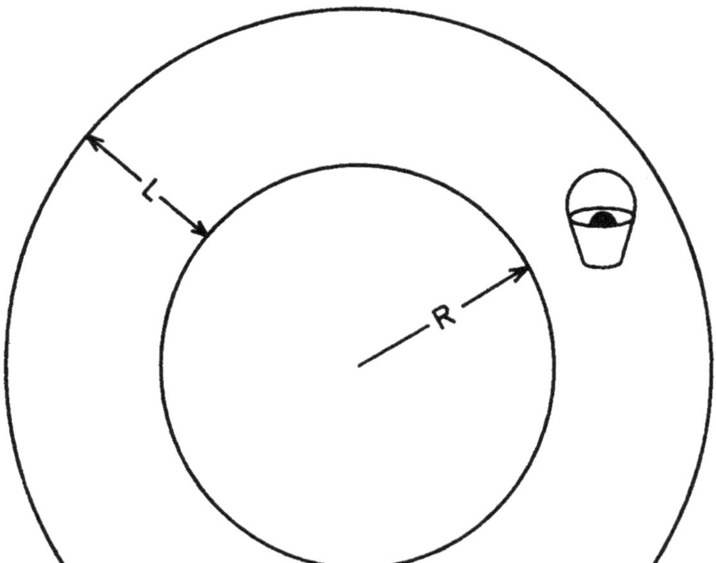

FIGURE 2

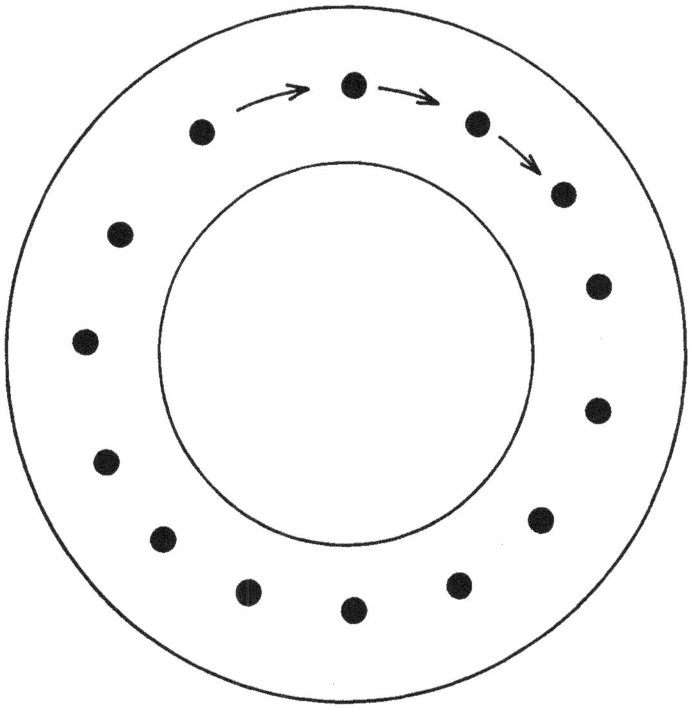

FIGURE 3

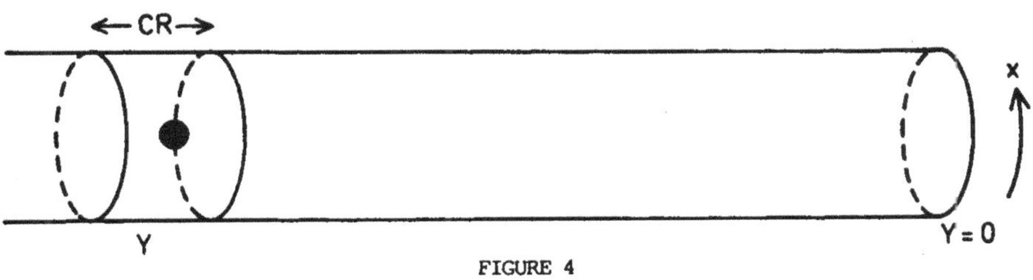

FIGURE 4

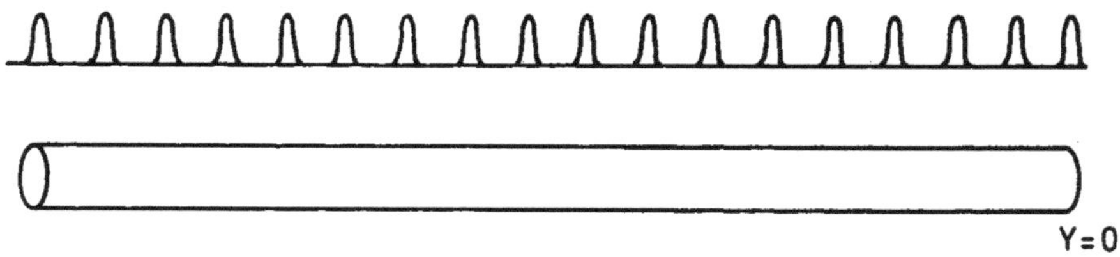

FIGURE 5

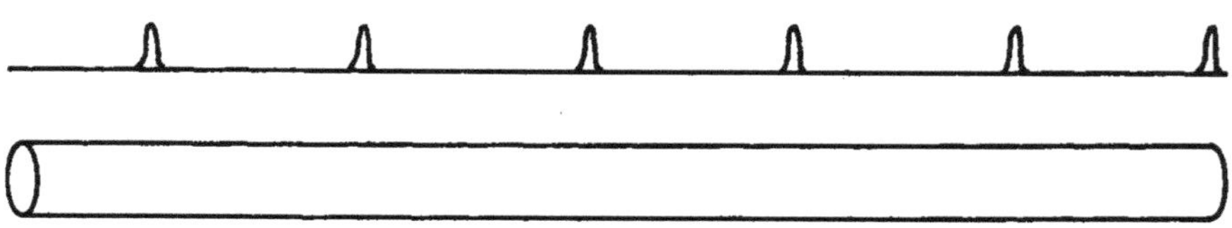

FIGURE 6

FIGURE 7

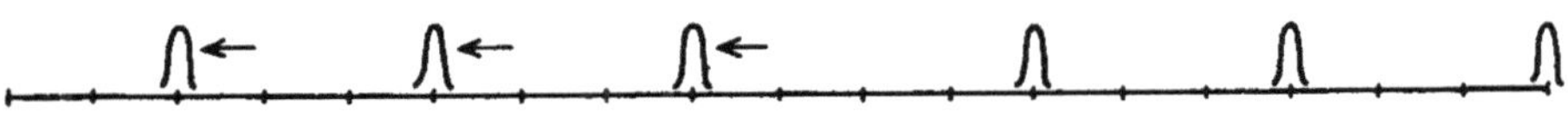

FIGURE 8

THE AHARONOV-BOHM EFFECT AND MASS OF THE PHOTON

DAVID G. BOULWARE
Physics Department, University of Washington
Seattle, Washington 98195, USA

and

S. DESER
Physics Department, Brandeis University
Waltham, Massachusetts 02254, USA

ABSTRACT

We show that the Aharonov-Bohm effect requires that the vector potential couple minimally to matter, not that it be a gauge field: the effect is present in massive (finite range) electrodynamics, reducing smoothly to the original result in the limit of infinite range. Indeed, it may be used to provide an experimental bound on the range which is much larger than the "table-top" apparatus, namely a lower limit of order 10^2 Km.

1. Introduction

The vast literature on the Aharonov-Bohm effect (see for example Ref. 1 is explicitly based on the gauge nature of the electromagnetic field. To our knowledge, there have been no discussions of whether it might exist for a non-gauge field such as finite-range electrodynamics (in quantum language, for finite photon mass). There is also a literature on massive electrodynamics (e.g., Ref. 2 and Ref. 3) which likewise does not include any analyses of the Aharonov-Bohm effect, let alone whether its infinite range limit is smooth.[4]

We will show that the effect indeed exists for finite mass, has a smooth limit and may even be exploited to provide a bound on the photon's mass: precision experiments could observe a range as high as 10^2 Km.

2. The Effect

We formulate the Aharonov-Bohm effect in the following terms. In the semiclassical limit, the wave function of a charged particle contains the path dependent phase $\exp(ie\int \vec{A}\cdot d\vec{x})$, which means that when a beam is split and recombined, there will be a phase shift of the form ($\hbar = 1 = c$)

$$\exp ie \oint_c \vec{A}\cdot d\vec{x} \equiv \exp i\Phi\,. \tag{1}$$

The flux

$$\Phi \equiv e\oint_c \vec{A}\cdot ds \equiv e\int \vec{\nabla}\times\vec{A}\cdot dS \equiv e\int \vec{B}\cdot dS \tag{2}$$

is through any surface bounded by the closed curve c defined by the two paths. Although this phase factor is usually invoked from gauge invariance of the theory, it is simply a consequence of minimal ($\partial_\mu \to \partial_\mu - ieA_\mu$) coupling to the vector field A_μ rather than of the specific action for A_μ. We will stay within the same minimal coupling framework here, (since Pauli terms are in any case irrelevant to the effect) so that the currents remain conserved. Thus, the value of Φ mod 2π will be observable in the interference pattern produced by the recombined beams. The transverse (non-local and gauge-invariant) vector potential $\vec{A}^T$ which contributes to Φ is uniquely determined by $\vec{B}$ according to $-\nabla^2\vec{A}^T = \vec{\nabla}\times\vec{B}$. Consequently, if there is a nonvanishing magnetic field $\vec{B} \equiv \vec{\nabla}\times\vec{A}$ anywhere in space (even one confined to a bounded region), its contribution to $\vec{A}^T = -\nabla^{-2}\vec{\nabla}\times\vec{B}$ in nonvanishing almost everywhere. Note that the above discussion is completely independent of the action governing A_μ.

We turn next to the equations which determine $(\vec{A}^T, \vec{B})$. If the field is of finite range, it obeys the Proca equation,

$$\partial_\nu F^{\mu\nu} + m^2 A^\mu = J^\mu \tag{3}$$

where $F_{\mu\nu} \equiv \partial_\mu A_\nu - \partial_\nu A_\mu$, and our metric is $(-+++)$. For stationary currents and magnetic fields, this means

$$\vec{\nabla}\times\vec{B} + m^2\vec{A} = \vec{J}\,, \tag{4}$$

$$(m^2 - \nabla^2)A_0 = 0\,. \tag{5}$$

Since $\vec{\nabla}\cdot\vec{J} = 0$, this implies vanishing longitudinal vector (and scalar) potentials* The relevant equation is then

$$(-\nabla^2 + m^2)\vec{A}^T = \vec{J}^T\,; \tag{6}$$

for the usual solenoidal configuration, it reads

$$(-\nabla^2 + m^2)\vec{A}^T = \hat{\phi} j\delta(\rho - a) = \vec{\nabla}\times j\hat{z}\theta(a-\rho)\,, \tag{7}$$

* Physical longitudinal photons, which are the hallmark of the $m \neq 0$ theory, are well-known to decouple from matter in the $m \to 0$ limit, provided that the current is conserved. If the current is not conserved, then the coupling to the longitudinal photons becomes strong as $m \to 0$ and only *extremely* weak coupling is allowed experimentally.[2]

where j is the current per unit length in the surface of an infinite cylinder of radius a, $(\hat{\phi}, \hat{z})$ are unit vectors, θ is the step function and ρ is the radius in the $x-y$ plane. Writing $\vec{A} = \hat{z} \times \vec{\nabla}\Pi(\rho)$, we obtain the scalar equation

$$(\partial_\rho^2 + \rho^{-1}\partial_\rho - m^2)\Pi(\rho) = j\theta(a-\rho)\,. \tag{8}$$

The relevant homogeneous solutions are the Bessel functions I_0, K_0 which are regular at the origin and at infinity respectively:

$$\begin{aligned} x \to 0: &\quad I_0(x) \sim 1 + x^2/4\,, \quad K_0 \sim -\ln(x/2)\,, \\ x \to \infty: &\quad I_0 \sim (2\pi x)^{-1/2} e^x\,, \quad K_0 \sim (\pi/2x)^{1/2} e^{-x}\,. \end{aligned} \tag{9}$$

Consequently the Green's function is just

$$G(\rho, \rho') = I_0(m\rho_<)K_0(m\rho_>) \tag{10}$$

and Π is given by

$$\begin{aligned} \Pi(\rho) = -j\Big\{&\theta(\rho - a)K_0(m\rho)\int_0^a \rho' d\rho'\, I_0(m\rho') \\ &+ \theta(a-\rho)\Big[K_0(m\rho)\int_0^\rho \rho' d\rho'\, I_0(m\rho') + I_0(m\rho)\int_\rho^a \rho' d\rho' K_0(m\rho')\Big]\Big\}\,. \end{aligned} \tag{11}$$

The magnetic field therefore reads

$$\vec{B} = \vec{\nabla} \times \vec{A} = \hat{z}\nabla^2\Pi = \hat{z}j\theta(a-\rho) + \hat{z}m^2\Pi(\rho)\,. \tag{12}$$

The first term on the right of (12) is of course just $\vec{B}_0$, the usual Maxwell field. It is clear from (11) that $m^2\Pi$ vanishes in the $m \to 0$ limit since I_0 is regular and K_0 is logarithmic at vanishing argument (for fixed ρ).

We have thus established continuity of the Aharonov-Bohm effect as the range increases (for fixed ρ, a): the magnetic field reduces to its Maxwell value B_0 in the region $\rho < a$ and to 0 outside. What about the corrections? Clearly, since there is now a length m^{-1} in the problem, there are three possible cases (we assume as usual that the wave function is excluded from the solenoid, *i.e.* that $\rho > a$), depending on whether the range is shorter than a, between a and ρ or greater than ρ. The first two possibilities are physically irrelevant since we know from the absence of an observed Yukawa decay of the terrestrial magnetic field[3] (let alone astrophysical considerations) that the range is much larger than the laboratory scales involved in ρ. Nevertheless, we record $m^2\Pi$ in these two cases:

$$\begin{aligned} m\rho \gg 1 \gg ma: &\quad m^2\Pi \sim (\pi/8m\rho)^{1/2}(ma)^2 e^{-m\rho} \\ m\rho \gg ma \gg 1: &\quad m^2\Pi \sim (a/\rho)^{1/2} e^{-m(\rho-a)}\,. \end{aligned} \tag{13}$$

We now concentrate on the interesting case of long range: $1 \gg m\rho > ma$. There are now corrections to $\vec{B}_0$ both inside and outside the solenoid:

$$\begin{aligned} \Delta\vec{B} &= \hat{z}m^2\Pi \sim \theta(\rho - a)C + \theta(a-\rho)D\,, \qquad\qquad 1 \gg m\rho \gg ma \\ C &\equiv -(j/2)(ma)^2 \ln(m\rho/2) \\ D &\equiv (j/2)\left[-(1/2)(ma)^2 + (1/2)(m\rho)^2 - (ma)^2\ln(ma/2)\right]\,. \end{aligned} \tag{14}$$

The magnetic field "leakage", C, outside the solenoid is an extremely small fraction of B_0, namely of order $(1/2)(ma)^2\ln(m\rho/2)$; it represents the closing of the field lines at finite ($\sim m^{-1}$) distance. The fractional change in the internal field is also small, of fractional order $(1/2)(m\rho)^2$. Of course the coupling is through $\vec{\nabla} - ie\vec{A}$ and the Hamiltonian is a function of $\vec{A}$ rather than $\vec{B}$. Although the local exterior $\vec{B}$ is in any case extremely small, it is the non-local $\vec{A}^T$ which carries the interaction effect and which is *not* small.

The contributions, $\Delta\Phi$, to the flux Φ inside the observation circle ρ come from both regions. The interior contribution is given by

$$\Delta\Phi_i = 2\pi\int_0^a \rho' d\rho' D(\rho') \sim -(\pi/2)ja^2(ma)^2\ln(ma/2) \tag{15}$$

where we have only kept the leading, logarithmic, part. The leading exterior correction,

$$\begin{aligned} \Delta\Phi_e &= 2\pi\int_a^\rho \rho' d\rho' C(\rho') \\ &\sim (\pi/2)j(ma)^2\left(\rho^2\ln(m\rho/2) - a^2\ln(ma/2)\right) \end{aligned} \tag{16}$$

is the dominant new effect ($\Delta\Phi_i \sim (a/\rho)^2\Delta\Phi_e$). Now the Maxwell flux Φ_0 is of course $\Phi_0 \equiv \int \vec{B}_0 \cdot d\vec{S} = \pi a^2 j$, so that the fractional correction is

$$\Delta\Phi/\Phi_0 \sim (-1/2)(m\rho)^2\ln(m\rho/2) = (m\rho/2)^2\ln(2/m\rho)^2\,, \tag{17}$$

which is essentially quadratic in the small parameter $m\rho$. Note that varying ρ is equivalent to "varying" m. Equation (17) is our basic result for the correction to the usual effect due to finite (but large) range.

In the case of a finite apparatus, this result is modified. The logarithmic dependence upon ρ remains, but it is scaled by some factor of order of the size of the apparatus. In general, the correction must be evaluated numerically; however, for a circular flux line of radius R, the correction to the flux through a small circle of radius $\rho \ll R$ enclosing the flux line may be written in terms of elliptic integrals. The approximate result is

$$\Delta\Phi/\Phi_0 \sim (m\rho/2)^2\left\{\ln(R^2/\rho^2) + \mathcal{O}(1)\right\}\,. \tag{18}$$

We have seen that the non-local effects of a magnetic field in normal electrodynamics are simply ascribed to the fact that the physical gauge-invariant transverse vector potential $\vec{A}^T$ is a non-local function of $\vec{B}$, and that it is $\vec{A}^T$ which couples to the wave function. These facts are independent of the dynamics governing the A-field, and persist whatever its range. For the standard solenoid configuration, finite range does lead to a small $\vec{B}$-field "leakage" outside the solenoid, as seen in (14); for the physical situation in which $1 \gg m\rho > ma$, the additional flux is dominated by the contribution from the leakage field, although the interior field corrections also contribute. This does not mean that the leakage field is the cause of the effect here; indeed, one could even arrange in principle for the leakage flux addition to be an integer, in which case only the interior field would contribute, making the picture just as nonlocal as in the massless case. Thus, the standard "paradoxical" features of Aharonov-Bohm are reproduced by the massive, non-gauge, field. We also noted that, as might be expected, the $m \to 0$ limit yields the $m = 0$ results.

3. Experimental Bounds on the Photon Mass

We now show that some non-trivial limits on the range of the transverse photon can be obtained from a tabletop experiment. (Experimentally, limits on electric field range from electrostatics are a quite different story, based on deviations from the Gauss law. The best of these limits based on the longitudinal field is $m^{-1} \geq 3.1 \times 10^4$ Km.[5]) In the typical experiment, the magnetic field is constrained to a small "whisker". If the whisker consists of a type II superconductor, then the flux quantization of the individual vortex lines will assure that Φ_0 will be an integer multiple of 2π, so that there is a "null" effect in Maxwell theory. A naive application of the Ginzburg-Landau theory modified to include finite range electrodynamics still yields flux quantization and we conjecture that a more sophisticated treatment will yield a fractional change in the shape of the free energy as a function of B of order $(m\ell)^2$ where ℓ is the size of the flux vortex (ℓ^{-1} is the effective London photon "mass").

To obtain an experimental limit on the range m^{-1}, one must detect a phase shift which is some fraction ϵ of 2π. Fortunately, since Φ_0 does not contribute when the "solenoid" is a finite superconducting whisker, this means that the correction $\Delta\Phi$ gives the whole effect, so we must have

$$\Delta\Phi \geq 2\pi\epsilon. \tag{19}$$

By (17), this means that the value of $m\rho$ at which the effect is just observable is

$$(m\rho/2)^2 = (2\pi\epsilon/\Phi_0 A) \tag{20}$$

where A depends upon the detailed experimental configuration but is of order 1. We consider a whisker of radius $a \sim 0.1$ cm, a magnetic field of 10 tesla and a resolution $\epsilon \sim 10^{-3}$. This yields, as the longest range which could be observed,

$$m^{-1} \simeq 14\rho \text{ Km}, \tag{21}$$

where ρ is measured in cm. For a coherent beam of $\rho \sim 10$ cm, the observable ranges are then $\lesssim 10^2$ Km. Naturally there will be uncertainties due to, *e.g.*, stray fields which vary with the flux in the whisker. Even if we have been overly optimistic in choosing our parameters, it is nonetheless interesting that a reasonably large lower bound on the range of electrodynamics can be obtained from a precision experiment of small dimensions. An analogous experiment has already been performed by Tonomura.[1] However,the geometry and field strengths obtained are not optimal for our purposes.

It should be emphasized that the proposed experiment probes the deviation of the magnetic field from its classical value at short distances. No measurements are proposed at a distance scale of m^{-1}. The precision is obtained only because the is a null experiment for $m = 0$ combined with the use of large magnetic fields and the sensitivity obtainable by detecting phase shifts.

4. Acknowledgements

We thank Ya. Smorodinskii for asking the right questions, R. Schrieffer for discussions on flux quantization in massive electrodynamics and B. Heckel for providing estimates on some of the parameters. This work was supported by the U.S. Department of Energy under Contract DE-AS06-88ER-40423 and by NSF grant PHY88-04561. S.D. thanks Boston University for its hospitality while this work was begun.

5. References

1. Y. Aharonov and D. Bohm, *Phys. Rev.* **115** (1959) 485; W. Ehrenberg and R.E. Siday, *Proc. Phys. Soc., London* **B62** (1949) 8; C.N. Yang, *Proc. Int. Symp. Found. of Quantum Mechanics, Tokyo*, (1983); S. Olariu and I. Popescu, *Revs. Mod. Phys.* **57** (1985) 339; A. Tonomura et al., *Phys. Rev. Letts.* **56** (1986) 792.
2. L. Bass and E. Schrödinger, *Proc. Rev. Soc.* **A232** (1955) 1; E.C.G. Stueckelberg, *Helv. Phys. Acta.* **30** (1957) 209; D.G. Boulware and W. Gilbert, *Phys. Rev.* **126** (1962) 1563; S. Deser, *Ann. Inst. Henri Poincare* **16** (1972) 79.
3. A.S. Goldhaber and M.M. Nieto, *Revs. Mod. Phys.* **43** (1971) 277.
4. H. Van Dam and M. Veltman, *Nucl. Phys.* **B22** (1970) 297; Y. Iwasaki, *Phys. Rev.* **D2** (1970) 2225; V.I. Zaharov, *J.E.T.P. Lett.* **12** (1970) 312; D.G. Boulware and S. Deser, *Phys. Lett.* **40**B (1972) 227; *Phys. Rev.* **D6** (1972) 3368.
5. E. R. Williams, J. E. Faller, and H. A. Hill, *Phys. Rev. Letters* **26** (1972) 721.

THE ELECTROSTATIC AHARONOV–BOHM EFFECT

GIORGIO MATTEUCCI
Department of Physics, University of Bologna
via Irnerio 46, 40126 Bologna, Italia

and

GIULIO POZZI
Department of Materials Science, University of Lecce
via Arnesano, 73100 Lecce, Italia

ABSTRACT

The quantum phase difference introduced between the two coherent split parts of an electron beam enclosing a line of electric dipoles has several analogies with the Aharonov–Bohm effect, where the phase difference is produced by the magnetic flux of a solenoid. In both cases the space is multiple–connected by the presence of the line of electric dipoles or the solenoid, the lateral deflection of the electron beam vanishes and the effect can be detected by means of electron interferometry methods. In this paper the experimental results obtained so far by our group concerning the line of electric dipoles associated either to a bimetallic wire or to two parallel wires having opposite charges are presented and discussed.

1.Introduction

In their classical paper [1] on the significance of the electromagnetic potentials in quantum theory, Aharonov and Bohm proposed two different electron interference experiments in order to test their conclusions. The first of them, concerning the effect of the vector potential associated to a static magnetic field, since 1959 has stimulated a wealth of experimental work (see [2,3] for reviews), which shows the attempts of the experimentalists to satisfy the ever more stringent conditions required by the theoreticians, especially by those who do not believe in the effect. The latest, and perhaps ultimate, experiment in this series is the one carried out by Tonomura and coworkers [4] using superconducting toroidal magnets (see also Tonomura in these proceedings).

Much less attention has been paid to the second experiment, which seemed (and perhaps still is) beyond the grasp of even the most advanced technology. In Aharonov and Bohm's [1] original paper, a coherent electron beam EB is chopped and split into two parts at A, Fig. 1(a). Each part is allowed to enter a cilindrical metal tube, the electric potential of which is varied only when the wavepackets are well inside the tubes and thus do not experience the action of the electric field. The beams are then recombined at B to give an interference pattern, in plane P, which should show a phase difference due to the time-dependent scalar potential, because no force is ever exerted on the electron wavepackets.

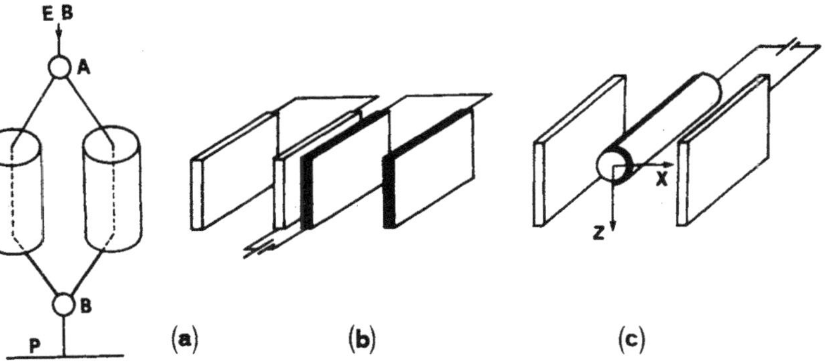

Fig.1 Set-ups for an electric phase shifting experiment: (a) two tubes; (b) parallel plates; (c) bimetallic wire.

In 1973 Boyer [5], in his considerations on the Aharonov-Bohm effect, noted that, if the experiment is carried out with *static* potentials, electrons experience retarding and accelerating fields at the entrance and exit of the two tubes, the net effect being that of a classical time lag between the two wavepackets, revealing itself again as constant phase difference introduced between them.

A demonstration of this electrostatic effect in an electron interference experiment has been carried out within a commercial electron microscope by using the electrostatic field associated to a bimetallic wire [6], whereas a realization closer to the original proposal was subsequently carried out by Schmid in a dedicated electron interferometer [7]. The electrostatic Aharonov–Bohm effect may also be responsible for the conductance fluctuations observed in a micrometric metal loop, where the transverse electric field has been used to tune the position, or phase, of the Aharonov–Bohm oscillations (see also Webb in these Proceedings) [8].

The aim of this paper is to illustrate the basic ideas underlying our experimental set-up and its following modifications, and to review the results obtained so far by our group using different experimental methods.

2. Experimental set-ups

If the two tubes of Fig. 1(a) are held at different electrostatic potential , the two beams experience a phase difference [5] given by:

$$\Delta\varphi = \frac{\pi \Delta V\, l}{\lambda E} \tag{1}$$

where ΔV is the potential difference between the two tubes of length l, λ is the electron wavelength, and E, in the nonrelativistic case, is the accelerating potential. As electrons do not suffer any lateral deflection, the outcome of the experiment should be a two-beam interference pattern, with interference fringes of spacing s, which are laterally displaced by an amount Δs given by $\Delta s = s\,\Delta\varphi/(2\pi)$ whereas the diffraction envelope is undeflected. However, in order to realize the experiment in this form, electron beams having a lateral coherence of at least a few hundreds of microns are required, i.e. about two orders of magnitude above the lateral coherence available in commercial electron microscopes, even when fitted with a field emission source.

In order to find experimental conditions for the realization of this experiment within an electron microscope, an alternative geometry could be used involving plates instead of tubes, Fig. 1(b). The main problem was to scale down the two inner plates to a micrometer scale, still maintaining a potential difference between them. The resulting solution was the following: a metallic wire of radius a was evaporated on one side with a thin layer of a different metal, Fig. 1(c). In this way, the contact potential difference ΔV between the two metals produces an electrostatic potential distribution around the wire given by

$$V(x,z) = \frac{\Delta V}{\pi} \arctan\left(\frac{2ax}{x^2+z^2-a^2}\right) \tag{2}$$

The following facts should be noted: if radius a of the wire tends to zero, while $a\Delta V = const$, the potential distribution becomes that of a line of dipoles and it can be easily verified [9,10] in this case that the electrons do not suffer any lateral deflection and leave the bimetallic wire with the same energy as when they enter its field. However, a constant phase difference given by

$$\Delta\varphi = \frac{4\pi a \Delta V}{\lambda E} \tag{3}$$

is introduced between the two split parts of the electron beam which travel on both sides of the wire. These same conclusions hold also for the wire with finite diameter. By comparing Eq. 3 with Eq. 1 it turns out that the device is equivalent to two tubes of length $4a$. For electrons accelerated at 100 kV, with a wire radius of 0.3 μm and a potential difference of 0.5 V, the phase difference amounts to 1.6 π.

This effect can be expected also on the basis of the following heuristic considerations: roughly speaking, the contact potential difference between the two metals causes a charge redistribution, Fig. 2 (a), in such a way that the resulting field is equivalent to that produced by two parallel linear charge densities of opposite sign (no net charge on the bimetallic wire) which are laterally displaced one with respect to the other, Fig. 2 (b). From the theory of the electron biprism [11,12] (whose electric field is equivalent to that of a single linear charge density) it is known that electrons passing on the same side of the biprism suffer a deflection proportional to the applied potential (or to the charge density) which is independent of their distance from the biprism. Moreover, from the wave point of view, the effect of the biprism on the electron wavefront is to introduce a phase shift proportional to $|x|$. Therefore, the bimetallic wire can be modelled by a system of two biprisms of opposite power.

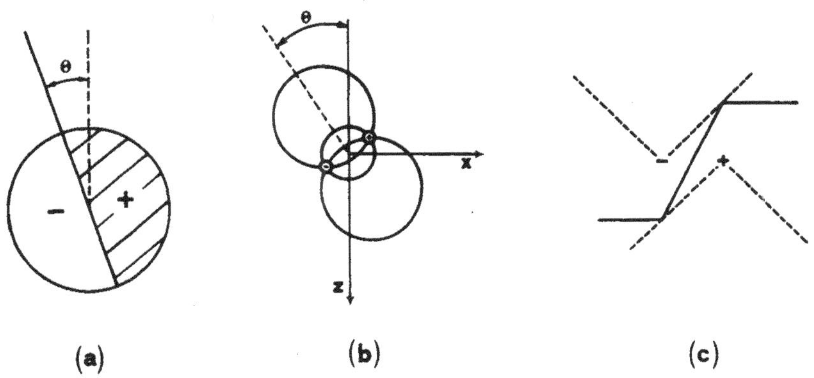

Fig. 2 (a) Schematic representation of the cross–section and charge distribution of the bimetallic wire; (b) electric field due to two lines having opposite charges; (c) phase shift due to the two equivalent configurations (a) and (b)

Fig. 2 (c) depicts the influence that each line of charge has separately on the phase of the impinging wavefront (dashed lines). The resulting phase shift (solid line) is simply given by the sum of the contributions of each biprism. Therefore the deflections produced on electrons passing on the left and right sides of the two lines of charge are exactly cancelled out, whereas a constant phase difference arises because of their separation. Taking into account the angle θ between the line of dipoles and the electron beam, Fig. 2 (a), the maximum phase difference given by Eq.3 is multiplied by the factor $\cos\theta$. In the region between the two lines of charge the phase shift varies linearly; this effect cannot be observed for the case of the bimetallic wire, because the wire itself acts as an impenetrable barrier for electrons, but can be observed in the case of two wires carrying opposite charges allowing thus the measurement of the total phase difference. This consideration has led us to develop our last device [9], namely two parallel conducting wires, held at opposite potentials by an external voltage supply. They act as a macroscopic line of dipoles with the additional advantage, with respect to the bimetallic wire, of controlling its strength.

3. Experimental methods and results.

A schematic drawing of the whole set-up of our first experiment [6] is shown in Fig. 3 (a). The coherent electron beam comimg from source S propagates to the biprism plane, located at the level of the selected area aperture of a Philips EM 400T electron microscope, equipped with a field emission gun. The electron biprism is a thin conducting wire W which can be charged with respect to the earthed plates P by means of an external voltage supply. The wire W was coated laterally for half of its length with a thin layer of gold, thus becoming a bimetallic biprism. The biprism wire splits the wavefront of the incoming beam and its electrostatic field produces a deflection and a subsequent overlapping in the plane OP below the wire, where a system of interference fringes will be observed.

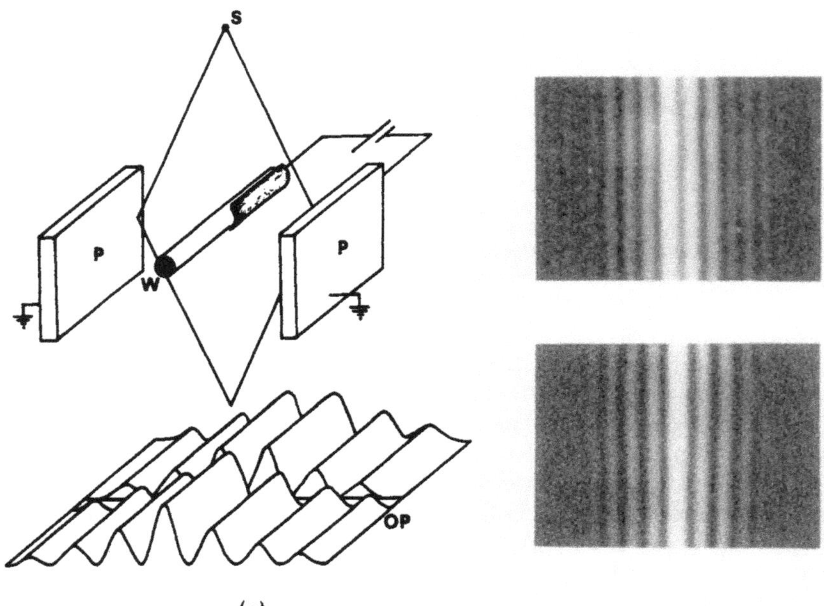

(a) (b)

Fig. 3 (a) Schematic set-up for the electron interference experiment. (b) Interference patterns corresponding to the coated (upper) and uncoated (lower) parts of the biprism wire.

The interference fringe systems of the wire recorded in correspondence with the coated and uncoated regions are shown in the upper and lower part of Fig. 3 (b). The displacement of the interference fringe system due to the constant phase difference with respect to the unperturbed diffraction envelope is clearly visible through a change of symmetry of the pattern: the central maximum of the uncoated part becomes nearly a minimum in the coated one, thus indicating a phase difference of about π.

In this experiment, the phase difference cannot be varied since the rotation of the wire around its axis is not allowed. The dependence of the effect on the angle θ cannot be revealed by this set-up. The second experiment has been carried out to exploit this dependence. A bimetallic wire was mounted in a standard microscope aperture, and inserted in the normal specimen position, where it could be rotated around its axis by means of the goniometer stage.

Observations were carried out both by interference electron microscopy [13] and diffraction methods [14]. In particular in the second case, even if the wire has no net charge, as it should have in the ideal case, it is, nonetheless, still possible to detect the effect by monitoring the changes in the faint interference image present in the region of the geometric shadow of the wire, see Fig. 4. These interference fringes are due to the coherent overlapping of the two tails of the Fresnel images concerning the two half-planes on both sides of the wire. It should be noted that, owing to a charging up of the wire due to secondary emission of electrons caused by the impinging electron beam, also biprism-like interference fringes are observable (see f.i. Fig. 3 (b)), which provide a striking evidence of the phase difference.

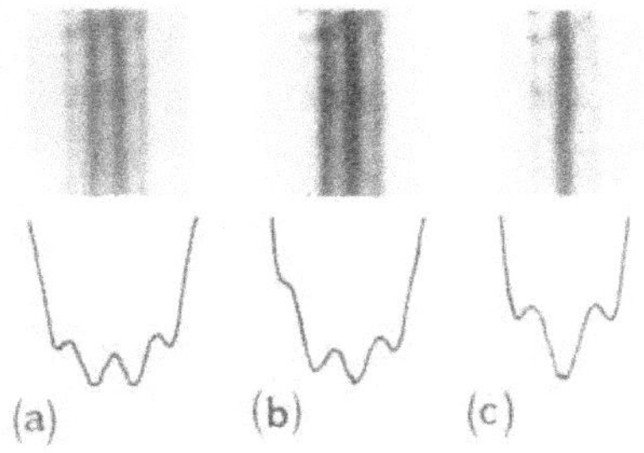

Fig: 4 Fresnel patterns showing the effect of the phase difference introduced by the bimetallic wire on the interference fringe system present within the shadow image. Rotation angle (a) $-24°$; (b) $0°$; (c) $24°$. Microdensitometer traces below the photographs are drawn at higher magnification. (Courtesy American Physical Society).

Our last experiment has been carried out by inserting in a special specimen holder, provided with electrical contacts, a macroscopic electric dipole D: two platinum wires soldered on platinum apertures, are superimposed and electrically insulated [9]. The wires were oriented parallelly and the whole assembly inserted in the microscope. The lower wire was earthed, whereas the upper one could be biased by means of an external voltage supply. Observations were carried out by means of interference electron microscopy, the instrument being equipped also with a conventional electron biprism, Fig. 5 (a). Quite surprisingly, by applying 24 V to the upper wire, the condition could be achieved when the two biprisms carried opposite charges. This fact was easily detectable by observing the relative alignment of the outer system of interference fringes.

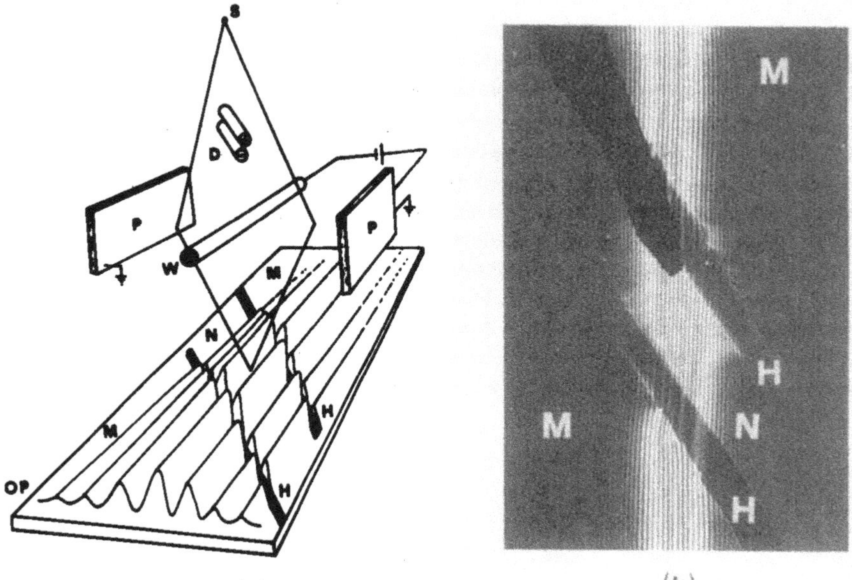

(a) (b)

Fig. 5 (a) Schematic set-up for the interference pattern formation by means of an additional electron biprism acting as interferometry device. The essential elements are the same as in Fig. 3 (a). (b) Interference pattern

The interference region in the OP plane due to the lower biprism is in fact in this arrangement divided into three parts by the projected shadows H of the two overlapped wires. In the outer regions M, provided the wires are carrying opposite charges, two wavefronts with the same phase shift overlap, and therefore a symmetric fringe system is expected, equal to that observable with the biprism alone. In the region N between the shadows, the waves coming from the opposite sides of the dipole overlap. Thus, electrons experience different phase shifts and a resulting phase difference arises. The net effect is a lateral displacement of the interference fringes with respect to the unperturbed system or to the diffraction envelope. Owing to the different geometry and to the value of the electrostatic potential of the dipole, the effect is much larger than in the case of the bimetallic wire. In our arrangement, by tilting the asssembly by an angle of 1°, it is possible to observe directly on the screen the changes in the interference fringe system between the two shadows, whereas the outer fringes are stationary.

The static recording on a photographic plate does not adequately render the observed phenomena, Fig. 5 (b), however, by tilting the macroscopic dipole in order to obtain an equivalent configuration of the two lines of charge as in Fig. 2 (b), electrons are allowed to travel also in the region between the two wires. It is expected to observe not a single shadow H as in Fig. 5, but the projected shadows of the two wires.

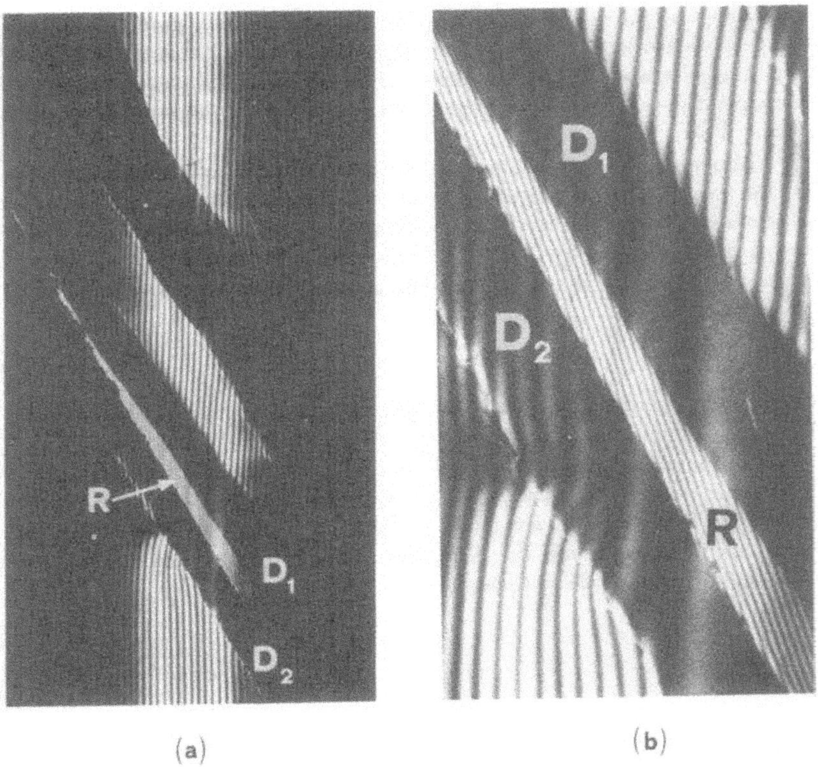

Fig. 6 (a) Interference pattern with the dipole rotated so that the two wires D_1 and D_2 are no longer overlapping. (b) Magnified view of the region R between D_1 and D_2 of (a).

In the interference pattern shown in Fig.6 (a) the shadows D_1 and D_2 of the two wires are clearly distinguishable together with the region R between them, shown enlarged in Fig. 6 (b). The linear increase of the phase is evident and it can be estimated that the phase difference is larger than 40π .

4. Conclusions.

The basic features and the obtained results of the experiments carried out by our group on the electrostatic Aharonov–Bohm effect have been briefly summarized.

Recently it has also been pointed out that asymmetry effects in the Fresnel patterns are due to the non–cylindrical shape of the wire [15]. A thorough theoretical and experimental analysis of the whole issue shows that by means of electron interference microscopy it is possible to ascertain that the contact potential difference is mainly responsible for the observed effects, which cannot be accounted for by the non–cylindrical shape of the wire alone [16]. In any case, the experiments with the macroscopic dipole are a definite demonstration of the existence of the electrostatic Aharonov–Bohm effect.

Although this effect has not the same paradoxical content as the magnetic effect, and can also be accounted for by means of a classical lag [5,10], it represents a clear demonstration of the fact that a change of phase for the quantum state of the electrons can be induced by their interaction with a particular class of fields, which have the peculiarity of changing the semiclassical action in a way independent of the trajectory [17].

5. Acknowledgements. Useful discussions with Prof. G. F. Missiroli and the skillful technical assistance of S. Patuelli are gratefully acknowledged. This work has been supported by funds from Ministero della Pubblica Istruzione, Italia. Both authors are members of the Center for Electron Microscopy, Department of Physics, University of Bologna, and are affiliated with the Consorzio Interuniversitario di Struttura della Materia and the Gruppo Nazionale di Struttura della Materia of the Consiglio Nazionale delle Ricerche.

6. References.

1. Y. Aharonov and D. Bohm, *Phys. Rev.* **115** (1959) 485.
2. S. Olariu and I. I. Popescu, *Rev. Mod. Phys.* **57** (1985) 339.
3. M. Peshkin and A. Tonomura, *The Aharonov–Bohm Effect* (Springer–Verlag, Berlin, 1989).

4. A. Tonomura, N. Osakabe, T. Matsuda, T. Kawasaki, J. Endo, S. Yano and H. Yamada, *Phys. Rev. Lett.* **56** (1986) 792.
5. T. H. Boyer,*Phys. Rev.* **D 8** (1973) 1679.
6. G. Matteucci, G. F. Missiroli and G. Pozzi, *Ultramicroscopy* **10** (1982) 247.
7. H. Schmid, in *Proceedings of the Eighth European Congress on Electron Microscopy, Budapest, Hungary, 1984* edited by Á. Csanády *et al.* (Programme Committee of the Congress, Budapest, 1984), p.285.
8. S. Washburn, H. Schmid, D.P. Kern and R. Webb, *Phys. Rev. Lett.* **59** 1791.
9. G. Matteucci and G. Pozzi, in *Quantum Uncertainties* edited by W. H. Honig *et al.* (Plenum Publishing Corporation,New York, 1987), p. 297.
10. T. H. Boyer, *Il Nuovo Cimento* **100B** (1987) 685.
11. G. Möllenstedt and H. Düker, *Z. Phys.* **145** (1956) 377.
12. G. F. Missiroli, G. Pozzi and U. Valdrè, *J. Phys. E* **14** (1981) 649.
13. G. Matteucci, G. F. Missiroli, M. Porrini and G. Pozzi, in *Proceedings of the International Symposium on Foundations of Quantum Mechanics,* edited by S. Kamefuchi *et al.* (Physical Society of Japan, Tokyo, 1984), p. 39.
14. G. Matteucci and G. Pozzi, *Phys. Rev. Lett.* **54** (1985) 2469.
15. A. Oshita, Y. Numata, Y. Mametani and H. Tomita, *J. Electron Microscopy* **37** (1988) 47.
16. G. Matteucci, F. F. Medina and G. Pozzi, *to be published.*
17. Y. Aharonov, in *Proceedings of the International Symposium on Foundations of Quantum Mechanics,* edited by S. Kamefuchi *et al.* (Physical Society of Japan, Tokyo, 1984), p. 10.

SPIN AND THE AHARONOV-BOHM EFFECT

C.R. Hagen
Department of Physics and Astronomy
University of Rochester
Rochester, New York 14627

Abstract

The Aharonov-Bohm scattering amplitude is calculated for a spin one-half particle. It is shown that the Zeeman interaction with the flux tube modifies the spin zero result in a way which can be experimentally detected by the use of polarized beams. States which are singular (but normalizable) at the position of the flux tube play an essential role in the calculation. They are also shown to have a remarkable effect upon the second virial coefficient of a gas of flux carrying fermions. In particular, the latter is found to have discontinuities at all even, nonzero values of the flux parameter.

I. Introduction

Over the past thirty years there has been amassed a significant body of experimental data which support the Aharonov-Bohm (AB) effect as a real physical phenomenon. This support has generally been in the nature of shifted interference fringes when a beam of particles is separated into two parts which then travel on opposite sides of a thin solenoid. The effect in this case is particularly simple - there is a phase difference between the two parts of the beam which is essentially just the flux in the solenoid. Since the beam is never allowed to penetrate into the region of nonzero magnetic field, there is no dependence on the spin of the particles.

On the other hand, the main part of the original AB paper[1] consisted of a calculation of the scattering of particles by a flux tube of arbitrarily small radius. Although it is natural that that calculation was carried out within the framework of a nonrelativistic spinless wave equation (i.e., the Schrödinger equation), the fact is that the only elementary particles generally available for such scattering experiments all have spin one-half! Furthermore, it is not clear (since the particles can in principle penetrate into the flux tube) that the effect of spin can safely be neglected. It will in fact be shown here that there are indeed observable effects associated with the spin degree of freedom. These can be observed either in the aforementioned scattering by a flux tube or by considering the equation of state of a gas of flux carrying fermions.

II. Spin and the AB Amplitude

Since it requires no additional effort to include relativistic kinematics (as well as spin), it is convenient to begin the calculation with the Dirac equation. There being no dependence on the direction of the flux tube, one is evidently dealing with the Dirac equation of a mass M particle

$$E\psi = [M\beta + \beta\gamma \cdot \Pi]\psi \tag{1}$$

in two spatial dimensions. In this case the matrices β and $\beta\gamma_i$ are conveniently defined in terms of the Pauli spin matrices by

$$\beta = \sigma_3$$
$$\beta\gamma_i = (\sigma_1, s\sigma_2)$$

and s is twice the spin value (+1 for spin "up" and -1 for spin "down"). This approach differs from the usual one which selects a particular value for s and has the advantage of making the results obtained more useful and transparent. The quantity Π_i is given by

$$\Pi_i = \frac{1}{i}\partial_i - eA_i$$

where the potential A_i is related to the magnetic field in the usual way

$$H = \nabla \times A.$$

Ideally, one is interested in the limit of a flux tube with finite flux but arbitrarily small dimensions, i.e.

$$eH = -\frac{\alpha}{r}\delta(r)$$

with r being the two dimensional radius vector. The corresponding form of A_i in the Coulomb gauge is

$$eA_i = \alpha\epsilon_{ij}r_j/r^2$$

where ϵ_{ij} is the usual Levi-Civita tensor. The trouble with this choice is that (1) implies

$$\begin{aligned}(E^2 - M^2)\psi &= -\gamma\Pi\gamma\Pi\psi \\ &= [\Pi^2 + \alpha s\sigma_3\frac{1}{r}\delta(r)]\psi.\end{aligned} \tag{2}$$

When (2) is then separated into radial and angular differential equations one has a Bessel equation for the radial coordinate with a delta function at $r = 0$. Since $r = 0$ is already a singular point of Bessel's equation, one is clearly faced with an ill-defined mathematical problem.

There have been attempts to deal with this complication. Alford and Wilczek[2] have chosen to impose the boundary condition that the upper component of ψ should be regular at the origin. Gerbert and Jackiw[3] have suggested a more general boundary condition which introduces a new parameter into the problem. This has been applied to the calculation under consideration here by Gerbert[4]. Unfortunately, none of these approaches is compelling. Regularity of a particular Dirac component is not an obvious thing, nor is it clear why there should be an undetermined parameter in the calculation as in refs. 3 and 4.

However, there is available a physical formulation which eliminates the mathematical difficulties inherent in Eq. (2)[5]. This is realized by allowing H to be independent of angle but otherwise quite arbitrary for $r < R$. For $r > R$ the H field is assumed to be zero. The Dirac equation is now mathematically well defined. The problem posed here is then considered to be the limit of this regularized version in the limit $R \to 0$ subject to the condition that

$$e \int H r dr = -\alpha.$$

In practice the simplest model is one in which one puts the H field on a ring of radius R, i.e.

$$eH = -\frac{\alpha}{R}\delta(r - R)$$

so that

$$\begin{aligned} eA_i &= \alpha \epsilon_{ij} r_j / r^2 \qquad & r > R \\ &= 0 \qquad & r < R. \end{aligned}$$

This may be thought to leave open the question of model independence. In particular does the result depend on the "hole" in the H field? Can other forms of H yield different results for $R \to 0$? Physically, it would seem that they should not. It is gratifying that the mathematics (Appendix of ref. 5) confirms this intuitive view.

The calculation can now readily be performed. Expansion of the upper component ψ_1 via

$$\psi_1 = \sum_{-\infty}^{\infty} a_m f_m(r) e^{im\phi}$$

yields the equations

$$\begin{aligned} &[\frac{1}{r}\frac{d}{dr}r\frac{d}{dr} - \frac{m^2}{r^2} + k^2] f_m(r) = 0 \qquad & r < R \\ &[\frac{1}{r}\frac{d}{dr}r\frac{d}{dr} - \frac{(m+\alpha)^2}{r^2} + k^2] f_m(r) = 0 \qquad & r > R \end{aligned}$$

where $k^2 = E^2 - M^2$. The delta function is then taken into account by means of the continuity relations

$$\begin{aligned} f_m(R - \epsilon) &= f_m(R + \epsilon) \\ R\frac{d}{dr} f_m |_{R-\epsilon}^{R+\epsilon} &= \alpha s f_m(R). \end{aligned} \tag{3}$$

In these terms the problem has been cast into a form which should be familiar to any graduate student. Namely, what is the solution of the Schrödinger equation for a delta function potential?

The solution is obtained by writing

$$f_m(r) = C_m J_{|m|}(kr) \qquad r < R$$

and

$$f_m(r) = A_m J_{|m+\alpha|}(kr) + B_m J_{-|m+\alpha|}(kr) \qquad r > R\ .$$

Equation (3) yields that the (unnormalized) $f_m(r)$ for $r > R$ is

$$\begin{aligned} f_m(r) = (kR)^{|m|}\{(kR)^{-|m+\alpha|}\left[\frac{1}{2} + \frac{|m| + \alpha s}{2|m+\alpha|}\right] J_{|m+\alpha|}(kr) \\ + (kR)^{|m+\alpha|}\left[\frac{1}{2} - \frac{|m| + \alpha s}{2|m+\alpha|}\right] J_{-|m+\alpha|}(kr)\}. \end{aligned} \tag{4}$$

Upon taking $R \to 0$ one clearly finds that the $J_{-|m+\alpha|}(kr)$ drops out unless

$$|m + \alpha| = -|m| - \alpha s. \tag{5}$$

Upon going to the next higher power of kR in the coefficient of the $J_{|m+\alpha|}(kr)$ term in (4) one finds (when (5) is satisfied) that

$$f_m(r) = (kR)^{|m|}\{D_m (kR)^{|m|+\alpha s+2} J_{|m+\alpha|}(kr) + (kR)^{-|m|-\alpha s} J_{-|m+\alpha|}(kr)\}$$

where $D_m \neq 0$. Thus for $R \to 0$ it follows that

$$f_m(r) \longrightarrow J_{-|m+\alpha|}(kr)$$

provided that

$$|m| + \alpha s + 1 > 0 \tag{6}$$

which upon using (5) is seen to imply

$$|m + \alpha| < 1. \tag{7}$$

This is a remarkable result! In all cases the correct solution is the regular one (reminiscent of ref. 2) except when (5) and (6) are simultaneously satisfied. In the latter case it is the irregular solution which obtains. But even in that case the solution is normalizable because of (7)! That the solution should remain normalizable without the necessity of imposing this condition *ab initio* is indeed gratifying.

To extract the significance of (5) and (6) it is convenient to write

$$\alpha = N + \beta$$

where N is an integer and

$$0 \leq \beta < 1\ .$$

Then one can show that $f_m(r)$ is always $J_{|m+\alpha|}(kr)$ unless

$$a)\ m = -N, N \geq 0, s = -1;\ or$$
$$b)\ m = -N-1, N+1 \leq 0, s = +1.$$

It is easily seen that singular states occur only when $\alpha s < 0$. This is intuitively very reasonable since in that case the delta function serves as an attractive potential which in turn gives it the capability of making the solution more concentrated at the origin than is the regular solution $J_{|m+\alpha|}(kr)$.

One can now construct the entire amplitude in the standard way[1,5]. For a wave incident from the right ψ has the asymptotic form

$$\psi = \begin{pmatrix} 1 \\ -\frac{k}{E+M} \end{pmatrix} e^{-ikr\cos\phi - i\alpha\phi} + \begin{pmatrix} 1 \\ \frac{k}{E+M} e^{is\phi} \end{pmatrix} r^{-1/2} e^{ikr} f(\phi)$$

where

$$f(\phi) = -\left(\frac{i}{2\pi k}\right)^{1/2} \frac{\sin \pi\alpha e^{-iN(\phi-\pi)}}{\cos\phi/2} e^{-i\phi/2} \left[\epsilon(\alpha s) e^{-i\phi\epsilon(s)\theta(-\alpha s)}\right] .$$

The step function $\theta(x)$ is given by

$$\theta(x) = \frac{1}{2}\left(1 + \frac{x}{|x|}\right)$$

and the alternating function by

$$\epsilon(x) = x/|x| .$$

It is to be noted that the function $f(\phi)$ differs from the spinless result only by the factor in square brackets, a remarkable fact which leads to the immediate conclusion that for unpolarized particles no effect is observed. In the case in which the beam is polarized in the direction $\vec{n}$, however, the cross section becomes

$$\frac{d\sigma}{d\phi} = \frac{\sin^2 \pi\alpha}{2\pi k \cos^2\phi/2}[1 - (n \times \hat{z})^2 \cos^2\phi/2]$$

where $\hat{z}$ is a unit vector parallel to the flux tube. In the optimal case of $\vec{n}$ and $\hat{z}$ being perpendicular there occurs a complete cancellation in the cross section for backward scattering (i.e., $\phi = 0$). Measurement of such effects and thus the confirmation of the existence of singular states may well be experimentally feasible.

III. The Second Virial Coefficient

Having demonstrated that the existence of irregular states can have an effect on the scattering of polarized beams, it is natural to raise the issue as to whether there exist other physical phenomena where their presence can be detected. Since the second virial coefficient of a system of flux carrying spinless particles is known[6] to be amenable to calculation, it would seem that this offers an ideal area in which to study the effects of spin (and thus irregular states)[7].

By way of introduction it should be remarked that the second virial coefficient is the second term in the equation of state of a system when expanded in powers of the density. Specifically

$$PV = NkT\left[1 + B_2(T)\frac{N}{V} + \ldots\right]$$

where P, V, T are the pressure, volume, and temperature respectively and N is the number of particles. One can show that[7]

$$B_2(T) = \frac{V}{2} - 2\lambda_T^2 Z_{2\mathrm{rel}}$$

where

$$\lambda_T^2 = 2\pi/MkT$$

and $Z_{2\,\mathrm{rel}}$ is the two body partition function (after the separation of the center of mass coordinate). In order to calculate this latter quantity it is helpful to make the spectrum discrete. Thus one includes an external harmonic oscillator potential to obtain the two body Hamiltonian

$$H = \left(\frac{P^2}{4M} + M\omega^2\right) + \frac{1}{M}\left(p - \frac{\alpha\bar{r}}{r^2}\right)^2 + \frac{1}{4}M\omega^2 r^2 + \frac{\alpha}{2M}(s_1 + s_2)\frac{1}{r}\,\delta(r) \ .$$

where (P, R) and (p, r) are the center of mass and relative coordinates respectively and $\bar{r}_i \equiv \epsilon_{ij} r_j$. The desired coefficient $B_2(\alpha, T)$ is then obtained in the $\omega = 0$ limit.

As in the case of AB scattering the magnetic field is placed on a ring of radius R and the limit $R \to 0$ taken. Not surprisingly, it is found that irregular states occur as seen in the formula for the energy levels

$$E^{\mathrm{rel}}_{n,m} = (2n + 1 \pm |m + \alpha|)\omega \tag{8}$$

where $n \geq 0$ is an integer. The + (-) sign refers to solutions which are regular (irregular) at the origin with the minus sign applying only for those cases enumerated in the preceding section, i.e.,

a) $m = -N,\ N \geq 0,\ s = -1$; or

b) $m = -N - 1,\ N + 1 \geq 0,\ s = +1$.

It is clear from (8) that there are an infinite number of states which are singular, these being spanned by the index n for that particular value of m for which irregular states are allowed.

Before summarizing the results some observations are in order. Clearly, $s_1 + s_2$ can assume the values $\pm 2, 0$. For $s_1 + s_2 = 0$ no singular states can occur. Also, it is convenient to examine separately the spin symmetric triplet states and the spin antisymmetric singlet state. Since these particles are fermions, the former set must have spatially antisymmetric wavefunctions while those of the latter are symmetric. The complete $B_2(\alpha, T)$ is obtained by averaging over the four states with the result[7]

$$B_2(\alpha,T) = \frac{1}{4}\,\lambda_T^2\left(\pm\frac{1}{2}+\beta-2\beta^2\right)\ \text{for}\ N\begin{pmatrix}\text{even}\\ \text{odd}\end{pmatrix} \tag{9}$$

for $N \geq 0$. The result for negative N is obtained from

$$B_2(\alpha,T) = B_2(-\alpha,T)\ .$$

It is of interest to note that had the irregular states been ignored one would have obtained B_2 (for odd N) $= \lambda_T^2/4(-\frac{1}{2}+3\beta-2\beta^2)$.

The most striking feature of these results is the fact that discontinuities occur for even values of N. This is in addition to the cusps which were already noted in ref. 6 in the context of the spinless boson case. Figure 1 shows the results obtained in Eq. (9) (solid curve) and the corresponding spin zero case (dotted curve). Figure 2 again displays the results (9) (solid curve) with the dotted curve in this case

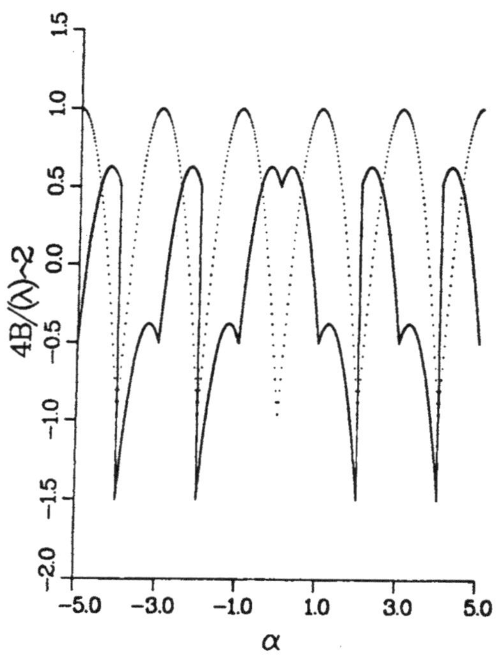

Figure 1. The fermionic $B_2(\alpha, T)$ (solid curve) along with the Arovas *et al.* result (dotted curve) for the spinless case.

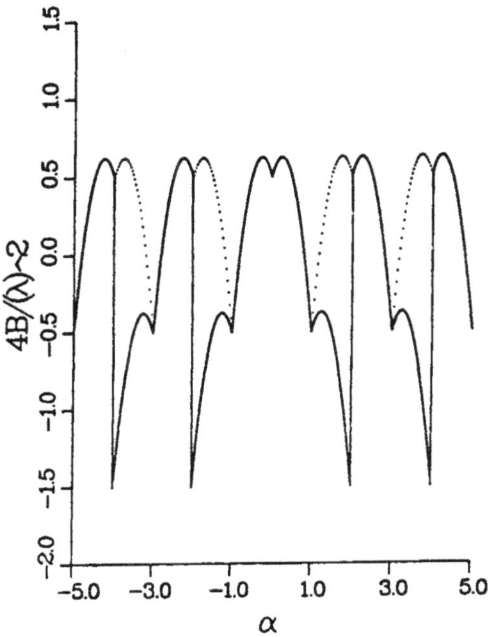

Figure 2. The fermionic second virial coefficient $B_2(\alpha, T)$ (solid curve) plotted as a function of α along with the result (dotted curve) when the irregular states are ignored.

giving the results when singular states are ignored. This clearly shows that the discontinuities in $B_2(\alpha, T)$ are a consequence of the presence of singular states.

A physical picture of what is going on at the discontinuities is easily obtained. Since singular states can only occur in the triplet configuration, one has to consider antisymmetric (i.e., m odd) spatial states. As α approaches a positive N value from below one clearly sees that m can refer to a singular state $m = -N$ (with m also being odd) only if N is even. As it approaches that even N the state becomes more and more singular. Thus the gas should compress (i.e., $B_2(\alpha, T)$ should decrease). As α exceeds that even integer N, the singular state becomes disallowed and the particles must make transitions to regular (i.e., less localized) states. A discontinuous expansion is the result.

Thus far it has been assumed here that the case of interest is one in which there is a mixture of equal numbers of positive ($s = 1$) and negative ($s = -1$) spins. This, of course, is the expected situation when the system under consideration can be idealized to a plane in a three-dimensional world of unpolarized spin-one-half objects. Since in two dimensions there is no compelling reason that both spin values must occur, one could perform the same calculation of the second virial coefficient assuming that s_1 and s_2 can only take a single value. In that case one readily obtains from the preceding analysis that for $s_1, s_2 < 0$,

$$B_2(\alpha,T) = \frac{1}{4}\lambda_T^2 \begin{cases} 1-2\beta^2, & \text{N even,} \\ 1-2(\beta-1)^2, & \text{N odd, N} < 0, \\ 1-2(\beta+1)^2, & \text{N odd, N} > 0. \end{cases} \qquad (10)$$

The corresponding result for $s_1, s_2 > 0$ is found from this expression by allowing $\alpha \to -\alpha$. The result (10) is plotted in Fig. 3 with the result being that discontinuities occur now only at the positive even integers. The discontinuities are 4 times as large as those observed in Figs. 1 and 2, an expected result in light of the fact that in the latter case only one-quarter of the particle pairs are in the spin state which can contribute to a discontinuity in $B_2(\alpha,T)$.

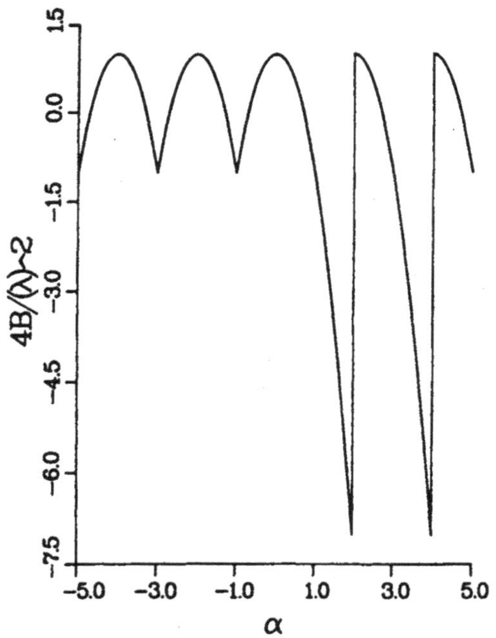

Figure 3. The fermionic $B_2(\alpha,T)$ for the case of polarized spins ($s = -1$).

IV. Conclusion

The calculations presented here can be viewed from different perspectives. At one level they are seen to fill a significant gap in the literature of the AB effect by clearly displaying the effect of spin. A notable bonus in that study has been the detection of an important role for the solutions of the wave equation which are singular at the origin. It is in fact very tempting to make the general observation that the rejection of singular wave functions solely on the basis of their singularity should be viewed with some suspicion. It is far more likely that such wave functions are simply not solutions of the differential equation at the singular point. In unusual circumstances, such as encountered here, they can, however, become relevant.

Another way in which the results presented here can be viewed is within the context of fractional statistics. There have been many attempts to apply the techniques of this approach to high temperature superconductivity since the latter is believed to be capable of being idealized to a field theory in two spatial dimensions. According to this view the α or flux parameter interpolates between bosons and fermions. While this is tenable at the level of spinless wave mechanics (not quantum field theory![8]) the work presented here shows that there is a huge difference between the equation of state of a spinless boson system and a fermionic one with spin (i.e., the latter has discontinuities in $B_2(\alpha, T)$). Of course, in two spatial dimensions fermions without spin can be postulated and the fractional statistics view given new meaning. The question must arise, however, as to whether such particles are meaningful since spin is unquestionably present in the three dimensional system from which this two dimensional idealization is derived. Certainly what is true is that considerably greater care is required in the statements which seek to link bosonic and fermionic systems through a flux parameter.

Acknowledgement

This work is supported by the U.S. Department of Energy Contract No. DE-AC02-76ER13065.

REFERENCES

1. Y. Aharonov and D. Bohm, *Phys. Rev.* **115**, 485 (1959).
2. M. G. Alford and F. Wilczek, *Phys. Rev. Lett.* **62**, 1971 (1989).
3. Ph. Gerbert and R. Jackiw, *MIT Report* No. CTP-1594, 1988.
4. Ph. Gerbert, *Phys. Rev.* **D40**, 1346 (1989).
5. C. R. Hagen, *Phys. Rev. Lett.* **64**, 503 (1990).
6. D. P. Arovas, P. Schrieffer, F. Wilczek, and A. Zee, *Nuc. Phys.* **B251**, 117 (1985).
7. This calculation of $B_2(\alpha, T)$ is from a recent paper by T. Blum, C. R. Hagen, and S. Ramaswamy, *Phys. Rev. Lett.* **64**, 709 (1990).
8. C. R. Hagen, *Phys. Rev. Lett.* **63**, 1025 (1989).

Analogue of the Aharonov-Bohm Effect for Black Holes and Strings

Steven B. Giddings[†‡]

Lyman Laboratory of Physics
Harvard University
Cambridge, MA 02138

Abstract

It has recently been proposed[1] that black holes may have quantum hair. This may occur in theories with an antisymmetric-tensor axion coupled to gravity, in which case the black hole carries an axion charge that can be measured by a string in analogy to the Aharonov-Bohm effect.

In this lecture I'll describe a new charge that can be carried by black holes; this charge is associated to axions described by a rank-three antisymmetric tensor field strength. I will also discuss a process by which this charge may be observed in a higher dimensional analogue of the Aharonov-Bohm effect. This process involves 'lassoing' an axion-charged black hole with a string, either fundamental or cosmic. This scenario was originally published in a paper by Mark Bowick, Jeff Harvey, Gary Horowitz, Andy Strominger, and me.[1]In this lecture I will begin by reviewing some of the basics of two-form axions, after which the existence of axion charge on black holes will be described. This is followed by a discussion of the observation of this charge by means of an Aharonov-Bohm-like effect using a string.

This new charge that black holes may carry is associated with axions described in unconventional fashion by a rank-two antisymmetric tensor potential, $B_{\mu\nu}$. Such axions generically arise in superstring theory and in some supergravity theories; *a priori* we cannot rule out the possibility that fundamental axions (if they exist) are of this type.

---

† Supported in part by the Harvard Society of Fellows, and by NSF contract PHY-87-14654.

‡ Address after July 1, 1990: Department of Physics, University of California, Santa Barbara, CA 93106.

The kinematics and dynamics of such axions are analogous to those of electromagnetism. In electromagnetism the potential $A_\mu(x)$ gives rise to a field strength $F_{\mu\nu} = \partial_\mu A_\nu - \partial_\nu A_\mu$ which is gauge invariant under transformations of the form $\delta A_\mu = \partial_\mu \Lambda$. Similarly the field strength associated to $B_{\mu\nu}$ is $H_{\lambda\mu\nu} = \partial_\lambda B_{\mu\nu} + \partial_\mu B_{\nu\lambda} + \partial_\nu B_{\lambda\mu}$, and $H_{\lambda\mu\nu}$ is unchanged by the gauge transformation $\delta B_{\mu\nu} = \partial_\mu \Lambda_\nu - \partial_\nu \Lambda_\mu$. Furthermore, in electromagnetism half of Maxwell's equations are given by the Bianchi identity $\partial_\lambda F_{\mu\nu} + \partial_\mu F_{\nu\lambda} + \partial_\nu F_{\lambda\mu} = 0$, to which there is an analogue,

$$\partial_\kappa H_{\lambda\mu\nu} + \partial_\lambda H_{\mu\nu\kappa} + \partial_\mu H_{\nu\kappa\lambda} + \partial_\nu H_{\kappa\lambda\mu} = 0 \; . \tag{1}$$

The dynamics of the theories arises from kinetic actions of the form $S_{\mathbf{Maxwell}} = \int d^4x \sqrt{g} F_{\mu\nu} F^{\mu\nu}$ for electromagnetism, and

$$S = f^2 \int d^4x \sqrt{g} H_{\lambda\mu\nu} H^{\lambda\mu\nu} \tag{2}$$

(where f is a constant) in the case of axions. The equations of motion derived from these actions are

$$0 = \frac{\delta S}{\delta A_\mu} \Rightarrow \nabla^\mu F_{\mu\nu} = 0 \tag{3}$$

for electromagnetism, and likewise

$$0 = \frac{\delta S}{\delta B_{\mu\nu}} \Rightarrow \nabla^\lambda H_{\lambda\mu\nu} = 0 \tag{4}$$

for axions.

Before proceeding, it is worthwhile reviewing why $H_{\lambda\mu\nu}$ is referred to as an axion, which is usually thought of as a (pseudo)-scalar field a with a Peccei-Quinn symmetry $a \to a + \text{const}$. To see the connection, define a current $j^\kappa = \epsilon^{\kappa\lambda\mu\nu} H_{\lambda\mu\nu}$, where $\epsilon^{\kappa\lambda\mu\nu}$ is the ordinary Levi-Civita (antisymmetric) tensor. The equation of motion (4) is then easily seen to imply that the curl of j, $\partial_\mu j_\nu - \partial_\nu j_\mu$, vanishes, so that (at least locally) j is a gradient, $j_\mu = \partial_\mu a$. Thus on shell the theory with potential $B_{\mu\nu}$ is equivalent to a pseudo-scalar field a, the axion. To further emphasize equivalence to the ordinary axion, note that the Peccei-Quinn preserving couplings of the axion to gauge theories and to matter may also be written in terms of H. The coupling to gauge potentials is

$$\int d^4x \sqrt{g} a \epsilon^{\kappa\lambda\mu\nu} F_{\kappa\lambda} F_{\mu\nu} \leftrightarrow \int d^4x \sqrt{g} H_{\lambda\mu\nu} \Omega_{\mathrm{CS}}^{\lambda\mu\nu} \tag{5}$$

where

$$\Omega_{\rm CS}^{\lambda\mu\nu} = A^\lambda F^{\mu\nu} + \frac{2}{3}A^\lambda A^\mu A^\nu + \text{cyclic permutations} \tag{6}$$

is the Chern-Simons form for the gauge field A_μ. Similarly, we may write the axial coupling to matter as

$$\int d^4x\sqrt{g}\partial_\mu a\bar{\psi}\gamma^\mu\gamma_5\psi \leftrightarrow \int d^4x\epsilon^{\kappa\lambda\mu\nu}H_{\kappa\lambda\mu}\bar{\psi}\gamma_\nu\gamma_5\psi \,. \tag{7}$$

In the scalar theory, to the Peccei-Quinn symmetry is associated the conserved Nöther current $j_\mu = \partial_\mu a$; this can be seen to be conserved as a result of the equation of motion $\nabla^\mu\nabla_\mu a = 0$. In the tensor theory the current j is also conserved, but now by virtue of the Bianchi identity (1). The conserved charge associated to these currents is written

$$A = \int_{V^3} d^3x\sqrt{g}j^0 = \int_{V^3} d^3x\sqrt{g}\epsilon^{ijk}H_{ijk} \tag{8}$$

where V^3 is a spacelike three-surface and i, j, k are indices tangent to it. The axion field carries this charge. An obvious question is whether a black hole can carry such a charge, and if so whether it is externally measurable.

Before addressing this issue, recall that generally speaking black holes don't support external fields and thus don't have externally measurable attributes; this is often stated as "a black hole has no hair." The reason for this general fact is basically that if one attempts to set up a field distribution outside a black hole, it will either fall down the hole or be radiated to infinity. One example is baryon number: to the best of our knowledge, baryon number will not be preserved in black hole processes. We can dump atoms, old refrigerators, or conference participants down the hole, and not only can't we get them back out but also it's impossible to tell how many of them were dropped in in the first place – the baryon number of a black hole is not meaningful.

Statements such as this are formalized by the black-hole uniqueness theorems, which basically state that there are few externally measurable attributes of a black hole. Of the few, three are the mass and angular momentum (which both leave their signature in the gravitational field of the hole), and the electric (or magnetic) charge (whose signature resides in the electromagnetic field of the hole). As we'll see shortly, to be added to this short list is the axion charge A. We can summarize the results by a table of allowed external fields:

spin	mass= 0	mass≠ 0
0	no*	no
$\frac{1}{2}$	no	no
1	yes	no
$\frac{3}{2}$	yes	no
2	yes	×

The positive result for spin 3/2 arises from the existence of supersymmetry; we can apply a supersymmetry transformation to a black hole and find a solution with non-zero external spin 3/2 (gravitino) field and with an associated spinor charge[2]. I am unaware of any result for massive spin two, but fundamental massive spin two fields are not believed to exist in nature anyway (outside string theory). Finally, the asterisk at spin zero, mass zero, indicates the exception that occurs in the case of axion charge. Proof of the uniqueness theorems typically proceeds in two steps. First one assumes that one is given a spacetime that is asymptotically flat, stationary (*i.e.* has a timelike symmetry generated by a Killing vector ξ), and contains a black hole with non-singular horizon. Then one takes a field ϕ and shows that the vanishing of the Lie derivative, $\mathcal{L}_\xi \phi = 0$ and the demand that $\phi \neq 0$ contradict if ϕ satisfies its equation of motion (see *e.g.* [1]). The second step is to show that the Kerr-Newman solutions are unique, and this is somewhat involved[3].

Naïvely one expects the same logic to apply here as well: if we attempt to set up a distribution of the H field outside the hole, it should either fall down the hole or be radiated to infinity, leaving no external signature. However, we know that if some axion field falls down the hole, then since the axion charge is conserved the hole should be ascribed an axion charge A. Indeed, it turns out that there is a non-trivial black hole solution to the coupled Einstein-axion field equations. This solution has $H_{\lambda\mu\nu} = 0$ everywhere except the origin, $r = 0$; thus the energy-momentum is zero and the geometry is that of the vacuum Schwarzschild solution,

$$ds^2 = -\left(1 - \frac{2M}{r}\right) dt^2 + \left(1 - \frac{2M}{r}\right)^{-1} dr^2 + r^2 d\Omega_2^2 \ . \tag{9}$$

The non-zero axion charge on the hole can be written as a boundary integral, using Gauss' law:

$$0 \neq A = \int_{V^3} d^3x \sqrt{g} \epsilon^{ijk} H_{ijk} = \int_{\partial V^3} dS \epsilon^{ij} B_{ij} \ . \tag{10}$$

Here V^3 is some volume including the black hole, and the surface integral is over the boundary of this volume. As a result of this, it is clear that the potential $B_{\mu\nu}$

must be non-zero outside the black hole; up to a gauge transformation it should take the form

$$B_{\mu\nu} = \frac{A\epsilon_{\mu\nu}}{4\pi r^2} \tag{11}$$

where $\epsilon_{\mu\nu}$ is the surface element on the two-sphere surrounding the hole. There is a uniqueness theorem for this solution which is discussed in [1].

We have established that a black hole can carry an axion charge. Next we consider the issue of whether this charge can be measured. Naïvely the answer is no, since $H = 0$ outside the hole and only H appears in the lagrangian for the coupling of the axion to other fields.

However, the situation looks suspiciously similar to that in the Aharonov-Bohm effect. In that context, one has a solenoid with a non-zero magnetic flux threading it, and the magnetic field outside the solenoid is zero although the potential is not. This non-zero potential is then measurable because it contributes a phase

$$e^{ie\oint A_\mu dl^\mu} \tag{12}$$

to the amplitude for an electron to propagate around the solenoid. This phase arises because in quantum mechanics particles couple to the *potential* A_μ rather than to the field strength $F_{\mu\nu}$.

Therefore if we hope to measure the charge A we should seek objects that couple directly to the potential $B_{\mu\nu}$. We do not have to look far: it is natural for *strings*, either fundamental or cosmic, to have a coupling directly to the potential $B_{\mu\nu}$. To see this, recall that the ordinary kinetic term for string propagation in a background spacetime metric $g_{\mu\nu}$ is

$$S_g = T\int_\Sigma d^2\sigma\sqrt{\gamma}g_{\mu\nu}\left(X^\lambda(\sigma)\right)\gamma^{ab}\partial_a X^\mu\partial_b X^\nu\ . \tag{13}$$

Here the integral is over the world sheet Σ swept out by the string; this is taken to have coordinates σ. The metric on the string world sheet is denoted γ^{ab}, and T is a constant giving the string tension. An obvious generalization of this is found by adding to (13)

$$S_B = T\int_\Sigma d^2\sigma\sqrt{\gamma}B_{\mu\nu}\left(X^\lambda(\sigma)\right)\epsilon^{ab}\partial_a X^\mu\partial_b X^\nu \tag{14}$$

where now ϵ^{ab} is the antisymmetric tensor on the string world sheet. Such a coupling occurs in the theory of fundamental strings, and also may appear in cosmic strings arising from pseudo-anomalous $U(1)$'s [4]. Eq. (14) is a direct generalization of the coupling of a particle to electromagnetism written in first quantized language,

$$S_A = e \int_L d\tau A_\mu (X^\nu(\tau))\, \partial_\tau X^\mu \ , \tag{15}$$

where τ is a coordinate along the particle world line L. (Likewise, the kinetic action for the particle can be written

$$S = \int_L d\tau \sqrt{\gamma_{\tau\tau}} g_{\mu\nu} \gamma^{\tau\tau} \partial_\tau X^\mu \partial_\tau X^\nu \tag{16}$$

where $\gamma_{\tau\tau}$ is a one-dimensional metric on L.) The action (15) directly leads to the phase (12).

Therefore in analogy to the Aharonov-Bohm effect, there can be phases in the string propagation amplitude due to the coupling (14). Consider the difference between the two world sheets shown in figs. 1(a),(b). In fig. 1(a) the string completely avoids the vicinity of the hole. Since $H = 0$ in the entire volume bounded by the string world sheet, by the Poincare lemma the potential $B_{\mu\nu}$ is gauge trivial in that volume, *i.e.*, we can write $B_{\mu\nu} = \partial_\mu \Lambda_\nu - \partial_\nu \Lambda_\mu$ throughout that volume. As a result, the action (14) can be rewritten

$$S_B = T \int_\Sigma d^2\sigma \sqrt{\gamma} \nabla_a \left(\epsilon^{ab} \partial_b X_\nu \Lambda_\nu \right) = 0 \ . \tag{17}$$

However, if we consider the world sheet in fig. 1(b), then $B_{\mu\nu}$ is not gauge-trivial inside the world sheet, and the action is

$$S_B = T \int_\Sigma d^2\sigma \epsilon^{ab} B_{ab} = AT \tag{18}$$

giving an extra phase

$$e^{iAT} \tag{19}$$

to the propagation amplitude. This phase could be measured in beam splitting experiments similar to in the Aharonov-Bohm effect; for example, one could consider propagation as shown in fig. 2 and find a relative phase shift between the strings that lasso the hole and those that don't.

Finally, it should be obvious that higher dimensional generalizations of this effect exist as well. We could consider a $(p+1)$-form potential B on an n-dimensional manifold M, and in direct analogy to electromagnetism and axions define a field strength

$$H_{\mu_1 \cdots \mu_{p+2}} = \partial_{\mu_1} B_{\mu_2 \cdots \mu_{p+2}} \pm \text{permutations} . \tag{20}$$

Then if $H^{p+1}(M)$, the $(p+1)^{st}$ cohomology class of the manifold M, were non-zero, or if $H \neq 0$ on some $(p+2)$-dimensional submanifold of M, then there is a higher analog of the Aharonov-Bohm effect for p-dimensional extended objects (p-branes).

To summarize, black holes may carry axion charge and that charge may in principle be measured using an analogue of the Aharonov-Bohm effect that involves passing a string around the black hole. Unfortunately, in contrast to the situation when the Aharonov-Bohm effect was proposed in 1959, actual observation of this effect is not likely to occur soon. First off, we require two-form axions rather than electromagnetism, and there has not yet been experimental evidence for any axions. Secondly, we need macroscopic objects carrying axion charge rather than solenoids in the Aharonov-Bohm effect (actually black holes are not necessary – any object carrying a macroscopic axion charge will do). Finally, instead of electrons we require either fundamental or cosmic strings, neither of which are known to exist. The Aharonov-Bohm effect was observed the year after it was proposed; unfortunately the effect described in this paper may not be observed for many years. Nonetheless I think that it is an interesting and amusing effect.

References

[1] M.J. Bowick, S.B. Giddings, J.A. Harvey, G.T. Horowitz, and A. Strominger, "Axionic Black Holes and an Aharonov-Bohm Effect for Strings," *Phys. Rev. Lett.* **61** (1988) 2823.

[2] P. Cordero and C. Teitelboim, "Remarks on Supersymmetric Black Holes," *Phys. Lett.* **78B** (1978) 80; "No-Hair Conjecture in Supergravity," in *Relativity and Gravitation, SILARG III*, S. Hojman et al eds. (Universidad Nacional Autonoma de Mexico, Mexico 1982).

[3] P.O. Mazur, "Proof of Uniqueness of the Kerr-Newman Black Hole Solution," *J. Phys.* **A15** (1982) 3173; "A Global Identity for Nonlinear σ-Models," *Phys. Lett.* **A100** (1984) 341.

[4] J.A. Harvey and S.G. Naculich, "Cosmic Strings From Pseudoanomalous U(1)s," *Phys. Lett.* **B217** (1989) 231.

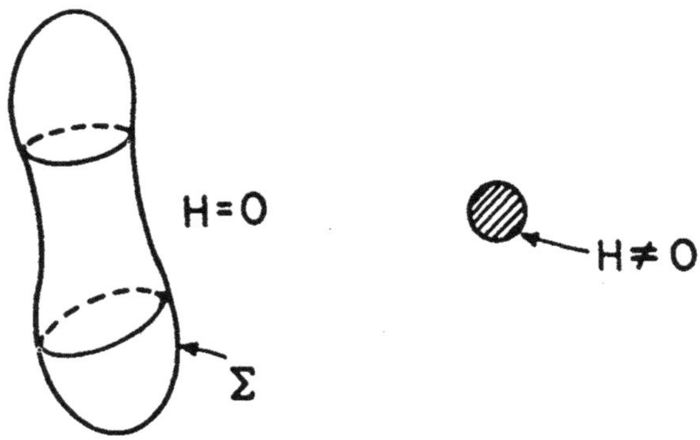

Fig. 1(a)

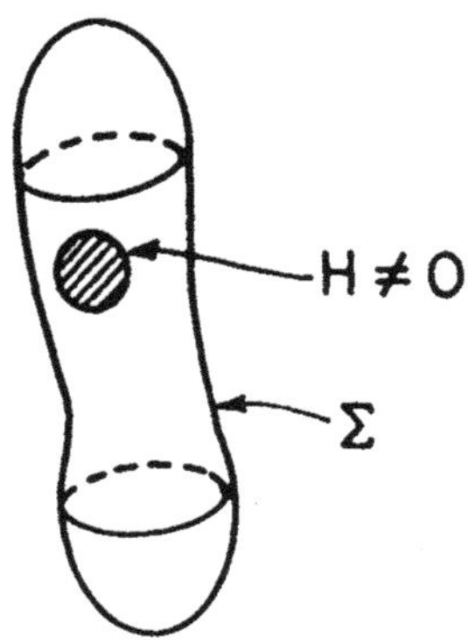

Fig. 1(b)

(a) A string world sheet corresponding to motion where the string avoids the black hole. (b) In this figure the string lassos the black hole.

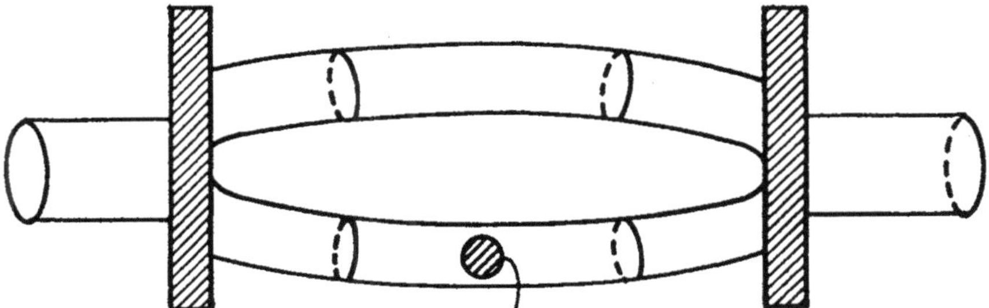

Fig. 2

Schematic diagram of a beam-splitting experiment that would allow measurement by strings of the phase shift produced by axion charge.

WAVE GEOMETRY: A PLURALITY OF SINGULARITIES

M.V.Berry
H.H.Wills Physics Laboratory, Tyndall Avenue, Bristol BS8 1TL, U.K.

ABSTRACT

Five interconnected wave singularities are discussed: *phase monopoles*, at eigenvalue degeneracies in parameter space, where the 2-form generating the geometric phase is singular; *phase dislocations*, at zeros of complex wavefunctions in position space, where different wavefronts (surfaces of constant phase) meet; *caustics*, that is envelopes (foci) of families of classical paths or geometrical rays, where real rays are born violently and which are complementary to dislocations; *Stokes sets*, at which a complex ray is born gently where it is maximally dominated by another ray; and *complex degeneracies*, which are the sources of adiabatic quantum transitions in analytic Hamiltonians.

1. Introduction

A rather old-fashioned view of quantum mechanics is that it is an application of the physics of waves. A not so old-fashioned way to get interesting physics out of mathematical objects is by studying their singularities. Here I will combine these two ideas and describe a variety of singularities (in several spaces) in waves, quantum and otherwise. This leads to an unorthodox way of thinking about waves, apparently very different from, but actually complementary to, the usual ones based on wave equations, Fourier analysis, etc. It is appropriate to be talking about it at a meeting celebrating the 30th birthday of the Aharonov-Bohm effect[1], for at least two reasons.First, their effect exemplifies several of the ideas I will describe. Second, it lies squarely in the geometrical tradition of the physics department at Bristol University where Aharonov and Bohm discovered their effect and where its existence was confirmed experimentally by Chambers[2].

All the material I will describe has been published, so this written version will be a summary of the talk that was given at the meeting, with references.

2. Phase monopoles

A vehicle to propel us into the unfamiliar world of wave singularities is the now-familiar geometric phase[3,4]. This phase is the flux of a 2-form in the space of Hamiltonians parameterized by $X=X_1, X_2,\ldots$. For the eigenstate $|n(X)\rangle$, the 2-form is

$$V_n(X) = \mathrm{Im}\langle \mathrm{d}n|\wedge|\mathrm{d}n\rangle \tag{1}$$

Phase monopoles are the singularities of this 2-form. They occur at parameters X^* for which the state $|n\rangle$ is degenerate. The generic case, where the degeneracy involves only two states (i.e. $|n\rangle$ and one other) and where the Hamiltonian has no symmetry, has codimension three. Thus a full exploration of the monopole singularities requires at least three parameters.

The simplest such family consists of the Hamiltonians for a spin-1/2 particle in a magnetic field $X=(X_1, X_2, X_3)$. Here the monopole[4] is at $X^*=0$. It is, however, more interesting to study the distorted monopoles in systems with more than two states. This has recently been done[5], choosing as family the *Aharonov-Bohm chaotic billiards*[6]. In these systems, a charged quantum particle moves freely in a bounded plane region D threaded by a line of magnetic flux of quantum strength α. (The purpose of the flux is to break time-reversal symmetry.) These Hamiltonians can be regarded as parameterized by the vector

$$X = \left(B_1, B_2, \alpha\right) \tag{2}$$

where B_1 and B_2 are parameters altering the shape of D. In terms of the wavefunctions

$$\psi_n(r;X) \equiv \langle r|n(X)\rangle \tag{3}$$

(depending on position r=x,y), the phase 2-form can be regarded as the vector

$$\begin{aligned} V_n(X) &= \mathrm{Im}\langle\nabla_X n|\wedge|\nabla_X n\rangle \\ &= \iint_{\text{inside D}} \mathrm{d}r\, \mathrm{Im}\, \nabla_X \psi_n^*(r;X) \wedge \nabla_X \psi_n(r;X) \end{aligned} \tag{4}$$

Degeneracies X^* can be determined by tracking the lines of V_n to their monopole sources or sinks. Several monopoles were studied in this way[5].

3. Phase dislocations

It is interesting to study the changing morphology of wavefunctions $\psi(r;X)$ in position space r, as they acquire their geometric phases during a circuit of X space. The crucial structural elements of the this process are the zeros of ψ. These are familiar as nodal lines (when r is two-dimensional) in the case where the Hamiltonian, and hence the wavefunctions, are real - because of time-reversal symmetry for example. In a study[7] of the family of triangular billiards (without flux) it was shown how in a circuit of a degeneracy (which in this case has codimension two) the nodal lines interact and change their topology so as to reverse the signs of all nodal cells and hence of ψ (for real wavefunctions the only possible phase change is π).

Things are different in the general case where there is no symmetry and the wavefunctions are complex. Then the zeros are points (in the plane), because the vanishing of a complex funtion is a codimension-two phenomenon. Writing

$$\psi \equiv \rho \exp\{\mathrm{i}\chi\} \tag{5}$$

we see that the zeros,where the modulus ρ vanishes, are also singularities of the phase χ. In three-dimensional space these singularities are lines. We[8,9] called them phase dislocations because of a far-reaching analogy between the wavefront surfaces χ=constant (mod π) and planes of atoms in a crystal: there exist exact solutions of wave equations in which phase dislocations are of edge, screw or mixed type, can be straight or curved, can have climb or glide motion, can collide and annihilate or bounce, and can grow in loops from points (punctured wavefronts).

In Aharonov-Bohm billiards the wavefunctions possess a rich dislocation structure[10]. During a circuit near a degeneracy in X space the dislocations can change their number by collision, either with the boundary or one another[5]. In the latter case we may have either annihilation or pair production, processes obeying two topological conservation laws[11]. It frequently happens that when a dislocation is moving whilst a parameter changes, a pair is created ahead of it and one member of the pair travels back and annihilates with the original dislocation, leaving its companion moving forward; this process, analogous to the zitterbewegung of electrons, was observed previously in acoustics[12] (the parameter was time).

Most commonly, phase dislocations are free, that is, they arise by interference and are not attached to any singularity of the propagation medium; for example, the air during speech is threaded with a forest of dislocation lines. They may, however, be attached to a point singularity and be movable, but not removable, by a gauge transformation; an example is the Dirac string in the wavefunction of an electron in the presence of a magnetic monopole[13,14]. Another possibility is that they may be fixed to a physical line singularity. This occurs in the ordinary Aharonov-Bohm scattering wavefunction[1], where the dislocation is pinned to the magnetic flux line; the strength S of the dislocation, that is $1/2\pi$ times the total change in χ during a circuit of it, is the integer closest to the magnetic flux α (measured in quantum units). Therefore S changes when α passes through half-integer values, the mechanism being that wavefronts unzip all the way from the flux line to infinity and reconnect with their 2π neighbours, This interesting phenomenon cannot be observed in quantum waves (nor can S)[14], but can be and has been observed in an analogue experiment with surface ripples on water[15].

Dislocations occur on a grand scale in the tide wave[9,16,17], as 'amphidromic point' singularities of the cotidal lines (wavefronts) connecing places where the tide is high at a given time. The dislocations are therefore places of no tide (the tide is 'always high').

4. Caustics

Dislocations involve phase and so are the characteristic spatial singularities of waves. It is interesting to enquire what are the corresponding singularities in the short-wave limit, that is of classical mechanics or geometrical optics, where there is no phase. These are the caustics, that is the focal singularities of the families of rays that represent short waves. The caustic of a family is its envelope, that is the surface (in space) touched by every ray in the family. Caustics are holistic phenomena, properties of the whole family and not inherent in any one ray.

Caustics are singular because across them the number of rays passing through a point jumps (usually by two), and the ray intensity diverges: caustics are the violent births of rays. The caustics that are geometrically stable can be classified according to their codimension by the Thom-Arnold theory of catastrophes[9,18,19]. The classification is more than geometric, because to each catastrophe singularity is associated a characteristic diffraction pattern that decorates the caustic when the wavelength is small but not zero.

In the fine detail of these wave patterns, dislocations reappear as the most delicate features[9]. Sometimes the association is amazingly complicated (although stable): in the elliptic umbilic diffraction catastrophe[20], for example, there are rows of dislocation rings; in the M'th row, the number of rings is Int[$512M/27 - 539/104$].

As singularities, ray caustics are complementary to phase dislocations, in the sense of Bohr. The complementarity is more than the fact that caustics are infinite in intensity (in the limit of zero wavelength), whereas disclocations have zero intensity. When the wavelength is so short that a caustic emerges sharply as a singularity, dislocation singularities are lost in fine detail and cannot be resolved; and under a magnification high enough to discern the structure of an individual dislocation, the caustic singularity is smoothed away.

5. Stokes sets

It is true that in the sense just described caustics are the violent births of rays. But in another sense they are not births but really transmutations, of a complex ray into (typically) two real rays. (This is most familiar in the stationary-phase approximation of diffraction integrals, when the change of a parameter causes two complex stationary points - only one of which contributes to the integral - to coalesce and separate into two real stationary points.) One can follow the complex ray and discover the moment and manner of its birth. In contrast to what happens at caustics, this is birth is a very gentle affair, known as *Stokes' phenomenon*.

Stokes discovered[21] (in the context of the asymptotics of certain Bessel functions in the complex plane) that the birth occurs when the complex ray is at its weakest in relation to another (real or complex) ray that dominates it. If the exponentials associated with the two rays are

$$\exp\{k\phi_+(X)\} \text{ (dominant) and } \exp\{k\phi_-(X)\} \text{ (subdominant)} \quad \text{where } \mathrm{Re}(\phi_+) > \mathrm{Re}(\phi_-) \tag{6}$$

in which k is a (large) asymptotic quantity and X are parameters, then the maximum disparity between the exponentials, and therefore the birth, occurs when X is on the Stokes set

$$\mathrm{Im}\{\phi_+(X) - \phi_-(X)\} = 0 \tag{7}$$

Stokes sets are familiar as lines in the complex plane (arising for example in the WKB approximation of functions defined by differential equations), but in diffraction catastrophes they can occur in real parameter spaces[22,23].

Recently[24,25], the details of the birth have been determined.The birth - that is, the change in the coefficient of the subdominant wave - is governed according to a universal scenario by the error function:

$$\begin{pmatrix}\text{change in} \\ \text{subdominant cofficient}\end{pmatrix} = \begin{pmatrix}\text{dominant} \\ \text{coefficient}\end{pmatrix} \times \frac{\mathrm{i}}{\sqrt{\pi}} \int_{-\infty}^{\sigma(X;k)} \mathrm{d}t \exp\{-t^2\}$$

$$\text{where } \sigma = \sqrt{\frac{k}{2\,\mathrm{Re}\{\phi_+ - \phi_-\}}} \times \mathrm{Im}\{\phi_+ - \phi_-\} \tag{8}$$

This result has generated some interest among mathmaticians[26,27,28], and applications to reflections[29] and quantum transition histories[30].

6. Complex degeneracies

For these final singularities I return to quantum adiabatics, and consider not the phase acquired by the original state during a slow parameter cycle but the probability of transition to a different state. It suffices to consider the simple 2-state Hamiltonian

$$\hat{H}(X(\delta t)) = X(\delta t) \cdot S \tag{9}$$

where $X=(X_1, X_2, X_3)$ are the parameters, S the vector of Pauli spin matrices and δ the (small) adiabatic parameter. If $\tau \equiv \delta t$ denotes 'slow time', then the history of any quantum state depends only on the *Hamiltonian curve* $X(\tau)$. We assume that there are no real degeneracies, i.e. that $X \equiv |X|$ does not vanish for any real τ, and that $X(\tau)$ is analytic in a sufficiently wide strip including the whole real τ axis. Then the probability of making a transition over infinite time is exponentially small in δ and dominated[31,32] by the *complex degeneracy* τ_c that is closest to the real axis. The degeneracy is a branch point singularity of the spectrum, determined by

$$X(\tau_c) = \sqrt{X_1^2 + X_2^2 + X_3^2} = 0 \tag{10}$$

I report two new results related to the complex degeneracies.

The first[33] concerns the final (infinite-time) transition probability. Of course this is dominated by the familiar 'dynamical' exponential whose exponent is is proportional to $1/\delta$. But in general there is another contribution which is purely geometric in that it is independent of δ (and $\hbar$) and depends only on the geometry of the Hamiltonian curve, analytically continued from the real axis to τ_c. The reason this geometric amplitude was not noticed before is probably that it vanishes for plane Hamiltonian curves (such as that for the familiar Landau-Zener problem) and for curves that up to a rotation are identical with their time-reverse (such a uniform helix). The simplest Hamiltonian curve giving a non-zero geometric amplitude is a helix that winds from $\tau=-\infty$ to $\tau=0$ and unwinds from $\tau=0$ to $\tau=+\infty$. It would be possible to measure the new amplitude in a spin experiment involving, say, neutrons in a magnetic field.

The second result[30] concerns the transition history, that is the manner in which the final exponentially small transition probability is reached. Of course this depends on the basis states with respect to which the evolving state is described. The usual choice is the 'adiabatic basis' of eigenstates of the instantaneous Hamiltonian. In this basis, the transition takes place via rapid system-dependent oscillations of order δ, i.e. far larger than the final probability. However, there exists a series of transformations to 'superadiabatic bases', close to the adiabatic basis, for which the transition history renormalizes onto a universal one in which the increase of the transition probability is maximally smooth and compact. The transition takes place according to an error function (analogous to (8)) near the instant when the complex Stokes line emanating from τ_c crosses the real τ axis. It is possible to define spin observables corresponding to the superadiabatic bases, so that in principle the universal transition history could be observed.

7. Conclusions

I have described a cycle in the space of singularities, with each singularity being related to the previous one(s) and with the end of the cycle being not quite the same as the beginning, as is familiar in situations of this type. Here is an outline of the connections. The first singularity, inhabiting parameter space, is the phase monopole associated with an eigenvalue degeneracy. Phase dislocations - the second class of singularity - provide the mechanism for wavefunctions in configuration space to acquire their geometric phases. Caustics - the third class - are the complementary ray singularities, the loci of violent births of real rays. Stokes sets - the fourth class - are the loci of the gentle births of complex rays, of small exponentials coming into existence whilst hidden behind large ones. And finally, completing the cycle, are the complex degeneracies, from which emanate Stokes lines near whose intersections with the real time axis quantum transitions occur.

8. Acknowledgment

I thank Professor J.A.Anandan and his colleagues for arranging, and generously supporting my participation in, the meeting, celebrating the Aharonov-Bohm effect, at which this talk was given.

9. References

1. Y.Aharonov and D.Bohm, *Phys.Rev.* **115** (1959) 485.
2. R.G.Chambers, *Phys.Rev. Lett.* **5** (1960) 3.
3. A.Shapere and F.Wilczek, *Geometric phases in physics*, (World Scientific, Singapore, 1989).
4. M.V.Berry, *Proc.Roy.Soc.Lond.* **A392** (1984) 45.
5. R.J.Mondragon and M.V.Berry, *Proc.Roy.Soc.Lond.* **A424** (1989) 263.
6. M,V.Berry and M.Robnik, *J.Phys A* **19** (1986) 649.
7. M.V.Berry and M.Wilkinson, *Proc.Roy.Soc.Lond.* **A392** (1984) 15.
8. J.F.Nye and M.V.Berry, *Proc.Roy.Soc.Lond.* **A336** (1974) 165.
9. M.V.Berry, in *Les Houches Lecture Series Session XXXV*, eds. R.Balian, M.Kléman and J-P.Poirier (North-Holland, Amsterdam 1981).
10. M,V.Berry and M.Robnik, *J.Phys A* **19** (1986) 1365.
11. J.F.Nye,J.V.Hajnal and J.H.Hannay, *Proc.Roy.Soc.Lond.* **A417** (1988) 7.
12. F.J.Wright and M.V.Berry *J.Acoust.Soc.Amer.* **75** (1984) 733.
13. P.A.M.Dirac, *Proc.Roy.Soc.Lond.***A133** (1931) 60.
14. T.T.Wu and C.N.Yang, *Phys.Rev.* **D12** (1975) 3845.
15. M.V.Berry, R.G.Chambers,M.D.Large,C.Upstill and J.C.Walmsley, *Eur.J.Phys* **1** (1980) 154.
16. W.Whewell *Phil.Trans.Roy.Soc.Lond.* **123** (1833) 147, (1836) 289.
17. M.V.Berry, *Proc.Roy.Soc.Lond.* **A413** (1987) 183.

18. T.Poston and I.N.Stewart, *Catastrophe theory and its applications* (Pitman, London, 1978)
19. M.V.Berry and C.Upstill, *Prog.in Optics* **18** (1980) 257.
20. M.V.Berry, J.F.Nye and F.J.Wright, *Phil.Trans.Roy.Soc.Lond.* **A291** (1979) 453.
21. G.G.Stokes, *Trans.Camb.Phil.Soc.* **10** (1864) 106
22. F.J.Wright, J.Phys.A. **13** (1980) 2913.
23. M.V.Berry and C.J.Howls *Nonlinearity* (in press, 1990).
24. M.V.Berry, *Proc.Roy.Soc.Lond.* **A422** (1989) 7.
25 M.V.Berry, *Publ.Math.I.H.E.S* **68** (1989) 211.
26 F.W.J.Olver, *Proceedings of International symposium on Asymptotic and Computational Analysis* (Manitoba, Winnipeg 1989) (Marcel Dekker, New York, 1990).
27 D.S.Jones, *Proceedings of International symposium on Asymptotic and Computational Analysis* (Manitoba, Winnipeg 1989) (Marcel Dekker, New York, 1990).
28. W.G.C.Boyd, *Proc.Roy.Soc.Lond.* (in press 1990).
29. M.V.Berry, *Proc.Roy.Soc.Lond.* (in press 1990).
30. M.V.Berry, *Proc.Roy.Soc.Lond.* (in press 1990).
31. J.P.Davis and P.Pechukas, *J.Chem.Phys.* **64** (1976) 3129.
32. J.T.Hwang and P.Pechukas, *J.Chem.Phys.* **67** (1976) 4640.
33. M.V.Berry, *Proc.Roy.Soc.Lond.* (submitted 1989).

ABELIAN AND NON-ABELIAN HOLONOMY IN NMR INTERFEROMETRY

A. Pines and K. T. Mueller

University of California
Berkeley, California 94720, U.S.A.

ABSTRACT

It is well known that a gauge ambiguity arises in physics when a partition is made, say, between "the system" (S) and "the rest" (R).[1]

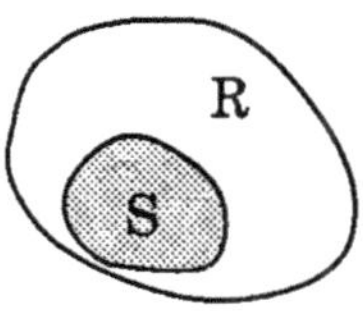

$$\begin{pmatrix} (\rho_S) & (\ \rho_{SR}\ ) \\ \begin{pmatrix} \rho_{RS} \end{pmatrix} & \begin{pmatrix} \rho_R \end{pmatrix} \end{pmatrix} \tag{1}$$

The geometric phase,[2,3,4] which appears as a change in ρ_{SR}, is the holonomy of the connection (gauge potential) over a circuit in the (projective) space of system density matrices (ρ_S).[5] The simplest case of (Abelian) holonomy arises in a projective two-level system for which ρ_S are 2×2 Hermitian matrices

$$\rho_S = \frac{1}{2}(1 + \vec{p} \cdot \vec{\sigma}) \tag{2}$$

where $\vec{p}$ is a polarization vector and $\vec{\sigma}$ is a vector of Pauli spin matrices. The gauge group is U(1):

$$\begin{pmatrix} U(1) & 0 \\ 0 & U(1) \end{pmatrix} \begin{pmatrix} \rho_a & 0 \\ 0 & \rho_b \end{pmatrix} \begin{pmatrix} U(1) & 0 \\ 0 & U(1) \end{pmatrix}^\dagger = \begin{pmatrix} \rho_a & 0 \\ 0 & \rho_b \end{pmatrix}. \tag{3}$$

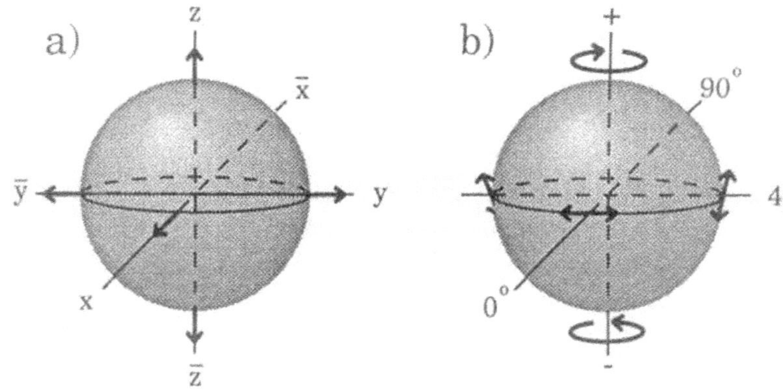

Figure 1. Projective space (S^2) for two-level systems. (a) For a spin-$\frac{1}{2}$ the poles (z, $\bar{z}$) correspond to "spin up" and "spin down" and the equator represents linear combinations of $x-$ and $y-$polarizations. (b) Space of light polarization states (Poincaré sphere) where the poles correspond to the two helicity states (circularly polarized light) and the equator describes states of linear polarization with plane angles from 0° to 180°.

The space of pure density matrices ($p = 1$) is S^2 (examples are spin-$\frac{1}{2}$ and photon polarization, Figure 1) and the connection one-form is[4]

$$A = -i\langle\tilde{\psi}|d\tilde{\psi}\rangle \tag{4}$$

where $|\tilde{\psi}\rangle$ is single valued on an appropriate patch of S^2. The geometric phase over a circuit C is

$$\gamma(C) = \oint_C A = \int_S dA = -\frac{1}{2}\Omega(C) \tag{5}$$

where S is the surface enclosed by C and $\Omega(C)$ is the solid angle subtended by C at the origin. This corresponds to the Aharonov–Bohm effect[6] arising from the circuit of a particle of unit charge near a magnetic monopole of strength-$\frac{1}{2}$.[7] The "spinor sign-change" under a 2π rotation (circuit 1 of Fig. 2)[8] is thus a special case of the geometric phase for $\Omega(C) = 2\pi$. It can be regarded as the Aharonov–Bohm effect around a flux tube of strength-$\frac{1}{2}$. A similar topology of real 2×2 matrices arises in "conical intersections" of degenerate electronic systems.[9,10]

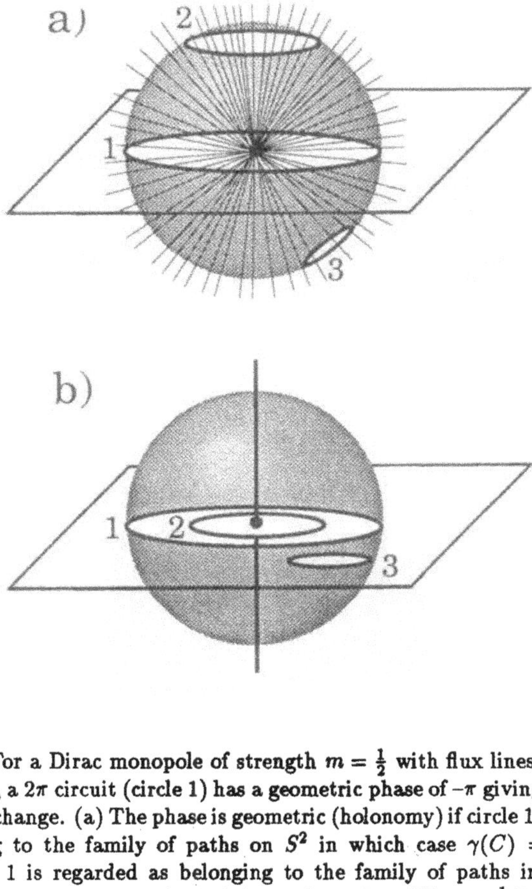

Figure 2. For a Dirac monopole of strength $m = \frac{1}{2}$ with flux lines emanating isotropically, a 2π circuit (circle 1) has a geometric phase of $-\pi$ giving rise to the spinor sign-change. (a) The phase is geometric (holonomy) if circle 1 is regarded as belonging to the family of paths on S^2 in which case $\gamma(C) = -\frac{1}{2}\Omega(C)$. (b) If circle 1 is regarded as belonging to the family of paths in the plane then the phase appears topological (monodromy): $\gamma(C) = -\frac{1}{2}(2\pi) = -\pi$ if the circuit includes the origin (1 and 2) and $\gamma(C) = 0$ if the circuit does not include the origin (3). In the planar system the lines of flux of the monopole can be considered as gathered together above and below the plane into a flux line (tube) of strength $m = \frac{1}{2}$.

The holonomy of a two–level system can be observed if it is imbedded in a three–level system[11] (the additional quantum level is the R of Eq. (1)) as in NMR interferometry[12,13] (normal two–beam interferometry[14,15] corresponds to a four–level system). An example of such an experiment is shown in Figs. 3 and 4.

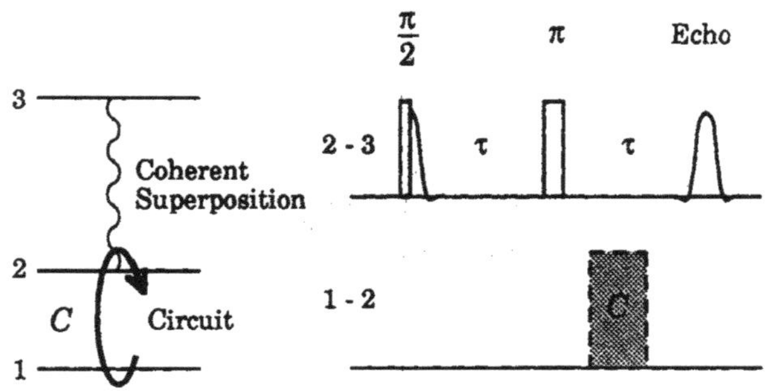

Figure 3. In NMR interferometry the holonomy of a circuit C in the 1–2 two-level system appears as a phase shift in the spin echo of a coherent superposition in the 2–3 system.

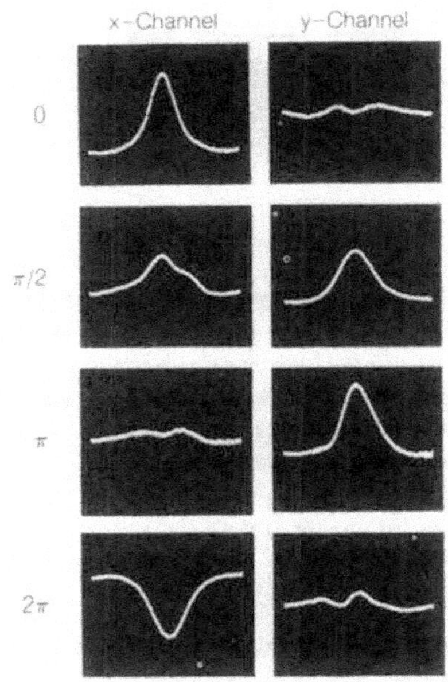

Figure 4. Holonomy in a two-level system (Fig. 3) observed in the phase shift of a spin-echo. The NMR echo is detected in orthogonal channels of a phase sensitive detector. The numbers to the left indicate the solid angle of a circuit on the sphere of polarizations for the 1–2 subsystem. The echo arising in the 2–3 superposition moves from one channel to the other as the geometric phase increases from 0° to 90°. (from Ref. 13, with permission)

An example of non-Abelian holonomy (the generalization of Berry's phase by Wilczek and Zee[16]) can be observed in systems with degeneracies, for example a time-reversal invariant fermion like a spin-$\frac{3}{2}$ quadrupole.[17] The gauge group is U(2):

$$\begin{pmatrix} \big(U(2)\big) & 0 \\ 0 & \big(U(2)\big) \end{pmatrix} \begin{pmatrix} \begin{pmatrix} \rho_a & 0 \\ 0 & \rho_a \end{pmatrix} & 0 \\ 0 & \begin{pmatrix} \rho_b & 0 \\ 0 & \rho_b \end{pmatrix} \end{pmatrix} \begin{pmatrix} \big(U(2)\big) & 0 \\ 0 & \big(U(2)\big) \end{pmatrix}^{\dagger} = \begin{pmatrix} \begin{pmatrix} \rho_a & 0 \\ 0 & \rho_a \end{pmatrix} & 0 \\ 0 & \begin{pmatrix} \rho_b & 0 \\ 0 & \rho_b \end{pmatrix} \end{pmatrix} \tag{6}$$

The first experiments on such a system were by Tycko[18] who rotated a crystal containing spin-$\frac{3}{2}$ chlorine-35 nuclei inside a coil and observed adiabatic shifts in the nuclear quadrupole frequencies. As Zee has pointed out[19] circuits other than circles are necessary in order for the non-Abelian character of the connection to express itself. Such circuits (in which the paths of the electric field gradient are not circles) have been implemented by rotating a sample about two axes.[20,21]

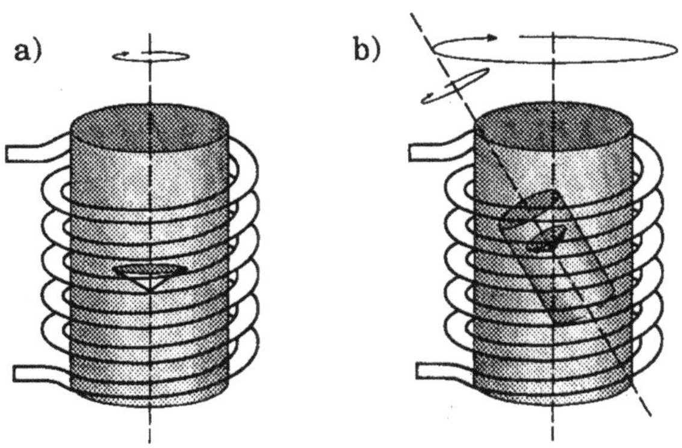

Figure 5. Experimental design for measuring geometric phases in a spin-$\frac{3}{2}$ quadrupole. (a) The sample is rotated about an axis and the geometric phase accumulated due to circuits of the electric field gradient appears as a frequency shift in the radiofrequency coil.[18] (b) Rotating the sample "about two axes" (where the spinning axis precesses about an axis parallel to the rf coil) allows a study of the non-Abelian Wilczek-Zee gauge.

Questions of holonomy also arise in quantum measurement or projection. In a normal Stern-Gerlach measurement of the components of spin (using an inhomogeneous magnetic field) the connection is Abelian. For example the sequence of spin–$\frac{1}{2}$ projections $z \to x \to y \to z$ gives rise to a geometric phase of $\frac{\pi}{4}$ (one-half of the solid angle of the geodesic triangle[22,23]). In the case of "incomplete measurement" of a spin–$\frac{3}{2}$ with a "quadrupole Stern–Gerlach" apparatus (using an inhomogeneous electric field gradient) the connection is non–Abelian.[24] Non–Abelian connections also arise naturally in the rotation of objects evolving through circuits in shape space.[1,25]

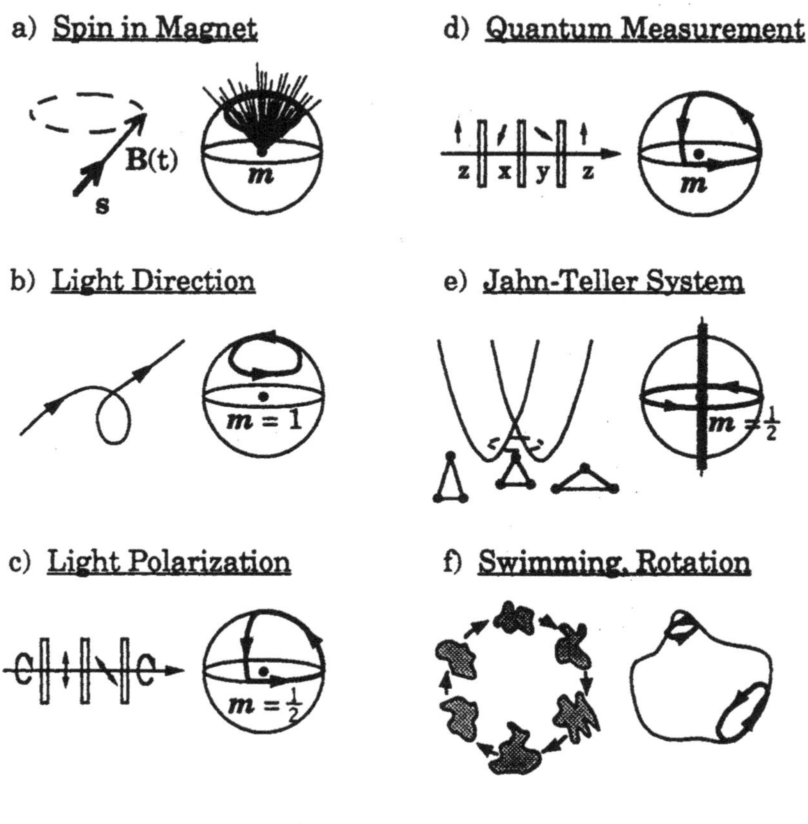

Figure 6. Summary of some physical systems exhibiting geometric phases.

In summary, we shall discuss a diverse range of phenomena in physics and chemistry unified by means of gauge symmetry, geometric phases, and the (now thirty–year old) Aharonov–Bohm effect. A review with an extended list of references has been published.[26]

References

1. A. Shapere and F. Wilczek, *Geometric Phases in Physics*, (World Scientific, Singapore, 1989).
2. M. V. Berry, *Proc. Roy. Soc. Lond.* **A392** (1984) 45.
3. B. Simon, *Phys. Rev. Lett.* **51** (1983) 2167.
4. Y. Aharonov and J. Anandan, *Phys. Rev. Lett.* **58** (1987) 1593.
5. R. R. Douglas and A. R. Rutherford, *Quantum Adiabatic Phase, Algebraic Topology, and Time–Reversal Invariance*, (Springer–Verlag, New York, 1990), in press.
6. Y. Aharonov and D. Bohm, *Phys. Rev.* **115** (1959) 485.
7. C. N. Yang, *Ann. NY Acad. Sciences* **294** (1977) 86.
8. Y. Aharonov and L. Susskind, *Phys. Rev.* **158** (1967) 1237.
9. G. Herzberg and H. C. Longuet–Higgins, *Disc. Farad. Soc.* **35** (1963) 77.
10. C. A. Mead and D. G. Truhlar, *J. Chem. Phys.* **70** (1979) 2284.
11. C. Bouchiat and G. W. Gibbons, *J. Phys. (Paris)* **49** (1988) 187.
12. M. E. Stoll, A. J. Vega, and R. W. Vaughan, *Phys. Rev.* **A16** (1977) 1521.
13. D. Suter, K. T. Mueller, and A. Pines, *Phys. Rev. Lett.* **60** (1988) 1218.
14. R. Y. Chiao *et al.*, *Phys. Rev. Lett.* **60** (1988) 1214.
15. R. Bhandari and J. Samuel, *Phys. Rev. Lett.* **60** (1988) 1211.
16. F. Wilczek and A. Zee, *Phys. Rev. Lett.* **52** (1984) 2111.
17. J. E. Avron, L. Sadun, J. Segert, and B. Simon, *Comm. Math. Phys.* **124** (1989) 595.
18. R. Tycko, *Phys. Rev. Lett.* **58** (1987) 2281.
19. A. Zee, *Phys. Rev.* **A38** (1988) 1.
20. A. Samoson and A. Pines, *Rev. Sci. Instrum.* **60** (1989) 3239.
21. J. W. Zwanziger, M. Koenig, and A. Pines, *Phys. Rev.*, in press.
22. S. Pancharatnam, *Proc. Ind. Acad. Sci.* **A44** (1956) 247.
23. M. V. Berry, *J. Mod. Optics* **34** (1987) 1401.
24. J. Anandan and A. Pines, *Phys. Lett.* **A141** (1989) 335.
25. R. Montgomery, *Comm. Math. Phys.* **128** (1990) 565.
26. J. W. Zwanziger, M. Koenig, and A. Pines, "Berry's Phase" in *Ann. Rev. Phys. Chem.* **41** ed. H. L. Strauss (Ann. Rev. Inc., Palo Alto, 1990).

GEOMETRICAL AND TOPOLOGICAL (AN)HOLONOMIES IN OPTICAL EXPERIMENTS

RAYMOND Y. CHIAO
Department of Physics, University of California, Berkeley, CA 94720, U. S. A.

ABSTRACT

The Aharonov-Bohm phase and the Berry phase possess important geometrical and topological characteristics. With these in mind, we review four recent optical manifestations of the Berry phase in optics: the spin-redirection phase, the Pancharatnam phase, the squeezed-states Berry phase, and the Kitano-Yabuzaki phase. These optical phenomena can be viewed as Aharonov-Bohm-like effects for the photon. A distinction is made between geometrical and topological effects. The third and fourth optical manifestations are shown to be topological ones closely analogous to the Aharonov-Bohm effect, in the limit of large squeezings and polarizations, respectively. These manifestations can be viewed as the result of parallel transport on the surface of a cone. Also, a distinction is made between classical and nonclassical Berry phases in optics. Nonclassical Berry phases can arise, for example, in an entangled (or EPR) state of two photons. Although these nonclassical phases have not yet been seen, as a preliminary step in this direction we have recently observed the dynamical phase for two entangled photons.

1. Introduction

The Aharonov-Bohm effect is one of the most profound effects in physics discovered in the past thirty years. It is at the heart of gauge field theories of the electrical, weak and strong forces in nature. Recently, the discovery of the closely related Berry phase has taught us that the geometrical and topological ideas involved in the Aharonov-Bohm effect are widespread in all of physics, not just in high-energy physics. For example, geometrical and topological aspects of low-energy physics, such as the appearance of fractional charge in condensed matter physics, and of fractional orbital angular momentum in molecular physics, can best be understood in terms of the Berry phase.

Here we review from the geometrical and topological viewpoint four recent optical manifestations of the Berry phase:

1) The spin-redirection phase (proposed by Chiao and Wu[1] and observed by Tomita and Chiao[2]).
2) The Pancharatnam phase (observed by Bhandari and Samuel,[3] Simon *et al.*[4,5]).
3) The squeezed-state phase (proposed by Chiao and Jordan[6] and Gerry[7]).
4) The Kitano-Yabuzaki phase (proposed and observed by Kitano and Yabuzaki[8]).

These manifestations can be viewed as Aharonov-Bohm-like effects for the photon.

All these phenomena can be understood in terms of what mathematicians call *holonomy* and what physicists, following Berry, call *anholonomy*. An *(an)holonomy* can arise whenever a system is processed through a sequence of changes such that it is returned to its original state. Most of the physical variables characterizing the system obviously return to their original values, but surprisingly certain ones may not. Since the most striking feature of the physics is the fact that certain anomalous variables may *not* return to their original values, I prefer the term *anholonomy*. An important example is the *phase anholonomy* which an electron exhibits when it returns to its original state after encircling a solenoid in the Aharonov-Bohm effect. A second important example is the *age anholonomy* which a travelling twin experiences relative to a nontravelling twin in the twin paradox. A third example is the *angle anholonomy* which a particle experiences after it is processed through a cycle of Lorentz boosts, which is at the heart of Thomas precession. When light is cycled through a sequence of states such that it returns to its original state, it can also acquire such anholonomies at both classical and quantum levels.

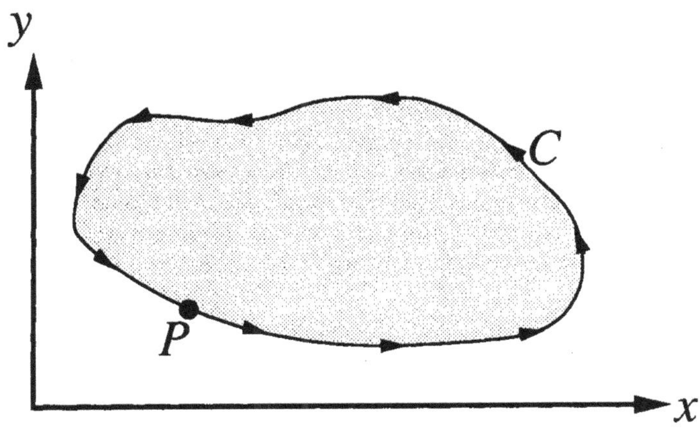

Fig. 1. *Anholonomies* may arise as a result of a circuit C of a system in its state space (x,y).

In Fig. 1, we sketch an abstract circuit C for a system which starts at point P and returns to point P, in a two-dimensional state space. Let x and y be two state variables. After such a circuit, it is obvious that most of the system's physical characteristics, such as x and y, return to their starting values. However, sometimes physical anholonomies may result after such a circuit.

One example is a *phase anholonomy*: Let an electron, starting at P, travel around C inside which is a solenoid. Let x and y represent the spatial position of the electron. After such a circuit, it can acquire an *extra phase* relative to a stationary electron at P, which is directly related to the amount of flux enclosed within C. This phase is also directly proportional to the winding number, or number of turns, around C. The phase anholonomy can be seen in the interference of a single electron with itself at a beam splitter placed at point P in the Aharanov-Bohm effect.

A second example is an *age anholonomy*: Let two identical twins be initially placed at a fixed point P on a merry-go-round, with x and y axes fixed onto the rotating platform. Let one twin stay at P; let the other walk around the circuit C on the rotating merry-go-round. When they meet again at P, although most of their physical characteristics remain identical, they can have *different ages*. For example, one twin may now be bald due to his extra age, whereas his otherwise identical twin brother, is hirsute. Their age difference is directly related to the area enclosed by the circuit C, a geometrical quantity. Again, it is directly proportional to the winding number, or number of turns, of the travelling twin around C. It is closely related to the fact that in general relativity, clocks cannot be globally synchronized. This age anholonomy becomes a phase anholonomy, when the twins are replaced by interfering particles. When a single particle interferes with itself at a beam splitter placed at point P, the Sagnac effect, which has been seen with photons, electrons and neutrons, results.

A third example is an *angle anholonomy*: Let us place a marble on a frictionless table. Let the marble be initially motionless at the origin of a Cartesian coordinate system, with x and y axes aligned with respect to the edges of the table, and the z axis perpendicular to the tabletop. Lorentz boost the marble in the x direction, and then boost it in the y direction. Then boost it in the $-x$ direction, and follow this by a boost in the $-y$ direction, such that it comes back to a motionless state. Thus a circuit is completed in the momentum space (p_x, p_y) of the marble. Now Lorentz boosts in different directions do not commute with each other. As a result, after this cycle of Lorentz boosts, the marble experiences an *angle anholonomy*: It comes back rotated around the z axis by the Wigner angle. Thomas precession is a physical manifestation of this angle anholonomy. This angle anholonomy becomes a phase anholonomy, if the marble is replaced by a quantum spinning particle whose spin is pointing in the z direction.

2. The Berry Phase

The phase anholonomy which Berry[9] discovered in the quantum adiabatic theorem can be expressed as an extra phase factor $\exp(i\gamma_n(C))$ which the wavefunction can acquire after adiabatically following around a circuit C the Hamiltonian $H(\boldsymbol{R})$, where $\boldsymbol{R}$ denotes some slowly varying parameters which return to their starting values. This *geometrical* and/or *topological* phase factor is accumulated by the wavefunction in addition to the usual *dynamical* phase factor, $\exp(-i\int E_n dt/\hbar)$. Explicitly, if $H(\boldsymbol{R})|n;\boldsymbol{R}\rangle = E_n(\boldsymbol{R})|n;\boldsymbol{R}\rangle$, then Berry's phase is

$$\gamma_n(C) = \oint_C \boldsymbol{A}_{\text{eff}} \cdot \mathrm{d}\boldsymbol{R} , \tag{1}$$

where $\boldsymbol{A}_{\text{eff}}(\boldsymbol{R}) = i\langle n;\boldsymbol{R}|\nabla_{\boldsymbol{R}}|n;\boldsymbol{R}\rangle$. This vector field over the parameter space (coordinatized by $\boldsymbol{R}$) can be viewed as the connection of a U(1) bundle over this space. This phase possesses the important property of *local gauge invariance*, viz., under the transformation

$$A_{\text{eff}} \rightarrow A_{\text{eff}} + \nabla_{R}\, \chi(R)\,, \tag{2}$$

where $\chi(\boldsymbol{R})$ is any single-valued function of $\boldsymbol{R}$, the phase $\gamma_n(C)$, and hence the phase factor $\exp(i\gamma_n(C))$, which is the *holonomy* of the U(1) bundle, is invariant. This phase shift is physical, and has been observed as a fringe shift in interference experiments by Chiao *et al.*[10] However, since one can choose $\chi(\boldsymbol{R})$ such that A_{eff} vanishes at each point $\boldsymbol{R}$ *locally*, this phase is identically zero except in *geometrically* and/or *topologically* nontrivial circumstances, just as in the Aharonov-Bohm effect.

In an important generalization, Aharonov and Anandan[11] removed the adiabatic restriction. In so doing, they replaced the notion of *parameter space*, by the notion of *state space*, which is the projective space of rays in Hilbert space. In Eq. 1, one now reinterprets $\boldsymbol{R}$ as coordinates on the state space manifold, rather than as coordinates on the parameter space manifold.

3. The Spin-Redirection Phase

The first recent optical manifestation of Berry's phase is the *angle anholonomy* which occurs when linearly polarized light enters a single-mode isotropic fiber which is wound into the shape of a helix, predicted by Chiao and Wu[1] and observed by Tomita and Chiao.[2] (See Figs. 2 and 3). One can arrange it so that the light returns to its original state in most of its variables, e.g., its frequency, its direction of propagation, and its state of polarization. However, the *axis* of linear polarization of this light does *not* return to its original direction, when referred to the axis of polarization at the entrance to the fiber. This can be understood at a classical level as arising from the *parallel transport* of the electric field vector of the light along the fiber.[12]

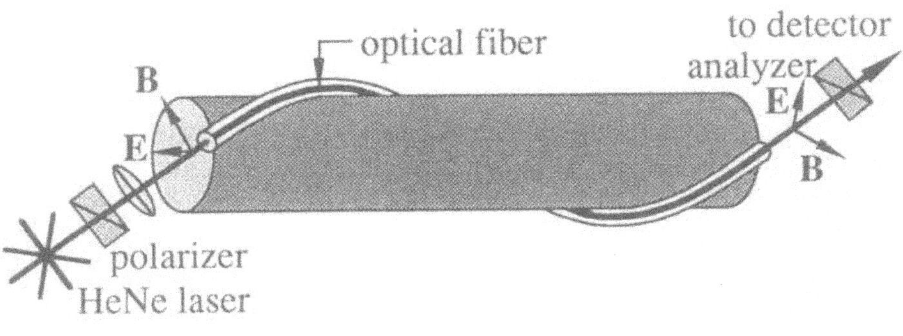

Fig. 2: Schematic of the optical fiber experiment of Tomita and Chiao.

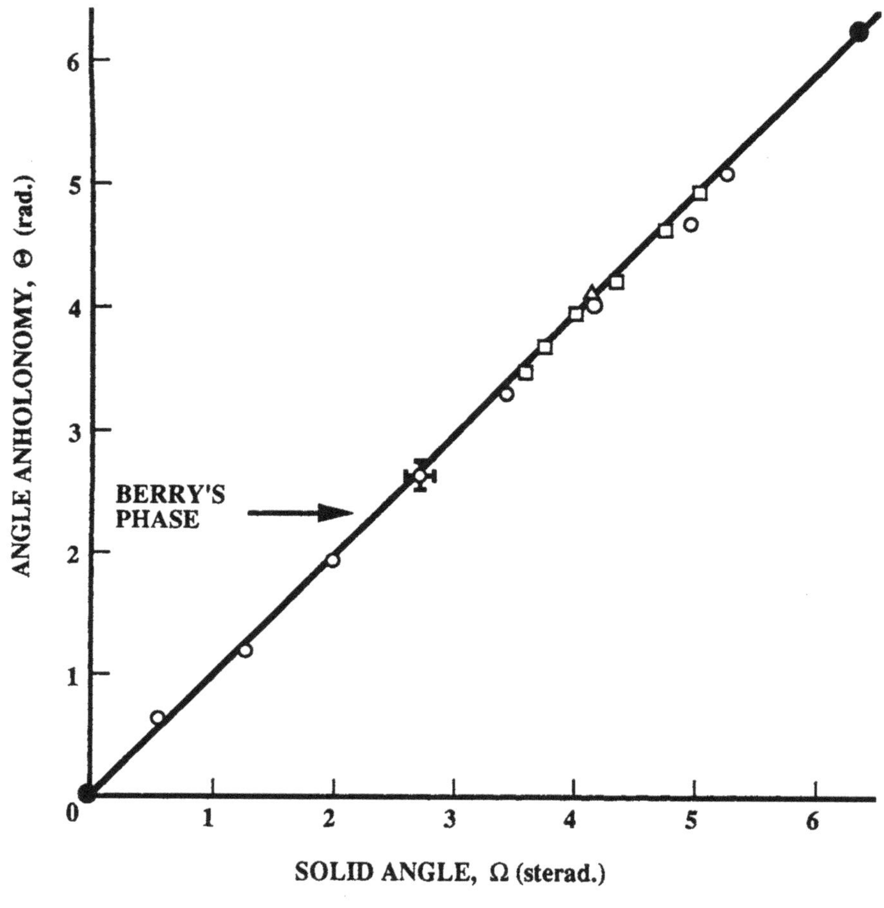

Fig. 3: Angle anholonomy vs. the solid angle $\Omega(C)$ in the experiment of Tomita and Chiao. Open circles represent uniform helices, closed circles planar curves, squares and triangle nonuniform helices.

A simple mechanical analogy helps us to understand this phenomenon:[13] Let one's extended right arm represent the instantaneous direction of propagation $\boldsymbol{k}$ of the light inside the fiber. Now extend the thumb such that it is always perpendicular to the arm. The direction of the thumb represents the electric field vector $\boldsymbol{E}$ of the light, which is always perpendicular to $\boldsymbol{k}$. Let us start with the extended right arm pointing straight in front of us, and the thumb pointing upwards. Let this represent the original state of the light entering the fiber.

Now move the arm upward toward the vertical, *keeping the thumb parallel to itself at all times*. This represents the changing direction of light propagation inside the fiber, with the electric field vector undergoing parallel transport during this process. The arm thus sweeps out an arc in a vertical plane. After the arm reaches the vertical position, move it downward along a perpendicular vertical arc, so that the arm extends to the right. The thumb now points backwards. Complete the cycle by moving the arm in a horizontal arc, so that the arm is brought back to its starting position pointing forwards. Surprisingly the thumb does not end up pointing in its starting direction. It now points towards the right, not upwards, as if it had been rotated by 90° around the axis of the arm. *Locally*, at each point in the motion, there has been no rotation of the thumb; *globally*, however, there exists such an effective rotation.

The above process represents parallel transport of a vector on the surface of a sphere. When generalized to curved spacetime manifolds, it becomes parallel transport in general relativity, where it embodies *the principle of equivalence*. Here, however, we should not invoke parallel transport as a principle; rather, it should be explicitly derivable from the underlying dynamical equations (i.e., Maxwell's equations) along with the symmetry (i.e., local isotropy) of the medium.[1] In fact, Berry has done this recently.[14]

The net result of the above cycle of changes in direction of light is an angle anholonomy: There is an effective rotation of the plane of polarization of the light by 90° in the above example. In general, the angle of rotation of the $\boldsymbol{E}$ vector (the thumb) is equal to the solid angle subtended by the closed curve swept out by the $\boldsymbol{k}$ vector (the arm). This is a consequence of the Gauss-Bonnet theorem, which states that the sum of the three interior angles of a spherical triangle is not 180°, but 180° plus the solid angle enclosed by that triangle, a result of noneuclidean geometry. The interior of any closed curve C can be broken up into a sum of infinitesimal triangles. Therefore the angle anholonomy is equal to the solid angle subtended by C.

This angle anholonomy can also be thought of as an *effective optical activity* induced by the helical path of the light. For instance, one can think of the helically wound single-mode optical fiber, composed of an isotropic glass medium without any intrinsic optical activity, as if it were an optically active medium. However, in contrast to ordinary optical activity, the magnitude of the geometrically induced optical rotation inside the helically wound fiber does not depend on the wavelength of the light, a point we have verified experimentally.[15]

Associated with this angle anholonomy is a phase anholonomy for circularly polarized light: The spin vector of a photon in a beam of circularly polarized light, after a cycle of changes in its direction, is rotated around its axis by this angle,[16,17] thus resulting in a phase shift between opposite senses of circularly polarized light,[1] which has been observed by Chiao *et al.* in a nonplanar Mach-Zehnder interferometer.[10] (See Figs. 4 and 5). This phase shift does not originate in the optical path length difference in the two arms of the interferometer, but rather in the difference in their handedness, a topological effect.

To understand the significance of these results, let us return to Young's original experiment, in which two closely spaced pinholes a distance d apart were inserted into a beam of sunlight. The resulting fringes at a distant screen were *chromatic*, but did not depend on the polarization of the light. Now insert into these pinholes two optical fibers cut to exactly the same length. Let these fibers be single-mode and nonpolarizing. Let us wind these fibers into two large helices of opposite handedness, and bring their output ends together so that they are again spaced a distance d apart, thus restoring Young's original geometry. *Locally*, unpolarized light is propagating down straight, infinitesimal

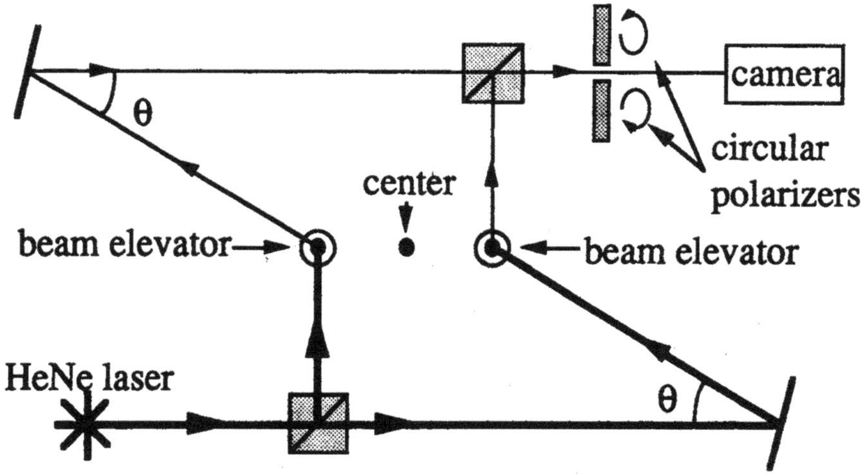

Fig. 4: Schematic of a nonplanar Mach-Zehnder interferometer. The beams in the upper half of the figure are on a higher plane than those in the lower half. Note the center of inversion symmetry. Hence the two arms have opposite handedness.

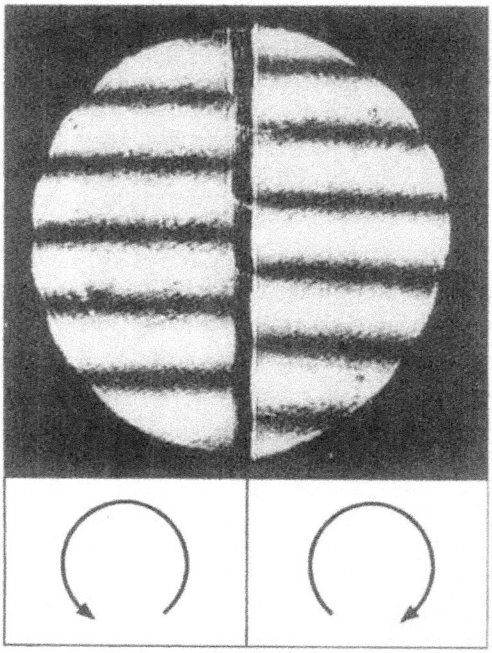

Circular polarizers of opposite senses

Fig. 5: Interferogram showing the phase anholonomy obtained with $\theta = 45^\circ$ in Fig. 4.

sections of isotropic glass fiber, and there seems to be nothing happening to the light. Naïvely, one might expect that there is therefore no change of Young's fringes at the distant screen. However, this is not correct. *Globally*, there exists a phase anholonomy. If this phase anholonomy is π, as in Fig. 5, Young's fringes actually disappear when viewed in natural (unpolarized) light. When viewed through circular polarization filters, Young's fringes reappear, but with a fringe shift which depends on the sense of the circular polarization. This fringe shift is *nonchromatic*, but depends on the polarization of the light. In our experiment (Fig. 4), we replaced for technical reasons these two optical fiber helices by two discrete helices formed out of mirrors, but the resulting fringe shift (Fig. 5) is the same in essence.

State spaces are the arenas in which these geometrical and topological phenomena associated with Berry's phases occur. A cycle of changes in the system corresponds to a closed curve in its state space, whose geometry and topology determine Berry's phases. Associated with the above example is a state space consisting of the surface of a sphere whose points correspond to all possible directions of the photon spin. This surface constitutes the state space manifold. The curvature of this manifold is nonzero. We shall call the associated Berry's phase *the spin-redirection phase*. This phase is given by[1]

$$\gamma(C) = -\sigma\Omega(C)\,, \tag{3}$$

where C is a circuit on the sphere, σ=±1 is the helicity of the photon, and $\Omega(C)$ is the solid angle subtended by C with respect to the center of the sphere. This phase is formally identical to the Aharonov-Bohm phase which an electron acquires after an analogous circuit on a sphere centered on a Dirac monopole. However, since the phase is not independent of arbitrary deformations of the path C, it is more appropriate to call this a *geometrical*, rather than a *topological* phase. The question naturally arises: Are there any truly *topological* Berry's phases? The answer is yes, as we shall see below.

4. The Pancharatnam Phase

In a second example of Berry's phase, the state space is the Poincaré sphere, which describes all possible polarization states of light. A cycle of changes in polarization states, *with the direction of the light kept fixed*, corresponds to a closed curve C on the Poincaré sphere. The associated Berry's phase is "Pancharatnam's phase".[3-5,18] This phase $\gamma(C)$ is given by

$$\gamma(C) = -\frac{1}{2}\Omega(C)\,. \tag{4}$$

Note that the fraction of one-half distinguishes this phase from that in Eq. 3. Since the state space is again a sphere, again, for the reasons given above, it is more appropriate to call this a *geometrical*, rather than a *topological* phase.

5. The Squeezed-States Phase

The generators of squeezed states of light are

$$K_1 = -i(aa-a^\dagger a^\dagger)/4 \text{ and } K_2 = (aa+a^\dagger a^\dagger)/4\,, \tag{5}$$

where $[a,a^\dagger]=1$ for a single mode.[19] These along with $J_3 = (aa^\dagger+a^\dagger a)/4$ satisfy the commutation relations

$$[K_1, K_2] = -iJ_3\,,\ [K_2, J_3] = +iK_1\,,\ [J_3, K_1] = +iK_2\,, \tag{6}$$

which generate the group SU(1,1).[20] This is isomorphic to the Lorentz group in two spatial dimensions.[21] The standard Pauli matrix representations of the Lorentz-group generators are $K_1 = i\sigma_z/2$, $K_2 = i\sigma_x/2$ and $J_3 = \sigma_y/2$. Apart from the minus sign in Eq. 6, the commutation relations are identical to those of the three angular momentum components which generate the group SU(2), whose manifold is a sphere. Starting with an arbitrary state vector, the generators of SU(2) can transport this state to all other possible state vectors of the system. Hence the group manifold is identical to the state space manifold. The state space for SU(2) is therefore a sphere. In the case of Pancharatnam's phase, it is the Poincaré sphere. Likewise, one can construct the state space for squeezed states by letting the elements of SU(1,1) act on any given ray in state space for squeezed states, e.g., the vacuum state $|0\rangle$.

Chiao and Jordan characterized the state space of squeezed states of light by three real numbers k_1, k_2, k_3 satisfying the relationships[6]

$$S_n \ldots S_2 S_1 J_3 S_1{}^\dagger S_2{}^\dagger \ldots S_n{}^\dagger = k_1 K_1 + k_2 K_2 + k_3 J_3\,, \tag{7}$$

$$-k_1{}^2 - k_2{}^2 + k_3{}^2 = 1\,, \tag{8}$$

where a sequence of squeezings associated with unitary operators $S_1, \ldots S_n$ takes J_3 to $k_1K_1+k_2K_2+k_3J_3$. Here each $S_i = \exp\{ir_i(K_1\cos\theta_i + K_2\sin\theta_i)\}$, where r_i is the ith squeezing parameter and θ_i is the ith phase of squeezing. The state space for squeezed states is therefore a *hyperboloid of revolution* given by Eq. 8. A mechanical analogy for a cycle of squeezings of light has already been mentioned in the Introduction. After a cycle of Lorentz boosts in two spatial dimensions, a particle experiences an angle anholonomy in which it is rotated by the Wigner angle ϕ.

The associated phase factor arises from the net rotation operator $\exp(i\phi J_3)$. Thus the resulting phase factor for a number state $|n\rangle$ so cycled is $\exp\{i\phi(n+1/2)/2\}$. For a coherent state $|\alpha\rangle$ so cycled, the result is that $|\alpha\rangle \to \exp(i\phi/4)\ |\alpha\exp(i\phi/2)\rangle$. One interesting feature of these results is that when the number state is processed through such a cycle of squeezings, the resulting Berry's phase is proportional to $n + 1/2$, which is proportional to the intensity of the light. However, when a coherent state is taken through the same cycle, there is no such proportionality.

The "mass hyperboloid" in the mechanical analogy, (with c=1),

$$-p_x^2 - p_y^2 + E^2 = m^2 , \tag{9}$$

corresponds to the hyperboloid of revolution given by Eq. 8. The mass hyperboloid becomes a cone in the ultrarelativistic limit $m \rightarrow 0$, or $|\boldsymbol{p}| \rightarrow \infty$. Since (p_x , p_y) correspond to the squeezing parameters (k_1 , k_2), the transition to a conical limit can be accomplished in the optical case by going to large squeezings $|\boldsymbol{k}| \rightarrow \infty$. But parallel transport on the surface of a cone gives rise to an angle anholonomy which is closely analogous to the Aharonov-Bohm effect.[22] (See Fig. 6). The curvature of a cone is zero

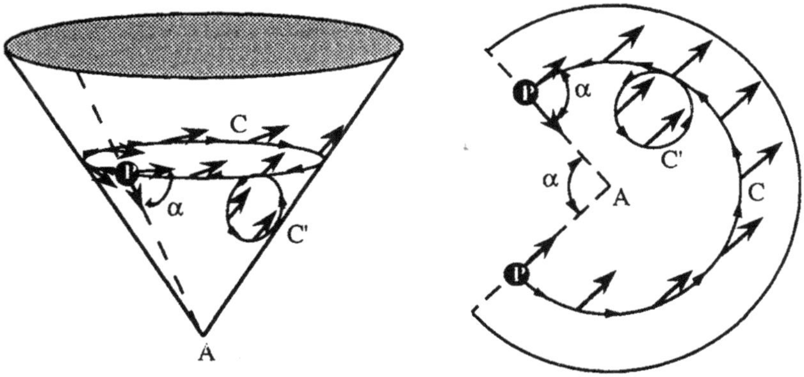

(a) 3D view of cone (b) 2D view of cut cone

Fig. 6: Parallel transport of a vector on a cone.

everywhere except at its apex, where it is infinite. Hence the surface of a cone can be cut along a dashed line which passes through its apex A, as shown in Fig. 6(a), and then be unfolded and laid flat onto a plane. The resulting flat surface is pie-shaped with a wedge angle α, as shown in Fig. 6(b). Parallel transport of a vector on this flat surface leads to no rotation of this vector. However, parallel transport of a vector on the *cone* around two different circuits C and C', where C encloses the singularity at the apex A, but C' does not, lead to different results. Parallel transport around C gives an angle anholonomy, whereas that around C' does not. To find out what the amount of the anholonomy is, consider a point P on the dashed line of the cone. P is a *single* point on the cone, but becomes, after cutting and unfolding, *two* points on the wedge. By plane geometry, we see that the angle anholonomy for circuit C is simply the wedge angle α.

Apex A corresponds to a thin solenoid in the Aharonov-Bohm effect. The wedge angle *α corresponds to the flux inside the solenoid. Circuit C* corresponds to a path of the electron which encircles the solenoid, and circuit C' corresponds to a path which does not. The angle anholonomy for path C corresponds to the phase anholonomy for this path. There are no anholonomies for path C'. There is no curvature at any point on C or C', and this corresponds to the fact that the electron experiences no forces. We should call these effects *topological*, since we can deform the circuit C into any shape as long as it does not cross the singularity at A, and still obtain the same anholonomies.

In the case of the squeezed-states Berry phase in the limit of large squeezing, we obtain a similar phase anholonomy. However, unlike the Aharonov-Bohm effect, where the flux, and hence α, is continuously variable, here α is quantized. Nevertheless, the effect is clearly *topological* here.

6. The Kitano-Yabuzaki Phase

Kitano and Yabuzaki[8] noticed that there is another optical situation where the Lorentz-group anholonomy appears. They found that if one cycles light through a sequence of *partial* linearly polarized states, the light acquires a phase which is formally identical to the above squeezed-states phase.

A partial linear polarizer can be represented by the 2x2 matrix

$$S = \mathrm{diag}(t_x, t_y) = t_a\, \mathrm{diag}(\kappa, \kappa^{-1})\,, \tag{10}$$

where t_i is the transmission coefficient for the ith component of the field, $t_a=(t_x t_y)^{1/2}$ and $\kappa=(t_x/t_y)^{1/2}$. Here the squeezing parameter r is replaced by $-2\ln\kappa$, which is a partial linear polarization parameter. The standard Pauli matrix representations of the Lorentz-group generators used above work here as well. The physical meaning of these generators is that they generate partial linear polarization along the x-axis, partial linear polarization along an axis 45° to the x-axis, and a rotation around the beam axis, respectively.

Associated with this phase anholonomy is an angle anholonomy, in which there is an extra rotation of the axis of partial linear polarization after it has passed through a sequence of four partial linear polarizers. This extra rotation is again analogous to the Wigner angle of the Lorentz group associated with Thomas precession. Kitano and Yabuzaki chose the axes of these polarizers to be at 0°, 45°, 90° and 135°, and observed this angle anholonomy by using tilted glass plates as partial linear polarizers.

Again, parallel transport on a cone is obtained in the limit of large polarizations. The resulting phase and angle anholonomies are again *topological*.

7. Nonclassical Berry Phases In Optics

Most of the Berry phases which have appeared in optical experiments so far can be viewed as phase shifts of a classical electromagnetic wave. The question arises: Are there any quantum mechanical Berry's phases in optics which have no classical explanation? The answer will shed light on uniquely quantum features of geometrical and topological phases. There are two situations in optics where nonclassical Berry's phases should appear. One involves a *single* beam of light; the second *two* beams. The first arises when the light in a single beam is in a nonclassical state, i.e., one that, unlike the coherent state, does not possess any correspondence principle limit in which it turns smoothly into a classical wave. Typical nonclassical states are number states and squeezed states. An example of the first kind of nonclassical Berry's phase is the phase resulting from a cycle of squeezings of a number state mentioned above. The second kind of nonclassical Berry's phase arises when the configuration space of two photons makes its appearance as the state space of the system. Then nonclassical, nonlocal effects can occur when two photons in two beams of light are in "entangled states".

As an example of the latter situation, consider the two-particle wavefunction $\psi(r_1,r_2;t)$ defined on the configuration space $(r_1,r_2;\mathrm{t})$ of the particles, which one can think of as a state space of the system. If these particles undergo a sequence of changes such that they return to their original state, then in general

$$\psi(r_1,r_2;t) \to \exp(i\gamma(C))\ \psi(r_1,r_2;t)\ , \tag{11}$$

where $\gamma(C)$ is a phase which depends on the circuit C in configuration space. A well-known special case arises when these two particles are identical, and their positions are interchanged by a transposition. In three spatial dimensions, the resulting phase factor gives rise to Bose or Fermi statistics, a nonclassical result. Now change the configuration space coodinates $(r_1,r_2;t)$ to angular variables $(\phi_1,\phi_2;t)$. Since the configuration space $(\phi_1,\phi_2;t)$ is doubly periodic with periodicity 2π in both of the variables ϕ_1 and ϕ_2, the state space in this case is a torus, whose geometry and topology is nontrivial. For example, consider a two-photon state prepared in an Einstein-Podolsky-Rosen-Bohm "singlet" combination of linear polarization states

$$|\psi\rangle = 2^{-1/2}\{|x\rangle_1|y\rangle_2 - |y\rangle_1|x\rangle_2\}\ . \tag{12}$$

This is an example of Schrödinger's "entangled states", i.e., superpositions of product states. These states have been the subject of much recent interest, since they are the source of nonlocal and nonclassical phenomena.[23] The wavefunction $\psi(\phi_1,\phi_2;t) = \langle\phi_1,\phi_2;t|\psi\rangle$ is the probability amplitude of finding the two photons linearly polarized along the ϕ_1 and ϕ_2 directions simultaneously, i.e., in coincidence detection at time t. As a simple example of a nonclassical Berry's phase, consider placing in the path of photon 1 an optically active medium which rotates its plane of polarization through 180°. The polarization of photon 1 is thus turned through space by 180°, which results in the sign change

$$\psi(\phi_1,\phi_2;t) \to \exp(i\pi)\psi(\phi_1,\phi_2;t) = -\psi(\phi_1,\phi_2;t), \tag{13}$$

which affects the behavior of photon 2. This is observable in two-photon interferometry.

Although we have not seen these *geometrical* and/or *topological* phases yet, we have recently taken an important step in this direction. Kwiat *et al.* have just completed an experiment in which we have seen the *dynamical* phase of correlated photon pairs in a dual-beam Michelson interferometer.[24] (See Fig. 7.) An entangled state of two photons, a "signal" photon at ω_s and an "idler" photon at $\omega_i=\omega_p-\omega_s$,

$$|\psi\rangle_{\text{in}} = \int d\omega_s\, \varphi(\omega_s)\,|\omega_s\rangle\ |\omega_p-\omega_s\rangle\ , \tag{14}$$

was produced by spontaneous parametric down-conversion in a KDP crystal of a UV argon ion laser "pump" photon at ω_p. In this state, the measurement of the energy of one photon, which had a broad spectrum before measurement, would cause a "collapse of the wavefunction", in which both its energy and the energy of the other photon at a distance would be instantly determined, so that the total energy of the two photons is sharp and conserved.

The two resulting beams of signal and idler photons were injected in a parallel-beam configuration into a single Michelson interferometer, and were detected separately by two photomultipliers, while its travelling mirror was slowly scanned. (After traversing the crystal, the UV beam was absorbed in a beam dump, and did not enter the interferometer.) We took data with sufficiently unequal arm lengths, so that the usual, single-photon interference was not visible. This was easy to do, since the coherence length of the signal and idler photons was quite short (50 μm); we made the optical path length difference much longer (240 μm) than this. However, the *coincidence* count rate between these two spatially separated photomultipliers exhibited fringes with a visibility of nearly 50%, and a period determined by half the UV wavelength. (See Fig. 8. There was no detectable leakage of UV into the interferometer, since the singles rate exhibited no UV fringes.)

This *two-photon* interference occurs because a photon pair could have both gone along the short arm (the "short-short" process), or both along the long arm (the "long-long" process): The probability ***amplitudes*** of these two processes interfere. This interference pattern shows that the dynamical phase of an entangled pair of photons, whose total energy equals the UV laser photon energy, governs the fringe periodicity. We are presently improving our apparatus both in terms of electronic resolution time and increased arm length difference, so that the visibility should approach 100%. This will allow fundamental tests of quantum mechanics, e.g., the violation of Bell's inequalities for time. Also, this type of interferometer should be well suited for observing nonclassical and nonlocal Berry's phases.

8. Acknowledgments

I would like to thank all my collaborators for their valuable contributions to this work. This work was supported by the National Science Foundation under grant ECS 86-13773, and by the Office of Naval Research under contract N00014-88-K-0126.

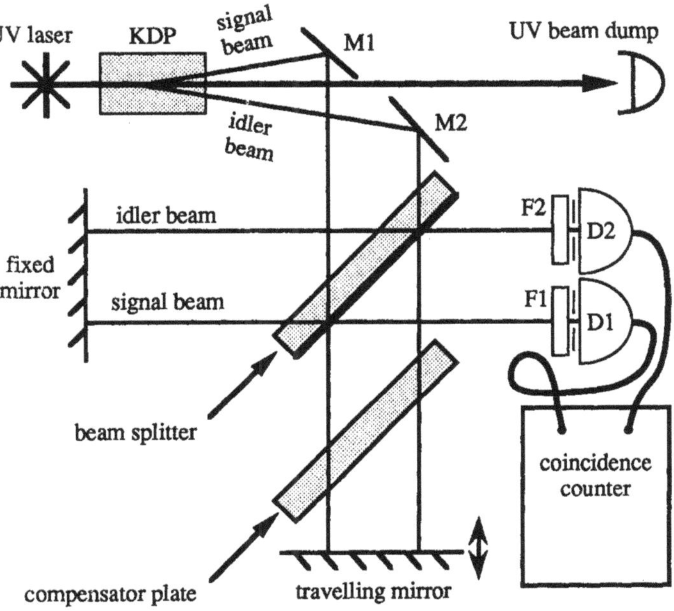

Fig. 7: Schematic of a dual-beam Michelson experiment to see the interference of two correlated photons produced in the parametric down-conversion of a UV laser photon in a KDP crystal.

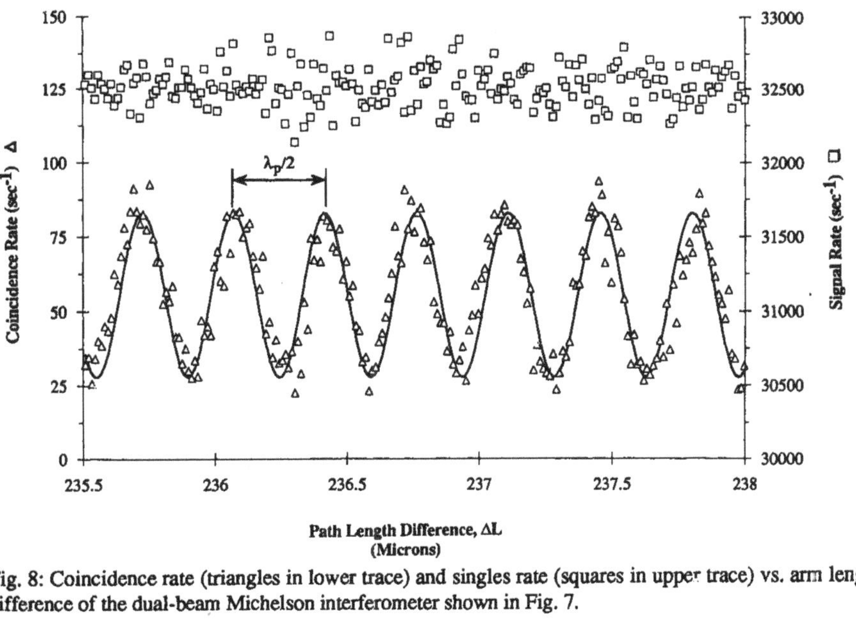

Fig. 8: Coincidence rate (triangles in lower trace) and singles rate (squares in upper trace) vs. arm length difference of the dual-beam Michelson interferometer shown in Fig. 7.

9. References

1. R. Y. Chiao and Y. S. Wu, *Phys. Rev. Lett.* **57** (1986) 933.
2. A. Tomita and R. Y. Chiao, *Phys. Rev. Lett.* **57** (1986) 937.
3. R. Bhandari and J. Samuel, *Phys. Rev. Lett.* **60** (1988) 1210.
4. R. D. Simon, H. J. Kimble, and E. C. G. Sudarshan, *Phys. Rev. Lett.* **61** (1988) 19.
5. T. H. Chyba, L. J. Wang, L. Mandel, and R. Simon, *Opt. Lett.* **13** (1988) 562.
6. R. Y. Chiao and T. F. Jordan, *Phys. Lett.* **A132** (1988) 77.
7. C. C. Gerry, *Phys. Rev.* **A39** (1989) 3204.
8. M. Kitano and T. Yabuzaki (preprint).
9. M. V. Berry, *Proc. Roy. Soc.* (London) **A392** (1984) 45.
10. R. Y. Chiao, A. Antaramian, K. M. Ganga, H. Jiao, S. R. Wilkinson, and H. Nathel, *Phys. Rev. Lett.* **60** (1988) 1214.
11. Y. Aharonov and J. Anandan, *Phys. Rev. Lett.* **58** (1987) 1593.
12. J. N. Ross, *Opt. Quantum Electron.* **16** (1984) 445; F. D. M. Haldane, *Opt. Lett.* **11** (1986) 730.
13. R. Cherry, *Sci. Am.* **260**(3) (1989) 9; I thank C. Wohl for pointing this analogy out to me.
14. M. V. Berry, "Quantum Adiabatic Anholonomy", Lectures at the Ferrara School of Theoretical Physics, June 1989, (to be published).
15. H. Jiao, A. Tomita and R. Y. Chiao (unpublished).
16. J. Anandan and L. Stodolsky, *Phys. Rev.* **D35** (1987) 2597.
17. T. F. Jordan, *J. Math. Phys.* **28** (1987) 1759; *ibid.* **29** (1988) 2042.
18. H. Jiao, S. R. Wilkinson, R. Y. Chiao, and H. Nathel, *Phys. Rev.* **A39** (1989) 3475.
19. See the special issue on squeezed states in *J. Opt. Soc. Am.* **B4** (1987) 1450.
20. R. Gilmore, *Lie Groups, Lie Algebras, and Some of their Applications*, (John Wiley and Sons, New York, 1974).
21. H. Han, Y. S. Kim, and M. E. Noz, *Phys. Rev.* **A37** (1988) 807.
22. B. Felsager and J. M. Leinaas, *Nucl. Phys.* **B166** (1980) 162; J. M. Leinaas, *Gauge Fields and Geometry*, (Thesis, Oslo, 1980, unpublished).
23. J. D. Franson, *Phys. Rev. Lett.* **62** (1989) 2205; M. A. Horne, A. Shimony, and A. Zeilinger, *Phys. Rev. Lett.* **62** (1989) 2209.
24. P. G. Kwiat, W. A. Vareka, C. K. Hong, H. Nathel, and R. Y. Chiao (to be published in *Phys. Rev. A*).

The Berry Phase as an Appropriate Correspondence Limit of the Aharonov-Anandan Phase in a Simple Model

Joy CHRISTIAN

Department of Physics, Boston University
Boston, Massachusetts 02215 USA

and

Abner SHIMONY

Departments of Philosophy and Physics, Boston University
Boston, Massachusetts 02215 USA

ABSTRACT

The relation between the Aharonov-Anandan phase and the Berry phase is studied in a simple model with a time-independent Hamiltonian, in which two spin-$\frac{1}{2}$ systems (I+II) are coupled to a plane rotator (III), so that the parameter ϕ in the part of the Hamiltonian governing I+II is a dynamical variable of III. A family of solutions $\Psi(t)$ with period τ to the time-dependent Schrödinger equation for I+II+III is investigated, yielding the following information. (*i*) The Aharonov-Anandan phase is obtained for each $\Psi(t)$ over the period τ. (*ii*) By tracing out the variables of III a statistical operator W(t) is obtained for I+II. (*iii*) In an appropriate correspondence limit, ϕ acquires a sharply defined value $\phi(t)$, and then W(t) represents a pure state v(t) of I+II. (*iv*) The Aharonov-Anandan phase is calculated for v(t) over the period τ, and it is independent of the Aharonov-Anandan phase of (*i*). (*v*) In the adiabatic limit, when τ goes to infinity, the Aharonov-Anandan phase approaches the Berry phase for I+II. (*vi*) Finally, comparisons are made of the exact energy eigenvalues of the total system with the traditional Born-Oppenheimer approximation, and the improved Born-Oppenheimer approximation.

1. Introduction

Analysis of a quantum mechanical system is usually simplified by dividing the universe into two parts — the system of interest and the rest of the universe or the environment. If it is possible to distinguish between the sets of dynamical variables corresponding to the system and the rest of the universe as the set of 'fast' variables and the set of 'slow' variables respectively, then one may employ the Born-Oppenheimer approximation in the analysis, using the quantum adiabatic theorem. A canonical example of this prescription is the traditional Born-Oppenheimer treatment[1] of the entanglement of the state of the electronic and nuclear variables in molecules. The geometrical phase of Berry is a quantum phase associated with such 'fast' variables under the adiabatic assumption for the 'slow' variables.

In his pioneering work[2], Berry considered a quantal system in an initial eigenstate $|n(0)\rangle$ (with non-degenerate eigenvalues $E_n(0)$) of a Hamiltonian $\hat{H}(X(t))$, which describes the effects of the rest of the universe through the adiabatically changing parameters X(t) such that $X(\tau) = X(0)$. The state evolves under the Schrödinger equation into

$$|n(\tau)\rangle = exp(i\gamma_n)\,|n(0)\rangle \, , \tag{1}$$

remaining, at each instant t, an eigenstate of the Hamiltonian $\hat{H}(t)$ of that instant. Berry showed that the phase γ_n is the sum of a dynamical contribution γ_n^d and a geometrical contribution γ_n^B (where we use B for Berry), with

$$\gamma_n^d = -1/\hbar \int_0^\tau E_n(t)\ \ dt \ , \tag{2}$$

and

$$\gamma_n^B = i \oint_C \langle n(X)\,|\nabla_X|n(X)\rangle\ \ dX \ . \tag{3}$$

Here C is the circuit in parameter space traversed by X(t) as t goes from 0 to τ. When the composite quantal system consisting of the system of interest and the environment is treated in this manner, the adiabatic phase γ_n^B is compensated by the appearance of an effective 'gauge field' in the 'slow' Hamiltonian of the environment[3].

Aharonov and Anandan[4] reformulated and generalized Berry's result by disregarding the parameter space and considering an arbitrary, not necessarily adiabatic, cyclic path in the projective Hilbert space $\mathcal{P}(\mathcal{H})$, which is the space of one-dimensional subspaces or rays of an appropriate Hilbert space $\mathcal{H}$. They showed that the projective Hilbert space itself has a non-trivial topological structure[5]. The Schrödinger equation induces a connection which, in fact, is a natural abelian connection of the projective Hilbert space. The geometric phase is then just an anholonomy due to this abelian connection for a parallel transported statevector along an arbitrary path on this projective Hilbert space $\mathcal{P}(\mathcal{H})$. Thus the geometrical phase as defined by Aharonov and Anandan is truly geometrical, because it is determined by the circuit alone, regardless of the parameterization of the circuit and regardless of the means by which the transport about the path is achieved; i.e., the phase is independent of the coordinates on the projective space $\mathcal{P}(\mathcal{H})$ and the Hamiltonian by which the transport is governed. Consequently, the statevector does not have to be an eigenvector of any instantaneous Hamiltonian, unlike in Berry's treatment. If, however, the transport of the state is achieved by dynamical evolution with Hamiltonian $\hat{H}(t)$, and $|\psi(t)\rangle$ is a solution to the Schrödinger equation with this Hamiltonian, such that the ray determined by $|\psi(t)\rangle$ traverses the cyclical path C in the projective space $\mathcal{P}(\mathcal{H})$ in time τ, then

$$|\psi(\tau)\rangle = exp(i\gamma)\,|\psi(0)\rangle, \tag{4}$$

and the total phase change γ is the sum of a dynamical contribution γ^d and a geometrical contribution γ^{AA} (where we use AA for Aharonov-Anandan):

$$\gamma = \gamma^d + \gamma^{AA} \tag{5}$$

with

$$\gamma^d = -1/\hbar \int_0^\tau \langle\psi(t)\,|\hat{H}|\psi(t)\rangle\; dt. \tag{6}$$

Thus, the geometrical phase γ^{AA} may be computed from the total phase γ and the dynamical phase γ^d without an explicit treatment of the geometrical structure of the projective space $\mathcal{P}(\mathcal{H})$.

The crucial point which concerns us here is the fact that the Berry phase is obtained by using a *classical treatment of the environment.* The treatment of Aharonov and Anandan, on the other hand, is associated with a cyclic evolution of a *pure* quantum state *independent* of a Hamiltonian, and hence it is completely quantum mechanical. It would be instructive, then, to investigate a situation in which the environment is also treated quantum mechanically, in order to bridge the conceptual gap between the two treatments. A natural question that follows is: how can one treat quantum mechanically the varying parameters representing the environment and then recover smoothly, in some appropriate correspondence limit, the definite time-dependence X(t) assumed by Berry? A partial discussion of such a question is found in various treatments[3] (which we shell refer to as "traditional" and "improved") of the Born-Oppenheimer approximation to stationary states of composite molecular systems. In all treatments of this approximation the entangled eigenfunctions of the complete Hamiltonian of the composite molecules are taken in such a form that the *quantized* position variables of the 'slow' subsystems, the nuclei, may be taken as parameters in the instantaneous Hamiltonian of the 'fast' subsystems, the electrons. None of these treatments, however, actually recover quantum mechanically the definite time-dependence of the parameters assumed by Berry. The purpose of our study is to give a rigorous analysis of a model composite system in which — under suitable conditions — the dynamical variable of a larger subsystem has a sharply defined value at each instant of time t, so that it may serve as the parameter in the effective Hamiltonian governing a smaller subsystem. The Aharonov-Anandan phase can be calculated for a cyclical transport of the composite system and — under appropriate conditions — also for the smaller subsystem alone, and in the adiabatic limit it reduces to the Berry phase for that subsystem. Furthermore, we compare the *exact* energy spectrum for the model composite system with the traditional Born-Oppenheimer and the improved Born-Oppenheimer energy spectra, showing that the latter does indeed bring us closer to the exact solution. In the model system we consider below, the agreement between the improved Born-Oppenheimer energy spectrum and the exact energy spectrum is perfect.

2. The Model

In a previous publication[6] exploring the relation between the Aharonov- Anandan phase and the Berry phase, we reasoned as below, using a model system consisting two interacting parts: a spin-$\frac{1}{2}$ system and a plane rotator in an *unchanging* external magnetic field perpendicular to its plane of rotation, so that the *time-independent* total Hamiltonian is

$$H = -i\mu\frac{\partial}{\partial\phi} + \lambda\,\hat{\sigma}\cdot\hat{r} \ , \tag{7}$$

with μ, λ real and non-zero and $\hat{r} = \hat{x}\cos\phi + \hat{y}\sin\phi$.

In this paper we extend our previous model by adding another interacting spin-$\frac{1}{2}$ system. The model system now consists of three interacting parts: I and II are two interacting spin-$\frac{1}{2}$ systems, and III is a plane rotator coupled to I and also to an unchanging external magnetic field perpendicular to its plane of rotation as above. The *time-independent* total Hamiltonian for I+II+III with the dynamical variable ϕ of III directly governing I and indirectly governing II is:

$$H = -i\mu\frac{\partial}{\partial\phi} + \lambda\,\hat{\sigma}_{\rm I}\cdot\hat{r} + \eta\left(\sigma_{x_{\rm I}}\sigma_{x_{\rm II}} + \sigma_{y_{\rm I}}\sigma_{y_{\rm II}}\right) \ , \tag{8}$$

with μ, λ, η real and non-zero and $\hat{r} = \hat{x}\cos\phi + \hat{y}\sin\phi$. The omission of a kinetic energy term for the rotator will be physically justified if the rotator has a very large moment of inertia. The peculiar interaction term $\eta(\sigma_{x_{\rm I}}\sigma_{x_{\rm II}} + \sigma_{y_{\rm I}}\sigma_{y_{\rm II}})$ was chosen after a long struggle with various possibilities, including a term proportional to $\hat{\sigma}_{\rm II}\cdot\hat{r}$. The most difficult part in the sort of analysis that we are interested in is to find an *exact* quantum mechanical solution in closed form for a physically interesting model. In our search, we have found the considered interaction term to be the most suitable one. The advantages for including such an interaction term will be transparent as we go on with our analysis of the model. The motivation for including the interaction term in the first place is, clearly, to extend the previously considered model such that all the results obtained in the *Ref.* 6 become special cases as $\frac{\eta}{\lambda} \longrightarrow 0$. The time-dependent Schrödinger equation with the Hamiltonian given by Eq. (8) reads:

$$i\hbar\frac{\partial\Psi}{\partial t} = \left[-i\mu\frac{\partial}{\partial\phi} + \lambda\begin{pmatrix} 0 & e^{-i\phi} \\ e^{+i\phi} & 0 \end{pmatrix}_{I} + \eta\begin{pmatrix} 0 & 1 \\ 1 & 0 \end{pmatrix}_{I}\otimes\begin{pmatrix} 0 & 1 \\ 1 & 0 \end{pmatrix}_{II} + \eta\begin{pmatrix} 0 & -i \\ i & 0 \end{pmatrix}_{I}\otimes\begin{pmatrix} 0 & -i \\ i & 0 \end{pmatrix}_{II}\right]\Psi. \quad (9)$$

It will be convenient to use a 4×1 column matrix to represent a general ket in the Hilbert space $\mathcal{H}_{I} \otimes \mathcal{H}_{II}$,

$$\begin{pmatrix} a \\ b \\ c \\ d \end{pmatrix} \quad (10)$$

representing

$$a\left(|+\rangle_{I}\otimes|+\rangle_{II}\right) + b\left(|+\rangle_{I}\otimes|-\rangle_{II}\right) + c\left(|-\rangle_{I}\otimes|+\rangle_{II}\right) + d\left(|-\rangle_{I}\otimes|-\rangle_{II}\right), \quad (11)$$

where $|\pm\rangle_i$ are spin-up and spin-down eigenkets of $\sigma_{z,i}$ for i = I,II. The dynamical evolution given by Eq. (9) can then be rewritten as a differential equation in variables ϕ and t for a four-component wavefunction Ψ:

$$-i\hbar\frac{\partial\Psi}{\partial t} = \left[-i\mu\frac{\partial}{\partial\phi} + \lambda\begin{pmatrix} 0 & 0 & e^{-i\phi} & 0 \\ 0 & 0 & 0 & e^{-i\phi} \\ e^{+i\phi} & 0 & 0 & 0 \\ 0 & e^{+i\phi} & 0 & 0 \end{pmatrix} + 2\eta\begin{pmatrix} 0 & 0 & 0 & 0 \\ 0 & 0 & 1 & 0 \\ 0 & 1 & 0 & 0 \\ 0 & 0 & 0 & 0 \end{pmatrix}\right]\Psi. \quad (12)$$

This dynamical equation admits solutions of the form:

$$\Psi_j(\phi,t) = \psi(\phi - \mu t/\hbar)\,exp(-i\mathcal{K}_j t/\hbar)\begin{pmatrix} 1 \\ \frac{2\eta\mathcal{K}_j(\mathcal{K}_j-2\mu)}{\lambda\{(\mathcal{K}_j-\mu)(\mathcal{K}_j-2\mu)-\lambda^2\}}e^{i\phi} \\ \frac{\mathcal{K}_j}{\lambda}e^{i\phi} \\ \frac{2\eta\mathcal{K}_j}{(\mathcal{K}_j-\mu)(\mathcal{K}_j-2\mu)-\lambda^2}e^{2i\phi} \end{pmatrix}, \quad (13)$$

where $\psi(\phi + 2\pi) = \psi(\phi)$, and

$$\mathcal{K}_j = \mu \pm \sqrt{\lambda^2 + 2\eta^2 + \frac{\mu^2}{2} \pm \sqrt{4\eta^2\lambda^2 + 4\eta^4 - 2\mu^2\eta^2 + \mu^2\lambda^2 + \frac{\mu^4}{4}}} \quad (14)$$

with $j = ++, --, +-, or -+$. (From here on we drop the subscript j for simplicity.) It is noteworthy that, except for the periodicity requirement, the wave-packet $\psi(\phi - \mu t/\hbar)$ may be chosen quite arbitrarily. Furthermore, these solutions have the advantage of showing that if there is initially a narrow packet for $\psi(\phi)$, concentrated at $\phi = 0$, then the packet rotates with a velocity $\omega = \frac{\mu}{\hbar}$ without spreading.

This non-spreading of the wave-packets is the result of the absence of kinetic energy term in the Hamiltonian.

We now focus our attention on the subsystem I+II, the two interacting spin-$\frac{1}{2}$ systems. For any solution $\Psi(\phi,t)$ given by Eq. (13) we form the statistical operator using the Dirac notation:

$$|\Psi(t)\rangle\langle\Psi(t)| = \int_0^{2\pi} d\phi \int_0^{2\pi} d\phi' \,\Psi(\phi,t)\Psi^*(\phi',t)\; |\phi\rangle\langle\phi'| , \tag{15}$$

where $\Psi(\phi,t)\Psi^*(\phi',t)$ is a 4×4 matrix. Tracing out of the variable of the plane rotator III yields the statistical operator $W(t)$ for the subsystem I+II :

$$W(t) = \mathrm{Tr}\,(|\Psi(t)\rangle\langle\Psi(t)|) \equiv \int_0^{2\pi} d\phi''\langle\phi''|\Psi(t)\rangle\langle\Psi(t)|\phi''\rangle. \tag{16}$$

The normalization of $|\Psi(t)\rangle$ implies that

$$\int_0^{2\pi} d\phi \;\; |\psi(\phi)|^2 = \frac{1}{(1+A^2+B^2+C^2)} \equiv \frac{1}{N} , \tag{17}$$

where

$$A = \frac{2\eta\mathcal{K}(\mathcal{K}-2\mu)}{\lambda\{(\mathcal{K}-\mu)(\mathcal{K}-2\mu)-\lambda^2\}},\; B = \frac{\mathcal{K}}{\lambda} \text{ and } C = \frac{2\eta\mathcal{K}}{(\mathcal{K}-\mu)(\mathcal{K}-2\mu)-\lambda^2} . \tag{18}$$

Defining

$$J_1 = \int_0^{2\pi} d\phi \;\; |\psi(\phi)|^2 e^{-i\phi} \quad\text{and}\quad J_2 = \int_0^{2\pi} d\phi \;\; |\psi(\phi)|^2 e^{-2i\phi} , \tag{19}$$

we obtain

$$W(t) = \begin{bmatrix} \frac{1}{N} & AJ_1 e^{-i\frac{\mu}{\hbar}t} & BJ_1 e^{-i\frac{\mu}{\hbar}t} & CJ_2 e^{-2i\frac{\mu}{\hbar}t} \\ AJ_1^* e^{i\frac{\mu}{\hbar}t} & \frac{A^2}{N} & \frac{AB}{N} & ACJ_1 e^{-i\frac{\mu}{\hbar}t} \\ BJ_1^* e^{i\frac{\mu}{\hbar}t} & \frac{AB}{N} & \frac{B^2}{N} & BCJ_1 e^{-i\frac{\mu}{\hbar}t} \\ CJ_2^* e^{2i\frac{\mu}{\hbar}t} & ACJ_1^* e^{i\frac{\mu}{\hbar}t} & BCJ_1^* e^{i\frac{\mu}{\hbar}t} & \frac{C^2}{N} \end{bmatrix} . \tag{20}$$

In general $W(t)$ is a mixed state, and none of its eigenvalues (i.e., the statistical weights of the orthogonal pure states in the mixture) α_k equals unity, where α_k are the four roots of the polynomial equation

$$\alpha^4 - \alpha^3 + \alpha^2\left[(A^2+B^2+AC^2+BC^2)\left(\frac{1}{N^2}-|J_1|^2\right) + C^2\left(\frac{1}{N^2}-|J_2|^2\right)\right]$$

$$+\alpha\left(A^2+B^2\right)C^2\left[\frac{2}{N}\left|J_1\right|^2+\frac{1}{N}\left|J_2\right|^2-\frac{1}{N^3}-\left(J_1\right)^2 J_2^*-J_2\left(J_1^*\right)^2\right]=0. \quad (21)$$

As required by the general properties of the statistical operator, $\alpha_k \geq 0$ and $\alpha_1+\alpha_2+\alpha_3+\alpha_4=1$.

Now in order to obtain sharply defined time dependence of the dynamical variable of the rotator, so that it may serve as the parameter in the effective Hamiltonian governing the subsystem I+II, we consider the case when $\psi(\phi)$ is highly localized — effectively a δ-function. Then J_1 and J_2 are effectively the maxima:

$$J_1 \longrightarrow \frac{1}{N}\, e^{-i\phi_0} \qquad \text{and} \qquad J_2 \longrightarrow \frac{1}{N}\, e^{-2i\phi_0}\ , \quad (22)$$

where ϕ_0 is the point at which $\psi(\phi)$ is concentrated. The eigenvalues α_k in this limit become

$$\alpha_1,\ \alpha_2,\ \text{and}\ \alpha_3 \longrightarrow 0 \qquad \text{and} \qquad \alpha_4 \longrightarrow 1\ , \quad (23)$$

which can be readily seen from Eq. (21), and the statistical operator $W(t)$ effectively represents a pure case of the subsystem I+II. This result is not surprising, since the insertion of a highly localized $\psi(\phi)$ into the solution $\Psi(\phi, t)$ implies that ϕ has an effectively well defined time-dependent value:

$$\phi\,(t)=\phi_0+\frac{\mu}{\hbar}\,t. \quad (24)$$

With no loss of generality ϕ_0 can be taken to be zero, by shifting the origin of time. The corresponding pure state of the spinor part I+II can be expressed as

$$\mathrm{v}\,(t)=\frac{e^{i\beta(t)}}{\sqrt{N}}\begin{pmatrix} 1 \\ Ae^{i\frac{\mu}{\hbar}t} \\ Be^{i\frac{\mu}{\hbar}t} \\ Ce^{2i\frac{\mu}{\hbar}t} \end{pmatrix}, \quad (25)$$

where $\beta(t)$ is not determined by $W(t)$.

The ray to which $\mathrm{v}(t)$ belongs is independent of $\beta(t)$, and hence the computation of the Aharonov-Anandan phase $\gamma^{AA}_{\mathrm{I+II}}$ associated with a cycle in the projective Hilbert space $\mathcal{P}(\mathcal{H}_{\mathrm{I+II}})$ of the subsystem I+II is also independent of this choice. To calculate $\gamma^{AA}_{\mathrm{I+II}}$ it suffices to find a Hamiltonian $h(t)$ and a $\beta(t)$ such that $\mathrm{v}(t)$ is the solution to the Schrödinger equation with this $h(t)$. One convenient choice is

$$\beta\,(t)=-\frac{\mathcal{K}}{\hbar}\,t\,, \quad \text{and}$$

$$h\,(t)=\lambda\begin{pmatrix} 0 & e^{-i\phi(t)} \\ e^{i\phi(t)} & 0 \end{pmatrix}_{\mathrm{I}}+\eta\begin{pmatrix} 0 & 1 \\ 1 & 0 \end{pmatrix}_{\mathrm{I}}\otimes\begin{pmatrix} 0 & 1 \\ 1 & 0 \end{pmatrix}_{\mathrm{II}}+\eta\begin{pmatrix} 0 & -i \\ i & 0 \end{pmatrix}_{\mathrm{I}}\otimes\begin{pmatrix} 0 & -i \\ i & 0 \end{pmatrix}_{\mathrm{II}}, \quad (26)$$

with $\phi(t) = \frac{\mu}{\hbar}\, t$. It can be checked that v(t) of Eq. (25) with $\beta(t) = -\frac{\mathcal{K}}{\hbar}\, t$ satisfies the Schrödinger equation

$$i\hbar \frac{\partial\, \mathrm{v}(t)}{\partial t} = h(t)\ \mathrm{v}(t). \tag{27}$$

It is now clear that our model composite system I+II+III, together with the choice of a highly localized $\psi(\phi)$, provides a quantum mechanical basis for a parameter-dependent Hamiltonian for the spinor subsystem I+II, the parameter being a well defined cyclic function of time. We emphasize that this result was obtained with a *time-independent* Hamiltonian for the composite system I+II+III.

3. Geometrical Phases

We now calculate the various phase changes in the period $\tau = 2\pi\frac{\hbar}{\mu}$. The total phase change in the state of the composite system I+II+III for a cycle of period $\tau = 2\pi\frac{\hbar}{\mu}$ is:

$$\gamma_{\mathrm{I+II+III}} = -\,2\pi\frac{\mathcal{K}}{\mu}\ , \tag{28}$$

since

$$\Psi(\phi,\tau) = exp\left(-i2\pi\frac{\mathcal{K}}{\mu}\right)\Psi(\phi,0). \tag{29}$$

The dynamical phase acquired by the system in this period is

$$\gamma^{d}_{\mathrm{I+II+III}} = -\,1/\hbar \int_0^\tau \langle\Psi(t)\,|H|\Psi(t)\rangle\ dt\ , \tag{30}$$

which gives

$$\gamma^{d}_{\mathrm{I+II+III}} = -2\pi\frac{\mathcal{K}}{\mu} - 4\pi\left(1 + A^2 + B^2 + C^2\right)\int_0^{2\pi} (Re\,\psi(\phi))\,\frac{\partial}{\partial\phi}(Im\,\psi(\phi))\,d\phi. \tag{31}$$

Hence the geometric phase $\gamma^{AA}_{\mathrm{I+II+III}} = \gamma_{\mathrm{I+II+III}} - \gamma^{d}_{\mathrm{I+II+III}}$ is:

$$\gamma^{AA}_{\mathrm{I+II+III}} = 4\pi\left(1 + A^2 + B^2 + C^2\right)\int_0^{2\pi} (Re\,\psi(\phi))\,\frac{\partial}{\partial\phi}(Im\,\psi(\phi))\,d\phi. \tag{32}$$

The intimate inner structure of $\psi(\phi)$ — specifically, the exact relation between its real and imaginary parts — determines the Aharonov-Anandan phase for I+II+III. There is essentially no restriction on this phase, even when $\psi(\phi)$ is localized very narrowly about one value of its argument. However, when $\psi(\phi)$ is pure real, pure imaginary, or even or odd about $\phi = \pi$, then $\gamma^{AA}_{\mathrm{I+II+III}}$ is zero.

For the subsystem I+II alone, evolving according to Eq. (27), the total phase change in a period τ is

$$\gamma_{I+II} = -2\pi\frac{\mathcal{K}}{\mu}, \tag{33}$$

since

$$\mathrm{v}(\tau) = exp\left(-i2\pi\frac{\mathcal{K}}{\mu}\right)\mathrm{v}(0). \tag{34}$$

The dynamical phase acquired by this subsystem in the period τ is:

$$\gamma^{d}_{I+II} = -1/\hbar\int_{0}^{\tau}\langle \mathrm{v}(t)|h(t)|\mathrm{v}(t)\rangle\ dt, \tag{35}$$

which gives

$$\gamma^{d}_{I+II} = -2\pi\frac{\mathcal{K}}{\mu} + \pi\left(\frac{2A^2+2B^2+4C^2}{1+A^2+B^2+C^2}\right), \tag{36}$$

where A, B, and C are given by Eq. (18). Hence the Aharonov-Anandan phase for the subsystem I+II alone, under the condition that the wave-packet $\psi(\phi)$ is narrowly localized, is:

$$\gamma^{AA}_{I+II} = -\pi\left[\frac{2\left(\frac{2\eta\mathcal{K}(\mathcal{K}-2\mu)}{\lambda\{(\mathcal{K}-\mu)(\mathcal{K}-2\mu)-\lambda^2\}}\right)^2 + 2\left(\frac{\mathcal{K}}{\lambda}\right)^2 + 4\left(\frac{2\eta\mathcal{K}}{(\mathcal{K}-\mu)(\mathcal{K}-2\mu)-\lambda^2}\right)^2}{1+\left(\frac{2\eta\mathcal{K}(\mathcal{K}-2\mu)}{\lambda\{(\mathcal{K}-\mu)(\mathcal{K}-2\mu)-\lambda^2\}}\right)^2 + \left(\frac{\mathcal{K}}{\lambda}\right)^2 + \left(\frac{2\eta\mathcal{K}}{(\mathcal{K}-\mu)(\mathcal{K}-2\mu)-\lambda^2}\right)^2}\right]. \tag{37}$$

This phase is trivial *iff* $\left(\frac{2\eta\mathcal{K}}{(\mathcal{K}-\mu)(\mathcal{K}-2\mu)-\lambda^2}\right)^2 = 1$, which happens only when $\mu = 0$, and hence it is non-trivial in general.

It should be noted again that for our model composite system the geometric phase γ^{AA}_{I+II} is defined only in the situation sketched above, when $\psi(\phi)$ is highly localized, so that the dynamical variable ϕ of the plane rotator behaves like a classical parameter from the standpoint of the two interacting spin-$\frac{1}{2}$ systems. Also noteworthy is that the geometric phase γ^{AA}_{I+II} for the subsystem I+II alone, when it is defined, is completely independent of the geometric phase $\gamma^{AA}_{I+II+III}$ given by Eq. (32) for the composite system as a whole. This independence is not surprising when one considers that phase information concerning a composite system is usually lost when one traces out the variables of one subsystem[7]. In the case of the composite system I+II+III, the geometric phase depends upon the details of the function $\psi(\phi)$, whereas the geometric phase for the subsystem I+II alone, after tracing out the variables of the rotator, depends only on $\psi(\phi)$ approaching a highly localized wave-packet; this localization ensures that the parameter ϕ of the subsystem I+II achieves a definite value as a function of time, giving an unequivocal geometrical phase for the spinor part independent of the details of the function $\psi(\phi)$.

The geometrical phase of Berry can now be easily recovered from the non-trivial Aharonov-Anandan phase γ^{AA}_{I+II} of Eq. (37) by simply taking the adiabatic limit $\tau \longrightarrow \infty$. In this limit the parameter $\phi(t)$ in the Hamiltonian $h(t)$, expressed in Eq. (26) for the subsystem I+II, changes adiabatically. This limit is equivalent to the limit $\mu \longrightarrow 0$ since $\tau = 2\pi\frac{\hbar}{\mu}$. But then the $\mathcal{K}_j$ of the Eq. (14) become

$$\mathcal{K}_j \longrightarrow \pm\sqrt{\lambda^2 + 2\eta^2 \pm 2\eta\sqrt{\lambda^2+\eta^2}}. \tag{38}$$

Substituting this as well as $\mu \longrightarrow 0$ into the expression (37) for the Aharonov-Anandan phase $\gamma^{AA}_{\mathrm{I+II}}$ for the subsystem I+II, we finally obtain

$$\lim_{\tau\to\infty} \gamma^{AA}_{\mathrm{I+II}} = \gamma^{B}_{\mathrm{I+II}} = -2\pi \ , \tag{39}$$

where B is used for Berry as before. We shell see in the next section (*cf.* Eq. (64)) that the Berry phase is indeed trivial for the subsystem I+II (unlike the previously considered model[6], in which case it is $-\pi$, and hence non-trivial).

This completes our demonstration that — *under appropriate correspondence conditions* — the Aharonov-Anandan phase in the adiabatic limit reduces to the Berry phase.

4. Born-Oppenheimer Approximation and the Berry Phase

In the traditional Born-Oppenheimer approximation to stationary states of a composite system the quantized position variables of the 'slow' subsystem are taken as parameters in the *instantaneous* Hamiltonian of the 'fast' subsystem. For our composite system I+II+III, the time-independent Hamiltonian is:

$$H = -i\mu\frac{\partial}{\partial\phi} + h(\phi) \ , \tag{40}$$

with

$$h(\phi) \equiv \lambda\begin{pmatrix} 0 & e^{-i\phi} \\ e^{i\phi} & 0 \end{pmatrix}_{\mathrm{I}} + \eta\begin{pmatrix} 0 & 1 \\ 1 & 0 \end{pmatrix}_{\mathrm{I}} \otimes \begin{pmatrix} 0 & 1 \\ 1 & 0 \end{pmatrix}_{\mathrm{II}} + \eta\begin{pmatrix} 0 & -i \\ i & 0 \end{pmatrix}_{\mathrm{I}} \otimes \begin{pmatrix} 0 & -i \\ i & 0 \end{pmatrix}_{\mathrm{II}} . \tag{41}$$

As in Eq. (12), we may write the time-independent Schrödinger equation as a four-component differential equation which has the following *exact* eigenfunctions and eigenvalues:

$$\Psi_p(\phi) = \frac{e^{ip\phi}}{\sqrt{2\pi\,(1+A^2+B^2+C^2)}}\begin{pmatrix} 1 \\ Ae^{i\phi} \\ Be^{i\phi} \\ Ce^{2i\phi} \end{pmatrix} \tag{42}$$

except for an irrelevant phase factor, and

$$E_p^j(exact) = p\,\mu + \mathcal{K}_j \;; \tag{43}$$

here p is an integer, $\mathcal{K}_j$ is given by Eq. (14), and A, B, and C are given by Eq. (18) with readoption of the subscript j on $\mathcal{K}$. The correctness of Eq. (42) and Eq. (43) can be checked by substitution into $H\Psi(\phi) = E\Psi(\phi)$.

We, however, wish to compare this exact solution of the complete eigenvalue equation with the traditional Born-Oppenheimer approximation applied to the system. Therefore, we follow the usual procedure[3] and look for approximate eigenfunctions of the complete Hamiltonian of Eq. (40) in the form:

$$|\Psi\rangle = \int d\phi\; U_m(\phi)\; |m;\phi\rangle \otimes |\phi\rangle\,, \tag{44}$$

where $U_m(\phi)$ are the wavefunctions of the rotator, $|\phi\rangle \in \mathcal{H}_{\rm III}$ is an eigenket of the angular observable, and $|m;\phi\rangle \in \mathcal{H}_{\rm I} \otimes \mathcal{H}_{\rm II}$ satisfies the energy eigenvalue equation

$$h(\phi)\; |m;\phi\rangle = \varepsilon_m(\phi)\; |m;\phi\rangle\,, \tag{45}$$

with ϕ in the eigenkets of $h(\phi)$ treated as the slowly varying parameter. Of course, $|\Psi\rangle \in \mathcal{H} \equiv \mathcal{H}_{\rm I} \otimes \mathcal{H}_{\rm II} \otimes \mathcal{H}_{\rm III}$ can be represented by the four-component wavefunction

$$\Psi(\phi) = U_m(\phi)\; u_m^\phi\,, \tag{46}$$

where the 4×1 column matrix u_m^ϕ represents $|m;\phi\rangle$ by the convention of Eq. (10). The eigenvalue equation (45) then yields:

$$u_m^\phi = \frac{1}{\sqrt{1 + \left(\frac{2\eta\varepsilon_m^2}{\lambda(\varepsilon_m^2-\lambda^2)}\right)^2 + \left(\frac{\varepsilon_m}{\lambda}\right)^2 + \left(\frac{2\eta\varepsilon_m}{\varepsilon_m^2-\lambda^2}\right)^2}} \begin{pmatrix} 1 \\ \frac{2\eta\varepsilon_m^2}{\lambda(\varepsilon_m^2-\lambda^2)}\, e^{i\phi} \\ \frac{\varepsilon_m}{\lambda}\, e^{i\phi} \\ \frac{2\eta\varepsilon_m}{\varepsilon_m^2-\lambda^2}\, e^{2i\phi} \end{pmatrix}, \tag{47}$$

except for an irrelevant phase factor, and

$$\varepsilon_m = \pm\sqrt{\lambda^2 + 2\eta^2 \pm 2\eta\sqrt{\lambda^2+\eta^2}}\,, \tag{48}$$

where $m = ++,\ --,\ +-$, or $-+$. To obtain an equation for $U_m(\phi)$ in the Born-Oppenheimer approximation, we project the complete eigenvalue equation $H\Psi(\phi) = E\Psi(\phi)$ on the eigenkets $|m;\phi\rangle$ and, under the adiabatic condition, ignore the off-diagonal matrix elements, yielding the following eigenvalue equation for the rotator wavefunctions $U_m(\phi)$:

$$-i\mu\left[\frac{\partial}{\partial\phi}U_m(\phi) + U_m(\phi)\,\langle m;\phi|\frac{\partial}{\partial\phi}|m;\phi\rangle\right] = (E - \varepsilon_m)\,U_m(\phi)\,. \tag{49}$$

In the traditional treatment of the Born-Oppenheimer approximation applied to a composite quantal system, an assumption is usually made that the phases of the 'fast' wavefunctions may always be chosen such that the wavefunctions are real[8]. If this is true then the term proportional to $\langle m;\phi|\frac{\partial}{\partial\phi}|m;\phi\rangle$ in the Eq. (49) would vanish, and hence the rotator wavefunctions would admit solutions of the form:

$$U_m(\phi) = U_m(\phi=0)\, e^{i\left(\frac{E}{\mu}-\frac{\varepsilon_m}{\mu}\right)\phi}\,. \tag{50}$$

The total wavefunction $\Psi(\phi) = U_m(\phi)\, u_m^\phi$ of the complete Hamiltonian H has to be single-valued in $\phi : \Psi(2\pi) = \Psi(0)$. Hence, $\frac{E}{\mu} - \frac{\varepsilon_m}{\mu} = p$ giving

$$E_p^m\,(\textit{Traditional Born–Opp.}) = \; p\,\mu + \varepsilon_m\,, \tag{51}$$

where p is an integer as before.

However, after Berry's work, we know that in some cases the impossibility of removing all phases gives rise to an effective 'gauge field' which appears in the 'slow' Hamiltonian of the total system[3] as in Eq. (49) above. Hence, in general, the connection coefficients

$$\mathbf{A}_{mm} = i\langle m;\phi|\frac{\partial}{\partial\phi}|m;\phi\rangle \tag{52}$$

of the 'Berry's connection' $\mathbf{A}$ are non-zero; using Eq. (47) we can easily obtain:

$$\mathbf{A}_{mm} = -1\left[\frac{\left(\frac{2\eta\varepsilon_m^2}{\lambda(\varepsilon_m^2-\lambda^2)}\right)^2 + \left(\frac{\varepsilon_m}{\lambda}\right)^2 + 2\left(\frac{2\eta\varepsilon_m}{\varepsilon_m^2-\lambda^2}\right)^2}{1+\left(\frac{2\eta\varepsilon_m^2}{\lambda(\varepsilon_m^2-\lambda^2)}\right)^2 + \left(\frac{\varepsilon_m}{\lambda}\right)^2 + \left(\frac{2\eta\varepsilon_m}{\varepsilon_m^2-\lambda^2}\right)^2}\right] = -1\,. \tag{53}$$

With these non-zero connection coefficients $\mathbf{A}_{mm}$, the solutions of Eq. (49) now become:

$$U_m(\phi) = U_m(\phi=0)\; e^{i\left(\frac{E}{\mu}-\frac{\varepsilon_m}{\mu}+\mathbf{A}_{mm}\right)\phi}\,, \tag{54}$$

amending the energy spectrum with a correction term $-\mu\mathbf{A}_{mm}$:

$$E_p^m\,(\textit{Improved Born–Opp.}) = \; p\,\mu - \mu\mathbf{A}_{mm} + \varepsilon_m\,, \tag{55}$$

which upon using Eq. (53) becomes,

$$E_p^m\,(\textit{Improved Born–Opp.}) = \; p\,\mu + \mu + \varepsilon_m\,. \tag{56}$$

This improved Born-Oppenheimer energy spectrum should be compared to the *exact* energy spectrum given by the Eq. (43) in the limit $\frac{\mu}{\lambda} \longrightarrow 0$.

The *exact* energy spectrum, Eq. (43), of the Hamiltonian given by Eq. (40) in the limit $\frac{\mu}{\lambda} \longrightarrow 0$ becomes

$$\lim_{\frac{\mu}{\lambda}\to 0} E_p^m\,(\textit{exact}) = \; p\,\mu + \mu + \varepsilon_m\,, \tag{57}$$

since from Eq. (14),

$$\lim_{\frac{\mu}{\lambda}\to 0} \mathcal{K}_j = \mu + \varepsilon_m \,, \tag{58}$$

with ε_m given by Eq. (48). Comparing Eqs. (56) and (57) we see that the correction term $-\mu \mathbf{A}_{mm}$ not only brings us closer than the traditional Born-Oppenheimer method to the exact solution, but also gives us complete agreement with the exact energy spectrum; for without the connection coefficients $\mathbf{A}_{mm}$ there would not be an extra μ-dependent term in the Born-Oppenheimer spectrum.

The simpler special case studied in the *Ref.* 6, corresponds to the vanishing of the interaction: $\frac{\eta}{\lambda} \longrightarrow 0$, so that subsystem II becomes decoupled from I+III. In this limit the exact energy spectrum given by Eq. (43) becomes

$$\lim_{\frac{\eta}{\lambda}\to 0} E_p^m\,(exact) = \; p\,\mu + \frac{\mu}{2} \pm \lambda\sqrt{1+\frac{\mu^2}{4\lambda^2}}\,, \tag{59}$$

and the eigenvalues of $|m;\phi\rangle$ given by Eq. (48) become

$$\lim_{\frac{\eta}{\lambda}\to 0} \varepsilon_m = \pm\lambda\,, \tag{60}$$

agreeing with the results of *Ref.* 6. When the energy spectrum of Eq. (59) is considered in the further limit of $\frac{\mu}{\lambda} \longrightarrow 0$, we obtain

$$\lim_{\frac{\mu}{\lambda}\to 0}\lim_{\frac{\eta}{\lambda}\to 0} E_p^m\,(exact) = \; p\,\mu + \frac{\mu}{2} \pm \lambda\,. \tag{61}$$

Turning now to the Born-Oppenheimer approximation in the case of $\frac{\eta}{\lambda} \longrightarrow 0$, we find that the connection coefficients given by Eq. (53) in this limit become

$$\lim_{\frac{\eta}{\lambda}\to 0} \mathbf{A}_{mm} \longrightarrow -\frac{1}{2}\,. \tag{62}$$

Then the improved Born-Oppenheimer spectrum given by Eq. (55) becomes

$$\lim_{\frac{\eta}{\lambda}\to 0} E_p^m\,(\textit{Improved Born-Opp.}) = \; p\,\mu + \frac{\mu}{2} \pm \lambda\,, \tag{63}$$

due to Eq. (60). Comparing Eq. (61) and Eq.(63) we see that in this restricted case we get complete agreement between the exact and the improved Born-Oppenheimer energy spectra, just as when we do not assume $\frac{\eta}{\lambda} \longrightarrow 0$.

It should be noted, however, that when $\frac{\eta}{\lambda} \longrightarrow 0$, the alteration of the traditional Born-Oppenheimer energy spectrum due to the correction term $\frac{\mu}{2}$ is observationally quite significant, whereas for non-zero $\frac{\eta}{\lambda}$, the alteration amounts to a mere relabeling of the traditional Born-Oppenheimer energy spectrum by the *trivial change of integer*, $p \longrightarrow (p+1)$. This difference is merely another manifestation of facts about the Berry phase. It can be easily checked, by using the values of the connection coefficients $\mathbf{A}_{mm}$ from Eqs. (62) and (53), that the Berry phase in each respective case is:

$$\gamma^{B}_{\mathbf{I+I}} \equiv \int_{0}^{2\pi} \mathbf{A}_{mm}\, d\phi = \begin{cases} -\pi, & \text{for } \frac{\eta}{\lambda} \longrightarrow 0; \\ -2\pi, & \text{otherwise.} \end{cases} \tag{64}$$

In the latter case the Berry phase is trivial (*cf.* Eq. (39)) as expected, since we are dealing with *two* spin-$\frac{1}{2}$ systems, whereas when $\frac{\eta}{\lambda} \longrightarrow 0$ we are effectively dealing with a single spin-$\frac{1}{2}$ system, for which the geometric phase obtained by Berry[2,6] for our geometry is $-\pi$.

5. Conclusion

Quantum mechanics postulates a one-one correspondence between the set of states of a physical system and the projective space $\mathcal{P}(\mathcal{H})$ of one-dimensional subspaces or rays of an appropriate Hilbert space $\mathcal{H}$. Consequently, it seems *prima facie* arbitrary whether we choose to represent a given state by the vector $|\psi\rangle \in \mathcal{H}$ or by $e^{i\alpha}|\psi\rangle$, where $\alpha \in \mathbf{R}$, since both belongs to the same ray. Nevertheless, the determination of phase turns out to be a subtle matter, with rich implications for physical phenomena, especially when the state of the system undergoes a cyclical change. In such a case, the state of the quantum mechanical system acquires a *path dependent* geometrical contribution in the phase, over and above the usual dynamical contribution. This geometric phase, as defined by Aharonov and Anandan is a quantal phase associated with a cyclic evolution of a *pure* quantum state. The geometrical phase discovered by Berry, on the other hand, is a geometrical phase defined for a state of a physical system, evolving in a cycle, and interacting with its environment in such a way that the dynamical effects may be represented by slowly and cyclically varying parameters of the Hamiltonian. At first glance, these two approaches appear to be quite different in spirit. However, in this paper we have given a pedagogical exposition aimed to bridge the conceptual gap between the two treatments by rigorously analyzing a model composite quantal system, and have shown that — under appropriate correspondence conditions — the Berry phase can be recovered from the Aharonov-Anandan phase. In a previous paper we presented a model for which a *non-trivial* Berry phase is recovered from a non-trivial Aharonov-Anandan phase. In this paper we have extended our previous model and have exhibited a case in which a *trivial* Berry phase is obtained from a non-trivial Aharonov-Anandan phase. Nevertheless, the extended model incorporates the results of the previously considered model as special cases, if care is taken with the order of limiting processes as in Eq. (61). Furthermore, we have taken this opportunity to compare the traditional Born-Oppenheimer energy spectra and the improved Born-Oppenheimer energy spectra with the exact spectra obtained for the model composite systems. As far as we know, this is the first comparison of the improved Born-Oppenheimer spectra with the energy spectra of exactly solvable completely quantum mechanical models.

Acknowledgment

This work was supported in part by the National Science Foundation, Grant No. DIR-8810713.

References

1. M. Born and K. Huang, *Dynamical Theory of Crystal Lattices*, (Oxford University Press, London, 1954), Chap. IV.
2. M.V. Berry, *Proc. R. Soc. Lond.* A **392** (1984) 45. For further elaborations and generalizations see the delightful collection, *Geometric Phases in Physics*, eds. A. Shapere and F. Wilczek, (World Scientific, Singapore, 1989).
3. See, for example, C.A. Mead and D.G. Truhlar, *J. Chem. Phys.* **70** (1979) 2284; I.J.R. Aitchison, *Phys. Script.* **T23** (1988) 12; and R. Jackiw, *Comm. At. Mol. Phys.* **21** (1988) 71.
4. Y. Aharonov and J. Anandan, *Phys. Rev. Lett.* **58** (1987) 1593.
5. For a further explication of this point and for the differential geometric setting for the geometric phase see, for instance, D. Page, *Phys. Rev. A* **36** (1987) 3479; J. Samuel and R. Bhandari, *Phys. Rev. Lett.* **60** (1988) 2339; and C. Bouchiat and G.W. Gibbons, *J. Phys. France* **49** (1988) 187.
6. J. Christian and A. Shimony, "The Relation Between the Aharonov – Anandan Phase and the Berry Phase in a Simple Model", in the *Proceedings of The 3rd International Symposium on Foundations of Quantum Mechanics -Tokyo'89* (1990).
7. J.M. Jauch, *Foundations of Quantum Mechanics*, (Addison-Wesley, Reading, MA, 1968), Chap. 11.
8. L.I. Schiff, *Quantum Mechanics*, Third Edition (McGraw-Hill, New York, 1968), p. 445; H. Bethe and R. Jackiw, *Intermediate Quantum Mechanics*, Third Edition (Benjamin/Cummings, Menlo Park, CA, 1986), p. 179.

A GEOMETRIC VIEW OF QUANTUM MECHANICS

J. ANANDAN*

Department of Physics and Astronomy
University of South Carolina
Columbia, SC 29208, USA

and

Max-Planck-Institute for Physics and Astrophysics
Föhringer Ring 6, D-8000 Munich 40, FRG

ABSTRACT

It is argued that quantum mechanics is fundamentally a geometric theory. This is illustrated by means of the connection and symplectic structures associated with the projective Hilbert space, using which the geometric phase can be understood. Using the natural metric on this space, Schrödinger's equation for an isolated system is reformulated. This metric is generalized to the space of all density matrices and a physical meaning is proposed.

1. Introduction

This talk is based on the view that the world is ultimately geometrical. We know that in some fields of physics, such as general relativity and the study of fractals, geometric ideas have been very useful. But the deepest physical theory we have today is quantum theory where geometric ideas are not readily apparant. Some quantum theorists have therefore argued that we must regard gravity quantum mechanically as a spin-2 field and that the beautiful geometry in general relativity is really not needed. I wish to reverse this argument today. I shall argue that there are interesting geometric structures in quantum theory also, and that perhaps we should look at quantum theory as a geometric theory, like special relativity formulated using Minkowski space-time.

The important role played by geometric ideas in quantum theory may be illustrated by examples beginning with the creation of quantum theory itself. As is well known, an important role was played in this creation by Hamilton's reformulation of classical mechanics, which inspired De Broglie and Schrödinger to discover wave-particle duality and quantum mechanics. What was Hamilton's motivation for doing this even though Newton's original formulation of classical mechanics was very intuitive and physical?

The answer is that Hamilton originally studied geometric optics in which the central question is the following: Given a set of rays entering an optical system, what is the set of outgoing rays? Each set of rays may be parametrized by the coordinates x, y of the point at which a typical ray intersects a fixed plane perpendicular to the optical axis and by p_x, p_y that are the corresponding direction cosines of the ray at this point multiplied by the refractive index. (In the wave description, p_x and p_y are proportional to the components of the wave vector along the x and y axes). In modern terms, he found the transformation between the two

*Alexander von Humboldt fellow.

sets of rays to be symplectic[1], i. e. it preserves the symplectic structure in the space $\{(x, y, p_x, p_y)\}$. A few years later it occured to Hamilton that if the optical axis is taken to be the time axis, then classical mechanics in two spatial dimensions can be reformulated using the same symplectic geometry, which is easily extended to any number of dimensions of the configuration space. Using this symplectic structure, the Poisson bracket between any two observables is formed. And the well known Dirac prescription to replace Poisson brackets by commutators gives quantum mechanics.

The second example that I wish to give is the Aharonov-Bohm (AB) effect[2], which the present conference was convened to celebrate. The most important lesson to me from this effect is that the classical electromagnetic field should be described geometrically as a connection, which is discussed elsewhere[3]. Then the AB phase factor is the holonomy transformation when the value of the wave function at an event in the interference region is parallel transported with respect to the electromagnetic connection around a closed curve, which passes through the interfering beams.

Whereas the above two examples have been studied in great detail over the years, the third example which I wish to mention is of recent origin. This is the geometric phase. Suppose a quantum system undergoes cyclic evolution, by which I mean that its state vector $|\psi(t)\rangle$ obeys Schrödinger evolution in the interval $[0,\tau]$ such that $|\psi(\tau)\rangle = e^{i\phi}\,|\psi(0)\rangle$. It was shown[4,5] that there is a part β of the total phase ϕ acquired which is geometric in the sense that it depends only on the motion of the system given by a closed curve C in the set of rays of the Hilbert space $\mathcal{H}$, which is called the projective Hilbert space $\mathcal{P}$, and is independent of the particular Hamiltonian used to obtain this motion.

This β can be understood in two ways. First, $e^{i\beta}$ is the holonomy transformation for parallel transporting around C in $\mathcal{P}$, or around a closed curve γ in a parameter space defined in section 2, with respect to the natural connection given by the inner product in $\mathcal{H}$ [5,6]. From this view point, β is like the AB phase. Alternatively, β is the 'area' of any surface spanned by C with respect to the natural symplectic structure in $\mathcal{P}$ determined by this inner product[3,7]. Interestingly, this symplectic structure when restricted to the submanifold of Gaussian wave packets with constant width, which may be identified with the classical phase space, is the symplectic structure found by Hamilton in the latter space[3]. Therefore, what Hamilton found, which made his reformulation of Newtonian mechanics so powerful, was a "shadow" of a geometric structure in quantum mechanics. Similar to how Hamilton's geometric reformulation of classical mechanics paved the way for quantum mechanics, perhaps, a geometric reformulation of quantum mechanics may lead to a new physical theory which we need today for a proper description of gravity, consistent with quantum phenomena.

2. Geometric Phase and Connection

The geometric phase is a consequence of just the inner product and is independent of the particular equation of motion which is first order in time[8]. But for simplicity, we shall now obtain it as a consequence of the Schrödinger evolution

$$i\hbar \frac{d}{dt}|\psi(t)\rangle = H(t)\,|\psi(t)\rangle\,, \qquad (1)$$

where H is the Hamiltonian and $\langle\psi(t)|\psi(t)\rangle = 1$. Given the cyclic evolution described above, choose $|\tilde{\psi}(t)\rangle$ such that $|\psi(t)\rangle = \xi(t)\,|\tilde{\psi}(t)\rangle$, where the function[9] $\xi(t)$ has complex values with unit modulus, and $|\tilde{\psi}(\tau)\rangle = |\tilde{\psi}(0)\rangle$. Then (1) implies

$$i\hbar\left(\frac{d}{dt}+\langle\tilde{\psi}(t)|\frac{d}{dt}|\tilde{\psi}(t)\rangle\right)\xi(t) = \langle\psi(t)|H|\psi(t)\rangle\,\xi(t)\ . \qquad (2)$$

The left hand side of (2) does not depend on the Hamiltonian and suggests the introduction of the covariant derivative:

$$\frac{D}{dt}\equiv\frac{d}{dt}+\langle\tilde{\psi}(t)|\frac{d}{dt}|\tilde{\psi}(t)\rangle \equiv v\cdot\nabla\ ,$$

where $\nabla = d + \langle\tilde{\psi}|d|\tilde{\psi}\rangle$, d being the exterior differential operator on $\mathcal{P}$, and v is the "velocity" in $\mathcal{P}$, i. e. v is the tangent vector to the curve C in $\mathcal{P}$ parametrized by t, representing the evolution. Here $|\tilde{\psi}\rangle$ is a "section" defined in some neighborhood U of C and $\langle\tilde{\psi}|d|\tilde{\psi}\rangle$ represents the "connection form" with respect to this section. $\langle\tilde{\psi}|d|\tilde{\psi}\rangle$ is analogous to ieA, where A is the electromagnetic vector potential, which is a connection form over space-time. The normalization $\langle\tilde{\psi}|\tilde{\psi}\rangle = 1$ implies that $\langle\tilde{\psi}|d|\tilde{\psi}\rangle$ is pure imaginary. If a different section $|\tilde{\psi}'\rangle = e^{ig}|\tilde{\psi}\rangle$ is chosen, where g is a real differentiable function on U, then ξ transforms to $e^{-ig}\xi$ and $\langle\tilde{\psi}|d|\tilde{\psi}\rangle$ transforms to $\langle\tilde{\psi}|d|\tilde{\psi}\rangle + i\,dg$, which is analogous to the electromagnetic gauge transformation. Clearly, (2) is covariant under this "local gauge" transformation.

Eq. (2) can be integrated immediately to yield

$$\xi(\tau) = \exp\{-\oint_C\langle\tilde{\psi}|d|\tilde{\psi}\rangle-\frac{i}{\hbar}\int_0^{\tau}\langle\psi|H(t')|\psi\rangle dt'\}\ \xi(0)\ . \qquad (3)$$

The real phase

$$\beta \equiv i\oint_C\langle\tilde{\psi}|d\tilde{\psi}\rangle = i\int_{\Sigma}\langle d\tilde{\psi}|\wedge|d\tilde{\psi}\rangle\ , \qquad (4)$$

where Σ is a surface spanned by C and $|d\tilde{\psi}\rangle = d|\tilde{\psi}\rangle$, is called the geometric phase, because $e^{i\beta}$ is the holonomy transformation associated with C with respect to the connection mentioned above. i. e. if we parallel transport a function η around according to the rule $\frac{D\eta}{dt} = 0$ then it acquires the phase factor $e^{i\beta}$. Also, $e^{i\beta}$ is invariant under a change of section similar to how the Aharonov-Bohm[2] phase factor $\exp\left(\frac{ie}{\hbar c}\oint A_\mu dx^\mu\right)$ is invariant under gauge transformations.

It may appear at first sight that it would be necessary to solve the full quantum mechanical problem in order to determine the path C and thereby predict $e^{i\beta}$. But this is not the case. I shall show now that for an arbitrary H(t) it is possible to determine $e^{i\beta}$ by solving the classical problem and an eigenvector problem for a set of invariants defined below. Suppose we have a class of Hamiltonians of the form $H(R) = \hbar R^\alpha J_\alpha$ where R^α are real variables and iJ_α generate a subgroup of the unitary group G determined by the Lie algebra relations

$$[J_\alpha,J_\beta] = i\,C_{\alpha\beta}{}^{\gamma}J_\gamma\ . \qquad (5)$$

Therefore, $C_{\alpha\beta\gamma}$ is totally antisymmetric, with the index being lowered using the metric $\delta_{\alpha\beta}$. Let $I(S) = S^\alpha J_\alpha$, where S^α are a set of parameters. Each $S = (S^1, S^2,\ldots)$ is a point in a parameter space $\mathcal{S}$. Now, let $|n(S)\rangle$ be a complete set of orthonormal eigenstates of a Hermitian operator I(S):

$$I(S)|n(S)\rangle = I_n(S)\,|n(S)\rangle\ . \qquad (6)$$

Suppose $|\psi_n(t)\rangle$ evolves according to Schrödinger's equation, i. e. $|\psi_n(t)\rangle \equiv U(t)|\psi_n(0)\rangle$, where U(t) is the time evolution operator:

$$U(t) = T \exp\{-\frac{i}{\hbar}\int_0^t H(t')dt'\} \ .$$

where T denotes time ordering. Let S(t) be a curve in $\mathbf{S}$ in the interval $[0,\tau]$,, such that each $|\psi_n(t)\rangle$ is an eigenstate of I(S(t)) with a constant eigenvalue I_n . This curve is not unique. But on requiring that $I(S(t)) = U(t)I(S(o))U^\dagger(t) \equiv S^\alpha(t)\, J_\alpha$, S(t) is unique for a given S(0). Then,

$$\frac{\partial I}{\partial t} - \frac{i}{\hbar}[I,H] = 0 \ , \qquad (7)$$

i. e. I(S(t)) is an explicitly time dependent Schrödinger observable which is an invariant of the motion[10]. Conversely, if I satisfies (7), then by reversing the steps, it follows that if $|\psi(t)\rangle$ was initially an eigenstate of I(S(t)) then it continues to be in an eigenstate with the same eigenvalue.

Suppose that the curve is closed, i. e. $S(\tau) = S(0)$, and denote the unparametrized closed curve by γ. In general, $|\psi_n(0)\rangle$ and $|\psi_n(\tau)\rangle$ need not belong to the same eigensubspace of $I(S(0)) = I(S(\tau))$. For example, if I has an infinite spectrum then it is possible for the nth state to go into the (n+1)th state during the cyclic evolution of I. But this is very unusual and we shall suppose here the usual case of them belonging to the same eigensubspace. I assume for the present that I(S) is non degenerate for every S. Then $|\psi_n(\tau)\rangle = e^{i\phi_n}|\psi_n(0)\rangle$ for some ϕ_n. Then, as shown by (3), ϕ_n is the sum of the dynamical phase and the geometric phase β_n[11]. For the present, the eigenvalues in (6) are assumed to be simple, i. e. I(S) is non degenerate, on γ. Then, β_n may be obtained from (4) by replacing $|\tilde{\psi}\rangle$ by $|n(S)\rangle$:

$$\beta_n \equiv i\oint_\gamma \langle n(S)|dn(S)\rangle = i\int_\sigma \langle dn(S)|\wedge|dn(S)\rangle \ , \qquad (8)$$

where σ is a surface spanned by γ and d is now the exterior differential operator on *parameter space*. Geometrically, $e^{i\beta}$ may be regarded as the holonomy transformation with respect to a connection in a line bundle over the space $\mathbf{S}$, whose fiber over $S \in \mathbf{S}$ is the eigensubspace of I(S) with eigenvalue $I_n(S)$, with connection form $\langle n(S)|dn(S)\rangle$.

To determine γ, obtain from (5) and (7),

$$\frac{dS^\alpha}{dt} = - C_{\beta\gamma}{}^\alpha S^\beta R^\gamma \ . \qquad (9)$$

It follows that $S_\alpha S^\alpha$ is a constant of motion. Therefore, for inessential economy, we may restrict our parameter space to the sphere $\mathbf{S}'$ in $\mathbf{S}$ defined by the condition $S_\alpha S^\alpha = 1$. Eq. (9) can be given the following classical interpretation: Let $H^c \equiv R^\alpha T_\alpha$ be the classical Hamiltonian obtained by replacing $\hbar J_\alpha$ in H by a set of classical observables T_α which obey the Poisson bracket relations

$$\{T_\alpha, T_\beta\} = C_{\alpha\beta}{}^\gamma T_\gamma \ . \qquad (10)$$

The T_α may be regarded as the classical limit of the quantum observables $\hbar J_\alpha$. Also, replace I by the classical observable $I^c(t) = S^\alpha(t)\, T_\alpha$ which is required to be an invariant of the evolution under H_c:

$$\frac{\partial I^c}{\partial t} + \{I^c, H^c\} = 0 \ . \qquad (11)$$

On using (10), (11) yields (9). Hence (9) may be regarded as the classical equation of motion for S^α. By solving it, the curve γ and hence the geometric phase (8) can be obtained.

From (6),

$$(I_n(S) - I_m(S)) \langle m(S)|dn(S)\rangle = \langle m(S)|dI(S)|n(S)\rangle, \quad m \neq n. \qquad (12)$$

Therefore, eq. (8) may be rewritten, after inserting a complete set of states and using(12), as

$$\beta_n = \int_\sigma \sum_{m\neq n} \frac{\langle n|dI|m\rangle \wedge \langle m|dI|n\rangle}{(I_n - I_m)^2}. \qquad (13)$$

There is no contribution m = n in the summation over m in (13), because $\langle n(S)|dn(S)\rangle$ is pure imaginary, due to the normalization of $|n(S)\rangle$, and therefore its wedge product with its complex conjugate vanishes. The prescription (13) has the advantage that $|n(S)\rangle$ need not be chosen differentiably or even continuously over the parameter space, unlike the prescription (8). This is similar to the prescription given by Berry[4] to evaluate the geometric phase in the adiabatic limit using the parameter space $\mathfrak{R}$ consisting of possible values of R^α with the Hamiltonian H(R) playing the role of I(S). In this limit, the evolving state may also be taken as an approximate eigenstate of H(R). Then S^α may taken to be parallel to R^α. But we prefer, even in this limit, to regard $\mathfrak{S}$ to be the parameter space for the evaluation of the geometric phase instead of $\mathfrak{R}$. Because, even when the curve in $\mathfrak{R}$ defined by H(R(t)) is not closed, the evolution may still be cyclic because the corresponding curve in $\mathfrak{S}$ defined by I(S(t) satisfying (9) is closed. Also, the present approach gives the freedom to take more general states which are not eigenstates of H, even in the adiabatic limit. We can always construct an explicitly time dependent invariant operator I(S(t)) whose eigenstate $|\psi(t)\rangle$ is, and the geometric phase for a cyclic evolution may be obtained from the above prescription.

As an example, consider the precession of spin in an arbitrary magnetic field $\mathbf{B}(t)$[12]. Then the Hamiltonian may be written as $H = -\eta\hbar\,\mathbf{B}\cdot\mathbf{J}$. Therefore, here $R^\alpha = -\eta B^\alpha$ and (9) is

$$\frac{dS}{dt} = \eta\, \mathbf{S} \times \mathbf{B}. \qquad (14)$$

Here, the parameters S^α may be given a classical meaning as the components of a spin vector; then $\mu = \eta$ S is the magnetic moment. Eq. (14) is the classical equation of motion for spin. By solving it we obtain a curve in $\mathfrak{S}$. When this curve γ is closed the evolution is cyclic. The geometric phase is obtained now from the holonomy transformation associated with γ with respect to the connection defined over $\mathfrak{S}$ or it may be evalutated using (13).

We therefore have now the following general prescription for obtaining the geometric phase: Given an arbitrary cyclic quantum evolution of a state associate with it a motion of an explicitly time dependent *invariant*. This motion is obtained as a closed curve in the space $\mathfrak{S}$ of possible values of the corresponding classical quantity, by solving the *classical* equations of motion. There is a natural connection over $\mathfrak{S}$ for parallel transporting eigenstates of this invariant. The geometric phase factor is the same as the holonomy transformation associated with this curve with respect to this connection. There are infinite number of Hamiltonians which give rise to a given *motion* of the invariant, given by an unparametrized curve γ in $\mathfrak{S}$. An adiabatic Hamiltonian, for example, is one way of achieving this motion. But the geometric phases for the eignstates of I are a property of this motion and independent of the particular choice of the Hamiltonian.

More generally, if I(S) is degenerate then the cyclic evolution of I(S) may be associated with the cyclic evolution of an eigensubspace whose dimension m may be greater than one. The geometric phase factor is then generalized to an element of SU(m), which is described elsewhere[13].

3. Symplectic Stucture

Another geometric meaning can be given to β as the symplectic area enclosed by C determined by the natural symplectic structure on $\mathcal{P}$ inherited from the symplectic structure in $\mathcal{H}$ due to the inner product[3]. More specifically, define the canonical coordinates and momenta $Q^j = \psi_j$ and $P_j = i\hbar\bar{\psi}_j$, where ψ_j are the components of $|\psi\rangle$ in an arbitrary orthonormal basis and the overbar denotes complex conjugation. Then (4) may be rewritten as

$$\beta = \frac{1}{\hbar}\oint_C P_j dQ^j = \frac{1}{\hbar}\int_\Sigma dP_j \wedge dQ^j \quad . \qquad (15)$$

Here $\hbar$ is used in the definition of P_j only in order that the integral in (15) has the dimension of action.

On the other hand, the action for the Schrödinger field is

$$S = \frac{1}{2}\int dt(i\hbar\langle\psi|\frac{d}{dt}\psi\rangle - i\hbar\langle\frac{d}{dt}\psi|\psi\rangle - \langle\psi|H|\psi\rangle) = \int d^4x\,\mathcal{L} \qquad (16)$$

Then the momentum conjugate to the canonical coordinate Q_j is

$$\frac{\partial\mathcal{L}}{\partial\dot{Q}_j} = i\hbar\bar{\psi}_j = P_j\,.$$

Thus we obtain, using dynamical considerations, the same pairs of canonically conjugate coordinate and momentum obtained earlier by purely geometric considerations. i. e. the symplectic structure given by the Lagrangian density $\mathcal{L}$ is the same as the symplectic structure given by (15). Therefore, in some sense, *the geometry is prior to the dynamics* here. Also, the procedure of second quantizing the Schrödinger field by imposing canonical commutation relations between P_j and Q_j, regarded as operators, uses the above symplectic structure determined by the inner product on $\mathcal{H}$.

We can go in the opposite direction of taking the classical limit of the first quantized theory. This may be done by using isotropic Gaussian states centered around a classical phase space point $(\mathbf{q},\mathbf{p})$ with a constant width Δq in coordinate space:

$$\tilde{\psi}_{\mathbf{q},\mathbf{p}}(\mathbf{x}) = \{2\pi(\Delta q)^2\}^{-1/4}\exp\{-\frac{(\mathbf{x}-\mathbf{q})^2}{(2\Delta q)^2} + \frac{i}{\hbar}\mathbf{p}\cdot(\mathbf{x}-\mathbf{q})\}\,. \qquad (17)$$

In the classical limit, the quantum system has a large enough mass for the spreading of the wave packet to be negligible during a time interval of interest. Then its state may be assumed to remain in the form (17), up to phase, with $(\mathbf{q},\mathbf{p})$ changing with time, to a very good approximation. Therefore, the submanifold $\mathcal{C}$ of $\mathcal{P}$ consisting of the states (17), for a fixed Δq, may be identified with the classical phase space $\{(\mathbf{q},\mathbf{p})\}$.

Consider a cyclic motion $(\mathbf{q}(t),\mathbf{p}(t))$ that is a closed curve C in $\mathcal{C}$. If this motion is due to an evolution in the classical limit then λ may be identified with time t. Then

$$i\langle\tilde{\psi}_{q(t),p(t)}|\frac{d}{dt}|\tilde{\psi}_{q(t),p(t)}\rangle = \frac{1}{\hbar}\,p\cdot\frac{dq}{dt} \qquad (18)$$

Hence the geometric phase for this motion, using (3) and Stoke's theorem, is[3,7]

$$\beta = \frac{1}{\hbar}\int_S dp \wedge\cdot dq \quad , \qquad (19)$$

where S is a surface in $\mathcal{C}$ spanned by C. On comparing (19) with (15), it is clear that $\frac{1}{\hbar}\,dp \wedge\cdot dq$ is the restriction to $\mathcal{C}$ of the symplectic form in $\mathcal{P}$, defined above. Hence, β may be regarded, in this classical limit, as the symplectic 'area' in the classical phase space and is therefore a canonical invariant.

The above treatment of the geometric phase therefore suggests why Hamilton's reformulation of classical mechanics should be much closer to quantum mechanics than Newton's original formulation: The symplectic structure used by Hamilton is close to the inner product in $\mathcal{H}$ whereas there is no such geometric structure in Newton's formulation. Hamilton's equations of motion determine a family of trajectories in the classical phase space $\mathcal{C}$ for a given Hamiltonian, that generates a 1-parameter family of transformations along these trajectories which preserve the symplectic structure. When the theory is quantized, the classical states may be represented approximately by Gaussian wave packets; but it now becomes possible to take superpositions of them to form our Hilbert space $\mathcal{H}$. These wave packets can spread and therefore Hamiltonian classical mechanics should be regarded as the geometric optics limit of quantum mechanics. But in $\mathcal{H}$, there is a symplectic 2-form $\Omega = \frac{1}{\hbar}\,dP_j \wedge dQ^j$, which gives $\Omega(\psi,\phi) = -2\,\mathrm{Im}\langle\psi|\phi\rangle$ for any two tangent vectors ψ, ϕ of $\mathcal{H}$. Since Schrödinger's equation preserves the inner product, it must preserve Ω. As seen above, the classical limit of Ω is $\omega = dp_j \wedge dq^j$ which determines the symplectic structure in $\mathcal{C}$. Therefore we expect ω to be preserved in the classical limit, which is satisfied by Hamilton's equations.

4. Metric

The inner product also gives a metric on $\mathcal{P}$. It is convenient to use fiber bundles which gives in a natural manner this metric as well as the connection and the symplectic structure discussed in sections 2 and 3, respectively. The reader who is not interested in the fiber bundle description may go directly to eq. (20), (24) or (31), which may be regarded as alternative definitions of this metric.

Suppose $\mathcal{H}$ has dimension N+1 where N is a non negative integer. The unitary group U(N+1) acting on $\mathcal{H} = \mathbb{C}^{N+1}$ can be identified with the set $\mathcal{B}$ of orthonormal bases of $\mathcal{H}$ because each element of $\mathcal{B}$ can be obtained from a fixed element of $\mathcal{B}$ by the action of a unique element of U(N+1). On defining an equivalence relation between two orthonormal bases whenever the first element of both bases are the same, the corresponding quotient set of $\mathcal{B}$, which is the same as U(N+1)/U(N), may be identified with the set of unit vectors of $\mathcal{H}$ and is called a Stiefel manifold S_N. The equivalence relation on S_N identifying any two unit vectors related by multiplication by a phase factor, gives the quotient set U(N+1)/U(N)×U(1) which can be identified with $\mathcal{P} = P_N(\mathbb{C})$, i. e. the N dimensional complex projective space. Now, U(N+1) is a principal fiber bundle over S_N with projection map Φ (say) and structure group U(N). Also, S_N is a principal fiber bundle over $\mathcal{P}$ with projection map Π (say) and structure group U(1), which is sometimes called the Hopf bundle.

Also, U(N+1) may be regarded as a principal fiber bundle over $\mathcal{P}$ with projection map $\Pi\Phi$ and structure group $U(N)\times U(1)$.

There are natural metrics on the above bundles defined as follows: A tangent vector X of U(N+1) is a (N+1) dimensional Hermitian matrix. Define the metric h in $\mathcal{B}$ by the condition $h(X,Y) = 2\, tr(XY)$ where X and Y are tangent vectors at any point in $\mathcal{B}$ = U(N+1). It is easily verified that this metric is real and positive definite and when restricted to the SU(N+1) subgroup it is the Cartan-Killing metric. Let g be the metric in S_N such that Φ is a Rimannian submersion, i. e. $d\Phi$ is an isometry when restricted to the orthogonal complement of the kernel of $d\Phi$. Similarly, define a metric f on $\mathcal{P}$ such that Π is a Riemannian submersion. This f is the *Fubini-Study metric*. There are now natural connections on each of the three bundles defined as follows: the horizontal space at each point is orthogonal to the fiber at that point with respect to the metric in that bundle. The connection over the parameter space $\mathcal{S}$ or $\mathcal{S}'$ defined in section 3 may also be understood as follows. There is a map f from $\mathcal{S}$ into $\mathcal{P}$ which takes each $S \in \mathcal{S}$ into the eigensubspace of I(S) corresponding to the eigenvalue $I_n(S)$. The pullback of S_N with the above connection by the map f is a U(1) principal fiber bundle over $\mathcal{S}$ with a connection. The closed curve γ is mapped by f into the closed curve C. The geometric phase factor[5] $e^{i\beta}$ is obtained as the holonomy transformation, i. e. parallel transport around C in $\mathcal{P}$ with respect to the connection in S_N or around γ in $\mathcal{S}$ with respect to the last mentioned connection.

Suppose that $|\psi\rangle$ and $|\psi'\rangle$ are such that $\Pi(|\psi\rangle)$ and $\Pi(|\psi'\rangle)$ are seperated by an infinitesimal distance ds in $\mathcal{P}$. Then, it can be shown that[14]

$$ds^2 = 4\,(1 - |\langle\psi|\psi'\rangle|^2) \qquad (20)$$

To obtain the metric coefficients, consider a smooth curve $|\psi(t)\rangle$ in $\mathcal{B}$ such that $|\psi(t)\rangle$ and $|\psi(t+dt)\rangle$ project to $\Pi(|\psi\rangle)$ and $\Pi(|\psi'\rangle)$, respectively. Now,

$$\langle\psi(t)|\psi(t+dt)\rangle = 1 + dt\,\langle\psi|\frac{d}{dt}\psi\rangle + \frac{1}{2}dt^2\,\langle\psi|\frac{d^2}{dt^2}\psi\rangle + O(dt^3)\,. \qquad (21)$$

Also, differentiating $\langle\psi(t)|\psi(t)\rangle = 1$ twice,

$$\langle\psi|\frac{d}{dt}\psi\rangle + \langle\frac{d}{dt}\psi|\psi\rangle = 0 \quad \text{and} \quad \langle\psi|\frac{d^2}{dt^2}\psi\rangle + \langle\frac{d^2}{dt^2}\psi|\psi\rangle + 2\langle\frac{d}{dt}\psi|\frac{d}{dt}\psi\rangle = 0\,. \qquad (22)$$

From (20), (21) and (22),

$$\left(\frac{ds}{dt}\right)^2 = 4\,\langle\frac{d}{dt}\psi|\frac{d}{dt}\psi\rangle - 4\,\langle\frac{d}{dt}\psi|\psi\rangle\langle\psi|\frac{d}{dt}\psi\rangle\,. \qquad (23)$$

Therefore, in terms of an arbitrary section $|\tilde\psi\rangle$, the metric can be written as

$$ds^2 = 4\,(\langle d\tilde\psi|d\tilde\psi\rangle - \langle d\tilde\psi|\tilde\psi\rangle\langle\tilde\psi|d\tilde\psi\rangle) = 2\,g_{\bar\mu\nu}\,d\bar z^\mu\,dz^\nu \qquad (24)$$

where z^μ are complex coordinates in $\mathcal{P}$ and the metric coefficients

$$g_{\bar\mu\nu} = 2\,(\langle\frac{\partial}{\partial z^\mu}\tilde\psi|\frac{\partial}{\partial z^\nu}\tilde\psi\rangle - \langle\frac{\partial}{\partial z^\mu}\tilde\psi|\tilde\psi\rangle\langle\tilde\psi|\frac{\partial}{\partial z^\nu}\tilde\psi\rangle) \qquad (25)$$

constitute a Hermitian matrix[15]. Indeed it is a Kahler metric[16]. It can be verified that (24) is invariant under change of section $|\tilde\psi\rangle \to |\tilde\psi'\rangle = e^{ig}|\tilde\psi\rangle$. It may be regarded as an alternative definition of the Fubini-Study metric which is valid also for an infinite dimensional $\mathcal{H}$. A unitary or anti-unitary transformation leaves the inner product invariant and is therefore an isometry in $\mathcal{P}$.

Eq. (23) and hence the expression (24) for the metric may also be obtained in another way. The horizontal component of the tangent vector $\frac{d}{dt}|\psi\rangle$ is $\frac{d}{dt}|\psi\rangle -$

$<\psi|\frac{d}{dt}|\psi>|\psi>$. Since the Fubini-Study metric on $\mathcal{P}$ is obtained by requiring that Π is a Riemannian submersion, twice the norm of the latter vector with respect to the inner product in $\mathcal{H}$ is $\frac{ds}{dt}$. This gives (23). Also, the restriction of this metric to the submanifold $\mathcal{C}$ of $\mathcal{P}$ consisting of Gaussian wave packets of constant width, defined in section 3, yields a positive definite metric on the classical phase space[7].

5. Geometry of Quantum Evolution

I shall now give a geometric description of quantum evolution, which need not be cyclic, by means of the metric on $\mathcal{P}$ described in section 4. Suppose $|\psi(t)>$ undergoes Schrödinger evolution. Then from (1) and (23), the speed of the quantum system in $\mathcal{P}$ is[14,17]

$$\frac{ds}{dt} = 2\frac{\Delta E}{\hbar} \tag{26}$$

where ΔE is the uncertainty in energy: $\Delta E^2 = <\psi(t)|H^2|\psi(t)> - <\psi(t)|H|\psi(t)>^2$. Hence the Fubini-Study distance s traveled during the time interval (t_1, t_2) is

$$s = 2\int_{t_1}^{t_2} \frac{\Delta E(t)}{\hbar} dt \quad . \tag{26'}$$

Suppose s_0 is the distance along the shortest geodesic joining $\Pi(|\psi(t_1)>)$ and $\Pi(|\psi(t_2)>)$, which I shall call the displacement between the two states. Then,

$$s \geq s_0 , \tag{27}$$

which is the same as

$$\int_{t_1}^{t_2} \frac{\Delta E(t)}{\hbar} dt \;\geq\; \cos^{-1}(|<\psi(t_1)|\psi(t_2)>|) \; . \tag{27'}$$

Now, define the "uncertainty of time" $\Delta t = \frac{\pi}{s_0}(t_1 - t_2)$, assuming $s_0 \neq 0$. This amounts to rescaling the actual time interval by the ratio of the displacement π between two orthogonal states to the displacement s_0 during the actual evolution. Then, in the special case when the initial and final states are orthogonal, $\Delta t = (t_1 - t_2)$. Conversely, the latter condition may be used to constrain the rescaling factor. Then, (27) reads

$$<\Delta E> \Delta t \geq \frac{h}{4} \quad , \tag{28}$$

where h is Planck's constant and $<\Delta E>$ is the time averaged uncertainty of energy in the interval (t_1, t_2). This is in the form of a time-energy uncertainty relation, which was previously obtained for the special case when the initial and final states are orthogonal[14]. But (28) is valid more generally when the system does not pass through orthogonal states. The usual problem in deriving the time-energy uncertainty relation is that time is not an operator in quantum mechanics, as ordinarily formulated. This prevents a canonical definition of the "uncertainty of time" Δt. In the present geometric approach this problem is solved by using the displacement between orthogonal states as the "standard" for defining Δt. The important physical result here is that the left hand side of (27'), for arbitrary quantum evolutions between given initial and final states, must have a lower bound, and this may be interpreted geometrically as a distance or an angle.

Consider now the Schrödinger evolution of an orthonormal basis under the action of a constant Hamiltonian H which is given by a geodesic curve in the bundle $\mathcal{B}$ defined in section 4. The projection C of this curve in $\mathcal{P}$ is the Schrödinger evolution of one of the states of this basis. Let $F_{\mu\nu}$ be the curvature of the connection in $\mathcal{B}$, regarded as a principal fiber bundle over $\mathcal{P}$, which takes its value in the Lie algebra of the structure group U(1) × U(N). By writing the geodesic equation in $\mathcal{B}$ and projecting it to $\mathcal{P}$, the equation of the curve C is obtained to be

$$\frac{dv^{\mu}}{ds} + \Gamma^{\mu}_{\nu\rho}\, v^{\nu} v^{\rho} = \frac{1}{2\Delta E}\, \mathrm{Tr}(H\, F^{\mu}{}_{\nu})\, v^{\nu} \qquad (29)$$

where $v^{\mu} = \frac{dz^{\mu}}{ds} = \frac{\hbar}{2\Delta E}\frac{dz^{\mu}}{dt}$ is the unit tangent vector to C which is proportional to the covariant velocity $\frac{dz^{\mu}}{dt}$, and $\Gamma^{\mu}_{\nu\rho} = g^{\mu\bar{\alpha}}\frac{\partial g_{\bar{\alpha}\nu}}{\partial z^{\rho}}$ are the connection coefficients formed from the Fubini-Study metric.

Eq. (29), which replaces Schrödinger's equation (1) in $\mathcal{P}$ for an isolated system, has the advantage that it is formulated in terms of the geometric quantities $g_{\mu\nu}$ and $F_{\mu\nu}$, which are obtained from the inner product in $\mathcal{H}$ and independent of the Hamiltonian H. The evolution is a geodesic if and only if H is such that the right hand side of (29) vanishes. Eq. (29) is analogous to the equation of motion of a classical particle in curved space-time in an arbitrary gauge field[18,19], with H playing the role of the gauge field charge (color, isospin etc.). This is the generalization of the Lorentz force law for a charged particle in an electromagnetic field to an arbitrary gauge field. In this case also, the trajectory of the particle in space-time is the projection of a geodesic in the corresponding Kaluza-Klein space[18].

Neither Planck's constant nor the time t, which occurs in (1), appears explicitly in (29). From (20), the Fubini-Study distance measures how far the state of the system moves from the original state as determined by its diminishing overlap with the original state. Therefore if we measure time by the motion of a quantum system, we would directly measure the Fubini-Study distance. Time then represents the correlation between the distances measured by different systems. This is similar to how money was invented to replace what was originally a direct exchange of goods and services, which gives new meaning to the old statement that "time is money"! On the other hand, (29) is non linear, unlike (1). Hence, the introduction of time allows the evolution to be treated linearly.

6. Mixed States

There is the following profound difference between classical and quantum physics: In classical physics if a system is in a pure state then any part of that system is also in a pure state. But in quantum physics, the subsystem can be and usually is in a mixed state. This is because in classical physics a system is in a mixed state due to our own ignorance of the state of the system so that we need to give a probability distribution for the possible states, e. g. a gas of molecules in classical statistical mechanics. If the state is completely known (pure state) then the state of each subsystem is also completely known. But consider now the state of two quantum systems which can always be written as

$$\sum_{k=1}^{N} c_k\, |\psi_k\rangle\, |\phi_k\rangle$$

where $|\psi_i\rangle$ and $|\phi_j\rangle$ are the states of the two individual systems satisfying $\langle\psi_i|\psi_j\rangle = \delta_{ij}$ and $\langle\phi_i|\phi_j\rangle = \delta_{ij}$. If two or more coefficients c_k are non zero then the combined system is said to be in an entangled state and each subsystem is in a mixed state. This is different from the mixed state for a classical system because an experiment can be performed to verify, in principle, that the combined system is in a pure state and therefore we do have the maximum possible information of the combined system. This entanglement is of very great physical importance because different types of entanglements lead to Bose and Fermi statistics, the EPR paradox and the violation of Bell's inequalities which a local realistic system should obey.

The density matrices for the two subsystems are[20]

$$\rho_1 = \sum_{k=1}^{N} |c_k|^2\, |\psi_k\rangle \langle\psi_k| \qquad \text{and} \qquad \rho_2 = \sum_{k=1}^{N} |c_k|^2\, |\phi_k\rangle \langle\phi_k| .$$

In the special case when only one of the coefficients is non zero, each subsystem is in a pure state. For this to happen, it is necessary and sufficient that the density matrix of either subsystem obeys $\rho^2 = \rho$. But since no system is completely isolated from its environment, a quantum system would in general be in an entangled state. It follows therefore that a description of a quantum system as a pure state is grossly inadequate and we should extend our description of the geometry of pure states given above to mixed states.

It is known that the set $\mathfrak{D}$ of density matrices of a given quantum system has a topology and an affine structure with respect to which it is a convex set with the pure states contained in the boundary[21]. Since the set of pure states can be identified with the projective Hilbert space, this part of the boundary has a natural metric, namely the Fubini-Study metric. It is therefore reasonable to extend this metric to the rest of $\mathfrak{D}$. To this end, write (20) in terms of the pure state density matrices $\rho = |\psi\rangle\langle\psi|$ and $\rho' = |\psi'\rangle\langle\psi'|$ as

$$ds^2 = 4(1 - \mathrm{tr}\,(\rho\rho')) = 2\,\mathrm{tr}\,(\rho - \rho')^2 \qquad (30)$$

where the last equality follows from $\mathrm{tr}\,(\rho^2) = \mathrm{tr}\,(\rho'^2) = 1$. It is therefore reasonable to introduce the following metric on all of $\mathfrak{D}$

$$dS^2 = 2\,\mathrm{tr}\,(d\rho^2)\ , \qquad (31)$$

which when restricted to pure states is the Fubini-Study metric.

A special case of a density matrix is a 'total mixture' for which ρ is of the form $\rho = \frac{1}{n} P$ where P is the projection operator which projects to a subspace of $\mathcal{H}$ of dimension n. The set of all such ρ's, for a fixed n, may be identified with the Grassmannian $G_{n,m}$ that is defined as the set of n-dimensional subspaces of $\mathcal{H}$ whose dimension is n+m. This manifold has a natural metric[22], which generalizes the Fubini-Study metric corresponding to n = 1, that may be defined as follows: Let $\{|\psi_k\rangle\}$ and $\{|\psi_k'\rangle\}$ be orthonormal bases in two subspaces of dimension n which are seperated by an infinitesimal distance dS in $G_{n,m}$. Let $Z_{ij} = \langle\psi_i'|\psi_j\rangle$. Then

$$dS^2 = \frac{4}{n^2}(n - \mathrm{tr}\,Z^\dagger Z) = \frac{4}{n^2}(n - \mathrm{tr}\,PP')\ , \qquad (32)$$

where $P = \sum_{k=1}^{n} |\psi_k\rangle\langle\psi_k|$ and $P' = \sum_{k=1}^{n} |\psi_k'\rangle\langle\psi_k'|$ are the projection operators for the two subspaces. Hence, $dS^2 = 2\,\mathrm{tr}\,(\rho - \rho')^2$, where $\rho' = \frac{1}{n} P'$, which shows that (31) restricted to the submanifold of total mixtures of rank n is the usual metric (32) on $G_{n,m}$. By means of an argument similar to the one which gave (24),

$$dS^2 = 4 \left(\sum_{k=1}^{n} <d\tilde{\psi}_k|d\tilde{\psi}_k> - \sum_{i=1}^{n} \sum_{j=1}^{n} <d\tilde{\psi}_i|\tilde{\psi}_j><\tilde{\psi}_j|d\tilde{\psi}_i> \right) = 2\, G_{\bar{\mu}\nu}\, d\bar{Z}^{\mu}\, dZ^{\nu}, \quad (33)$$

where $\{|\tilde{\psi}_k>\}$ is a smooth choice of orthonormal frame field in a neighborhood of $G_{n,m}$, Z^{μ} are complex coordinates in $G_{n,m}$ and the metric coefficients

$$G_{\bar{\mu}\nu} = 2 \left(\sum_{k=1}^{n} <\frac{\partial}{\partial Z^{\mu}}\tilde{\psi}_k|\frac{\partial}{\partial Z^{\nu}}\tilde{\psi}_k> - \sum_{i=1}^{n} \sum_{j=1}^{n} <\frac{\partial}{\partial Z^{\mu}}\tilde{\psi}_i|\tilde{\psi}_j><\tilde{\psi}_j|\frac{\partial}{\partial Z^{\nu}}\tilde{\psi}_i> \right). \quad (34)$$

It follows from (1) that the density matrix ρ satisfies the evolution equation

$$i\hbar \frac{d\rho}{dt} = [\, H, \rho\,]. \quad (35)$$

Therefore the velocity in $\mathcal{D}$ satisfies, from (31),

$$\left(\frac{dS}{dt}\right)^2 = 2\, \mathrm{tr} \left(\frac{d\rho}{dt}\right)^2 = \frac{4}{\hbar^2}\, \mathrm{tr}\{(\rho^2 H^2) - (\rho H)^2\} \quad (36)$$

For a pure state, (36) is the same as (26); indeed this is a simpler derivation of this result if (31) is used as the definition of this metric. To understand the physical meaning of this result, note that the uncertainty of energy for quantum mixed state as two parts to it. There is the "classical part" which exists even when $[\, H, \rho\,] = 0$, because of the energy spread of the common eigenstates of H and ρ. There is in addition a purely quantum mechanical uncertainty of energy ΔE_Q which exists merely because $[\, H, \rho\,]$ is non zero. It is therefore reasonable to define ΔE_Q by

$$\Delta E_Q{}^2 = -\, \mathrm{tr}\, \{[\, H, \rho\,]^2\} = 2\, \mathrm{tr}\{(\rho^2 H^2) - (\rho H)^2\}\ . \quad (37)$$

Then

$$\frac{dS}{dt} = 2 \frac{\Delta E_Q}{\hbar}, \quad (38)$$

which generalizes (26) to mixed states.

7. Discussion

There are two fundamental conceptions of geometry. One is Riemannian by which I mean broadly a geometry defined on a manifold by local structures such as a metric, connection and torsion. The other is Kleinian which is the conception of Felix Klein in his Erlanger program according to which a geometry is a set of properties that are invariant under a group of transformations. The Riemannian conception of geometry is very useful in physics basically because the fundamental laws of classical physics are local. Gravity and gauge fields, classically, are described using the Riemannian conception. On the other hand, quantum mechanics is formulated in the Hilbert space of states of a quantum system on which a symmetry group acts. The state of a single particle may be regarded as a wave function on physical space, which reflects a fundamental non locality of quantum mechanics. Therefore, it is the Kleinian conception of geometry which seems to be appropriate for quantum theory[23]. The task of constructing a quantum theory of gravity appears to be grounded in the problem of bringing these two conceptions of geometry together.

But the geometric structures in quantum theory described above from a Riemannian point of view may also be understood from a Kleinian point of view: The projective geometry on $\mathcal{P}$ is invariant under the group of linear transformations in $\mathcal{H}$ and may therefore be regarded physically as due to the principle of superposition. The metric on $\mathcal{P}$ is invariant under unitary and anti-unitary transformations in $\mathcal{H}$. The connection and the symplectic structure are invariant under transformations in $\mathcal{H}$ which leave invariant the imaginary part of the inner product, including unitary transformations. But they are not invariant under anti-unitary transformations, e. g. time reversal. Consequently, the geometric phase is not invariant under time reversal (it changes sign). However, all four geometric structures are invariant under the group of unitary transformations. Therefore the quantum geometry described by them is analogous to the Minkowski space-time geometry, which may also be regarded as a Riemannian geometry with zero curvature or as a Kleinian geometry invariant under the group of Poincare transformations on space-time. Unlike for Minkowski geometry, time reversal transformation is not a symmetry of this quantum geometry, which is interesting because time reversal symmetry is violated in any case by weak interaction.

However, the geometric structures in $\mathcal{P}$, regarded as Riemannian structures, are "local" in $\mathcal{P}$, but not local in the sense of space-time locality which gives rise to the Riemannian space-time. Since $\mathcal{P}$ represents the physical states of the system it is more closely related to the classical phase space rather than classical space-time. On the other hand the classical phase space can be constructed from particle states represented on space-time which has no analog in quantum theory. That the geometry of $\mathcal{P}$ is fundamentally different from space-time geometry may also be seen from the following observation: A free classical particle moves along a geodesic in space-time. But it is neither necessary nor sufficient for a quantum particle to be free in order to move along a geodesic in $\mathcal{P}$. Therefore, it seems that there must exist a more general quantum geometry which reduces to space-time geometry when Planck's constant tends to zero. Also, from the quntum states in this geometry it should be possible to construct the above geometry in $\mathcal{P}$ when the gravitational field is negligible. Such a geometry may enable us to obtain in a natural manner a quantum theory of gravity.

Acknowledgements

I thank R. Penrose for useful discussions concerning the geometry of mixed states. Part of this work was done at the University of Oxford. Also, this work was partially supported by NSF grant nos. PHY-8807812 and INT-8912968.

References

1. V. Guillemin and S. Sternberg, Symplectic Techniques in Physics (Cambridge Univ. Press 1984).
2. Y. Aharonov and D. Bohm, Phys. Rev. 115 (1959) 485.
3. J. Anandan, Ann. Inst. Henri Poincare, 49 (1988) 271.
4. M. V. Berry, Proc. Roy. Soc. Ser. A 392 (1984) 45 .
5. Y. Aharonov and J. Anandan, Phys. Rev. Lett. 58 (1987)1593; J. Anandan and Y. Aharonov, Phys. Rev. D 38 (1988) 1863.
6. See also B. Simon, Phys. Rev. Lett. 51 (1983) 2167; E. Kiritsis, Commun. Math. Phys. 111 (1987) 417; D. A. Page, Phys. Rev. A 36 (1987) 3479.
7. J. Anandan, Phys. Lett. A 147 (1990) 3.
8. J. C. Garrison and R. Y. Chiao, Phys. Rev. Lett. 60 (1988) 165; J. Anandan, Phys. Rev. Lett. 60 (1988) 2555.
9. A similar function was introduced in the adiabatic limit by I. J. R. Aitchison, Acta Physica Polonica, B18 (1987) 207.

10. This is because all matrix elements of I are invariant under the Schrödinger evolution. Such invariants were studied by H. R. Lewis and W. B. Riesenfeld, J. Math. Phys. 10 (1969) 1458. But since they did not consider a cyclic evolution, the geometric phase was not obtained.
11. Geometric phases for eigenstates of explicitly time dependent invariants were obtained by J. Anandan, Phys. Lett. A 129 (1988) 201, by an approach closely related to the present one.
12. This special case was treated using a time dependent invariant also by X-C. Gao, J-B. Xu and T-Z. Qian, to be published.
13. F. Wilczek and A. Zee, Phys. Rev. Lett. 52 (1984) 2111; J. Anandan, Phys. Lett. A 133 (1988) 171.
14. J. Anandan and Y. Aharonov, Phys. Rev. Lett. 65 (1990) 1697.
15. J. P. Provost and G. Vallee, Commun. Math. Phys. 76 (1980) 289; M. V. Berry in *Geometric Phases in Physics* by A. Shapere and F. Wilczek (World Scientific 1989), p. 14.
16. S. Kobayashi and K. Nomizu, Foundations of Differential Geometry (Interscience, New York, 1969).
17. This result was obtained independently by R. Montgomery, Comm. Math. Phys. 128 (1990) 565.
18. R. Kerner, Ann. Inst. Henri Poincare, 9 (1968) 143.
19. See also, S. K. Wong, Nuov. Cim. 65A (1970) 689; R. Montgomery, Lett. Math. Phys. 8 (1984) 59.
20. Since ρ_1 and ρ_2 are isomorphic, all properties of the two subsystems which depend only on the density matrix alone will be the same. For example, the entropy tr $(\rho \ln \rho)$ is the same for both subsystems even though their individual Hilbert spaces may have very different dimensions. This was realized during a conversation I had with Y. Aharonov and L. Stodolsky.
21. See, for example, B. Mielnik, Commun. Math. Phys. 37 (1974) 221.
22. This metric has been used in the study of a non abelian geometric phase[13] arising due to incomplete measurements by J. Anandan and A. Pines, Phys. Lett. A 141 (1989) 335.
23. J. Anandan, Found. of Phys. 10 (1980) 601.

EXPERIMENTAL VERIFICATION OF THE AHARONOV-CASHER EFFECT FOR NEUTRONS WITH A CRYSTAL INTERFEROMETER

G.I. OPAT, A. CIMMINO, A.G. KLEIN

School of Physics
The University of Melbourne
Parkville, Victoria 3052
Australia

and

H. KAISER, S.A. WERNER, M. ARIF† and R. CLOTHIER

Research Reactor and Department of Physics
University of Missouri-Columbia
Columbia, MO 65211, U.S.A.

ABSTRACT

The phase shift predicted by Aharonov and Casher for a magnetic dipole diffracting around a charged electrode, has been verified for the case of thermal neutrons, using a perfect crystal neutron interferometer. The judicious use of the earth's gravitational field introduces a spin independent phase shift which enables unpolarized neutrons to be used. A supplementary magnetic field of the correct magnitude enables first order sensitivity to be achieved. The ratio of the experimental phase shift to that predicted theoretically is: 1.46 ± 0.35.

1. Introduction

We review the experiment[1] carried out at the Missouri University Research Reactor (MURR) to determine a topological phase shift predicted by Aharonov and Casher.[2]

In 1959 Aharonov and Bohm,[3] in whose honor this conference has taken place, predicted the existence of a topological phase shift which affects the wave-function of a charged particle diffracting around a remote tube of magnetic flux. This effect has been verified many times, initially by Chambers[4] and more recently by the beautiful experiments of Tonomura et al.[5]

In 1984 Aharonov and Casher[2] reasoned that if the diffraction of a charged particle around a tube of magnetic flux produced a topological phase shift, then the diffraction of a particle with magnetic dipole around a straight charged line would also produce a topological phase shift. The situation is shown in Fig. 1. We note that Ampére's circuital law ensures that a tube of flux can only be produced by a line of magnetic dipoles. The Aharonov-Casher theory of this phase shift is presented in the next section. In a less direct manner Anandan foreshadowed this same effect.[6,7] A review of topological effects in particle interferometry has been given by Klein.[8]

We shall see below, that the value of the Aharonov-Casher phase shift attainable with strong electric fields acting on the neutrons within the space permitted by a single crystal neutron interferometer amounts to only a few milli-radian at best. For this reason, the observation of the phase shift requires great care.

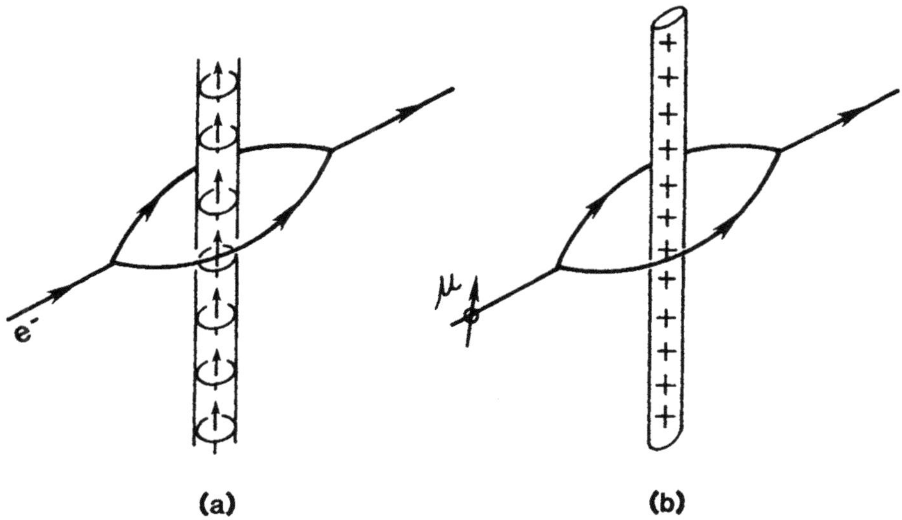

Fig. 1

The Aharonov-Bohm configuration with a charged particle diffracting around a magnetic flux tube (dipole chain) is shown in (a). The Aharonov-Casher configuration of a magnetic dipole diffracting around a line of charge is shown in (b).

2. Theory of the Aharonov-Casher Phase Shift

The Lagrangian of two electromagnetically interacting particles, one bearing only a charge, the other bearing only a magnetic dipole, is given by,

$$\mathcal{L} = \frac{1}{2}mv^2 + \frac{1}{2}MV^2 + e\mathbf{A}(\mathbf{r} - \mathbf{R}) \cdot (\mathbf{v} - \mathbf{V})/c \quad , \tag{1}$$

where the charged particle has position $\mathbf{r}$, velocity $\mathbf{v}$, mass m, and charge e. The other particle has position $\mathbf{R}$, velocity $\mathbf{V}$, mass M, and magnetic moment $\boldsymbol{\mu}$. We have assumed that retardation effects are unimportant.

In Eq. (1) $\mathbf{A}(\mathbf{r}-\mathbf{R})$ is the vector potential produced by the magnetic dipole at the electric charge. If $\mathbf{E}(\mathbf{R}-\mathbf{r})$ is the electric field produced by the electric charge at the magnetic dipole, we find that

$$\mathbf{E}(\mathbf{R}-\mathbf{r}) \times \boldsymbol{\mu} \equiv e\mathbf{A}(\mathbf{r}-\mathbf{R}) \quad , \tag{2}$$

where the electric field for a point charge is given by

$$\mathbf{E} = e(\mathbf{R}-\mathbf{r})/\mid \mathbf{R}-\mathbf{r}\mid^3 \quad . \tag{3}$$

Eq. (2) embodies the dual nature of the two effects.

Using the Lagrangian of Eq. (1) the Hamiltonian, $\mathcal{H}$, for the motion of the magnetic dipole alone can be shown to be:

$$\mathcal{H} = \frac{1}{2M}(\mathbf{P} - \boldsymbol{\mu} \times \mathbf{E}/c)^2 - \boldsymbol{\mu}^2\mathbf{E}^2/Mc^2 \quad . \tag{4}$$

We see that $\boldsymbol{\mu} \times \mathbf{E}$ is the analogue of a magnetic potential, $\mathbf{A}$. If we restrict our attention to motion of a neutral particle with magnetic moment $\boldsymbol{\mu}$ polarized along the axis of a long charged cylinder, then for any path encircling the cylinder we may evaluate an effective "flux":

$$\text{"flux"} = \oint \boldsymbol{\mu} \times \mathbf{E} \cdot d\boldsymbol{\ell} = \sigma 4\pi\mu\Lambda \quad , \tag{5}$$

where $\sigma = \pm 1$ depending on whether the spin is quantised "up" or "down" (parallel or antiparallel to the axis of the line charge) and Λ is the charge per unit length along the cylinder.

By analogy with the Aharonov-Bohm effect we may determine the Aharonov-Casher phase shift, $\Delta\Phi_{AC}$ by substituting the effective "flux" for the magnetic flux. We find

$$\Delta\Phi_{AC} = \sigma 4\pi\mu\Lambda/\hbar c \quad . \tag{6}$$

This phase shift is topological in the following sense:

(i) it is valid for any path of encirclement of the charged cylinder (as long as $\boldsymbol{\mu}$ is parallel to the axis of the cylinder).
(ii) the cross-sectional shape of the charged cylinder is irrelevant, and
(iii) it is independent of the speed of the particle or the orientation of the "orbit" encircling the charged cylinder.

The Aharonov-Casher phase-shift may also be viewed as a manifestation of a spin-orbit interaction $[\boldsymbol{\mu} \cdot (\mathbf{E} \times \mathbf{p})/mc]$.

3. The Neutron Interferometer

In this section we shall show that by judiciously mixing the nuclear (offset), gravitational, magnetic, and electric (Aharonov-Casher) phases in an interferometer, first order sensititivy to the Aharonov-Casher effect can be achieved, and moreover, the neutrons need not be polarized.

A Bonse and Hart, single-crystal silicon neutron interferometer was employed to make manifest the Aharonov-Casher phase shift. Werner and Klein[9], who have reviewed such interferometers, may be consulted for the details. Figs. 2 and 3 show disposition of the elements of the interferometer. The interferometer contains the Aharonov-Casher electrode in the first section and a biassing magnetic field assembly in the second section.

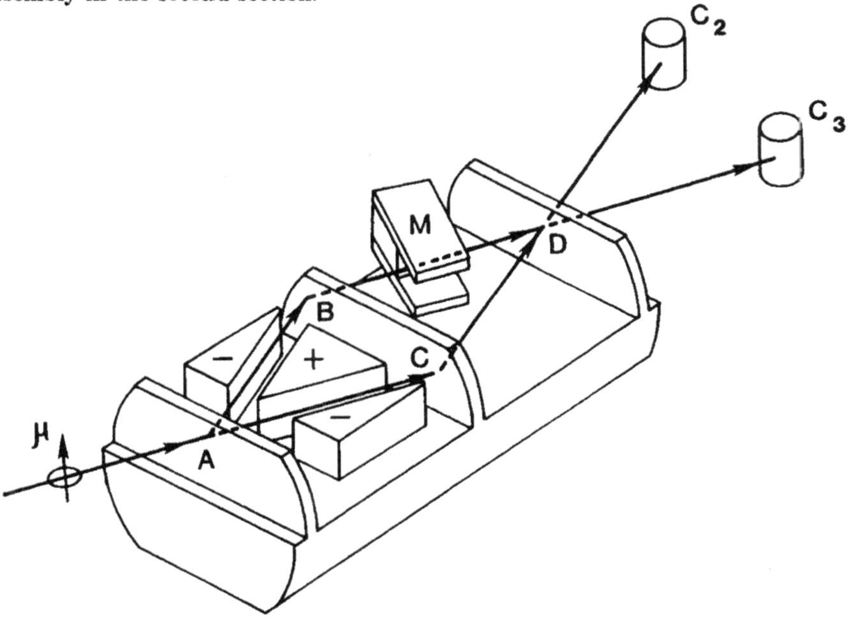

Fig. 2

A perspective view of the neutron beams in the interferometer, showing the electrode assembly and the bias magnet.

The charged electrode structure seen in Fig. 3 is housed in a vacuum box. Care with the electrode surfaces enabled potential differences of more than 50 kV to be maintained across the 1.54 mm gap, corresponding to fields of 30 MV/m. As the experiment had to be stable for months on end, slightly lower potential differences of $\Delta V = \pm 45$ kV were employed. The electrode assembly and the high tension lead were made rigid enough to avoid appreciable deformation under the action of the electrical stress which depends on $(\Delta V)^2$. However, even if there was some residual electrostatic deformation of the assembly, the experimental strategy, which involved reversing the potential difference periodically, would have eliminated the consequences of such a quadratic effect. A computation of the electrostatic field map, showed that the effective length of the neutron path in the electric field had to be increased by 8% over the naive value based on parallel plate theory.

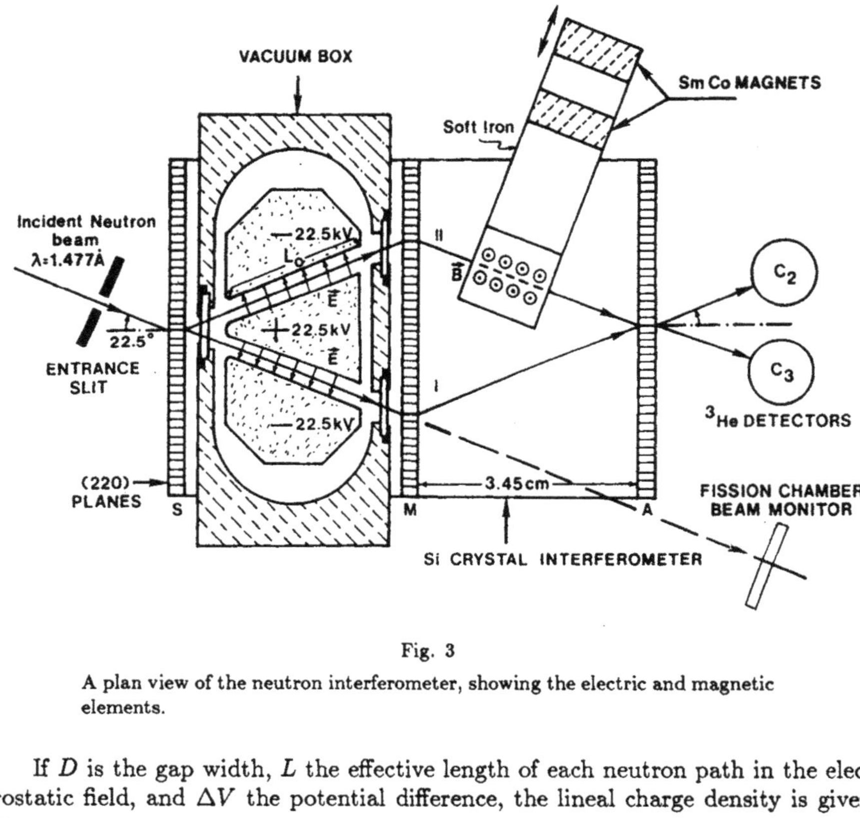

Fig. 3

A plan view of the neutron interferometer, showing the electric and magnetic elements.

If D is the gap width, L the effective length of each neutron path in the electrostatic field, and ΔV the potential difference, the lineal charge density is given by

$$\Lambda = 2\Delta V L/4\pi D \quad , \tag{7}$$

according to Gauss's Law.
For our experimental arrangement, we then find an Aharonov-Casher phase shift of

$$\Delta\Phi_{AC} = 1.52\sigma \text{ milliradian (theory)} \quad , \tag{8}$$

where $D = 1.54$mm, $2L = 2 \times 2.53$cm, and $\Delta V = 45$kV ($= 150$ statvolts). The magnetic field produces a spin-dependent bias phase shift $\Delta\Phi_M$ to be discussed below. The neutron counting rates in detectors C_2 and C_3 are given by

$$C_2 = a_2 - b_2 \cos \Delta\Phi \quad , \tag{9}$$

and

$$C_3 = a_3 + b_3 \cos \Delta\Phi \quad , \tag{10}$$

where $\Delta\Phi$ is the phase shift for the neutron wave on path II relative to path I. The constants a_2, a_3, b_2, b_3 characterize the actual interferometer setup, and $a_2/a_3 \approx 3$, while $b_2 - b_3$, such that $C_2 + C_3$ is independent of $\Delta\Phi$. The total phase difference consists of four terms

$$\Delta\Phi = \Delta\Phi_0 + \Delta\Phi_G + \sigma\{\mp \mid \Delta\Phi_{AC} \mid + \mid \Delta\Phi_M \mid\} \quad , \tag{11}$$

where the + sign is for negative polarity and the − sign is for positive polarity of the center electrode. $\Delta\Phi_0$ is a fixed phase of nuclear origin representing an inherent offset in the interferometer. $\Delta\Phi_G$ is a phase difference due to the earth's gravitational field. If the interferometer is rotated slightly about the incident beam direction, beam II is raised a little above beam I. As beam II now has a higher potential energy than beam I, its kinetic energy is lower, and its wavelength longer[10]. This gives rise to a gravitationally-induced phase shift $\Delta\Phi_G$, known as the COW effect.

We now assume that the incident beam is unpolarized, thus the total count rate in detector C_3 is the sum of the count rates for spin up and spin down incident neutrons,

$$C_3(\pm) = C_3^{\uparrow}(\pm) + C_3^{\downarrow}(\pm) = a_3 + b_3 \cos[\Delta\Phi_0 + \Delta\Phi_G] \cos[\mid \Delta\Phi_M \mid \mp \mid \Delta\Phi_{AC} \mid] \quad . \tag{12}$$

Similarly, the count rate in detector C_2 is given by

$$C_2(\pm) = a_2 - b_2 \cos[\Delta\Phi_0 + \Delta\Phi_G] \cos[\mid \Delta\Phi_M \mid \mp \mid \Delta\Phi_{AC} \mid] \quad . \tag{13}$$

4. Experimental Strategy

The gravitational phase shift $\Delta\Phi_G$ is adjusted by tilting the crystal so that $\Delta\Phi_0 + \Delta\Phi_G = 0 \pmod{2\pi}$, giving maximum sensitivity to $\Delta\Phi_{AC}$. Furthermore, adjusting the magnetic phase $\Delta\Phi_M$ to be $\pi/2$, we have

$$\begin{aligned} C_3(\pm) &= a_3 + b_3 \sin(\pm \mid \Delta\Phi_{AC} \mid) \\ &\approx a_3 \pm b_3 \mid \Delta\Phi_{AC} \mid \quad , \end{aligned} \tag{14}$$

and

$$\begin{aligned} C_2(\pm) &= a_2 - b_2 \sin(\pm \mid \Delta\Phi_{AC} \mid) \\ &\approx a_2 \mp b_2 \mid \Delta\Phi_{AC} \mid \quad . \end{aligned} \tag{15}$$

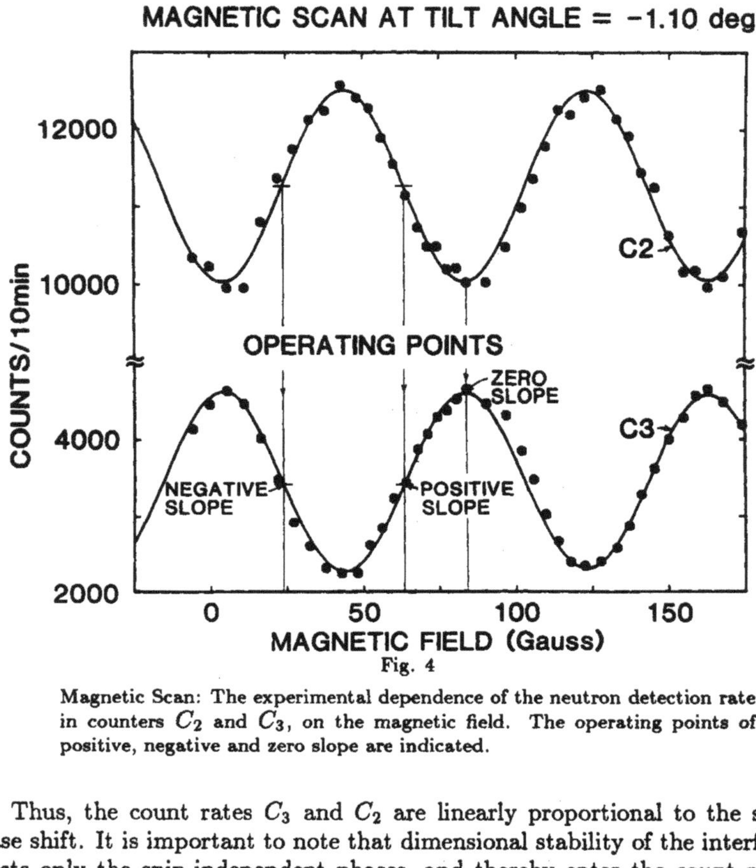

Fig. 4

Magnetic Scan: The experimental dependence of the neutron detection rate in counters C_2 and C_3, on the magnetic field. The operating points of positive, negative and zero slope are indicated.

Thus, the count rates C_3 and C_2 are linearly proportional to the small AC phase shift. It is important to note that dimensional stability of the interferometer affects only the spin-independent phases, and thereby enter the count rates only in second order. Had polarized neutrons been used, this would not have been the case, and typical drifts in $\Delta\Phi_0$ of order 50 milliradians/day would have precluded obtaining the sensitivity required.

The magnetic bias field, which was provided by *SmCo* magnets, was varied by changing the reluctance of a magnetic circuit. To find the correct operating point for $\Delta\Phi_G$ and $\Delta\Phi_M$, the following procedure was followed: With the magnetic field **B** set to zero, a gravity scan was carried out by tilting the interferometer in steps about the incident beam direction. The first maximum near zero tilt angle sets $\Delta\Phi_G + \Delta\Phi_0 = 0 \pmod{2\pi}$. Leaving the tilt fixed at this maximizing angle, operating points of negative slope ($\Delta\Phi_M = \pi/2$) and positive slope ($\Delta\Phi_M = 3\pi/2$) were selected by scanning the magnetic field as shown in Fig. 4. This procedure was repeated every week, since the spin-independent phase drifts 50 to 100 milliradians over this time period. The following additional procedure was also carried out periodically to fix the operating point more accurately: The magnetic field was changed in small increments around the value giving $\Delta\Phi_M \approx \pi/2$ (or $3\pi/2$). Gravity scans were executed for each value of the magnetic field and the optimum was chosen as that giving minimum oscillation. A representative gravity scan (for maximum oscillations) is shown in Fig. 5.

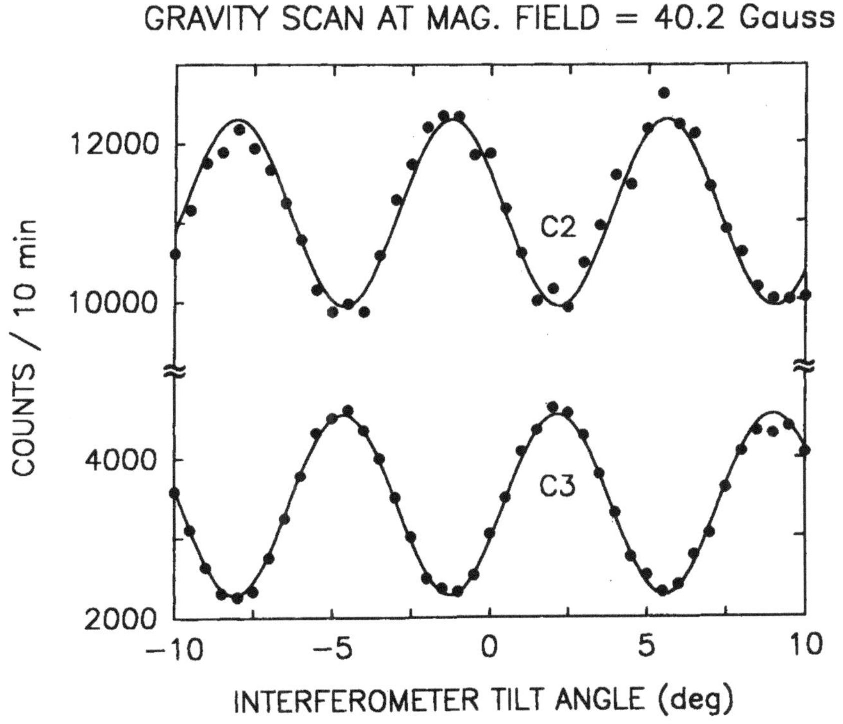

Fig. 5

Gravity Scan: The experimental dependence of the neutron detection rate in counters C_2 and C_3 on the angle of tilt about the incident neutron beam direction.

The voltage cycle was divided into three parts: positive, negative and zero. The signal for the voltage cycle was provided by the monitor counter reaching a preset value (35,000), taking about 10 minutes, for a total of 30 minutes per complete cycle. Data was accumulated over approximately 1700 voltage cycles, taking about 35 days of counting time each. The total accumulated counts were approximately 6,000,000 in detector C_3 and 19,000,000 in detector C_2 for each operating condition. Table I summarizes the difference counts for positive and negative electric fields, namely $C_2(+) - C_2(-)$ and $C_3(+) - C_3(-)$, for positive, negative and zero slope conditions. Statistically combining the results for positive and negative slope and normalizing each count to $C_2 + C_3 = 25,000,000$ (=1700 cycles), we find

$$\overline{\Delta C} \equiv \omega_2 \mid C_2(+)-C_2(-) \mid + \omega_3 \mid C_3(+)-C_3(-) \mid = 9010 \pm 2130 \text{ counts/1700 cycles} \tag{16}$$

Table I

This table summarizes the difference counts $C_2(+)-C_2(-)$ and $C_3(+)-C_3(-)$ for the three operating conditions of the bias magnetic field. There were approximately 1700 voltage cycles for each operating condition, taking approximately 35 days each.

Experimental Condition	$C_2(+)-C_2(-)$	$C_3(+)-C_3(-)$
(1) Positive Slope	$+13,462$ $\pm 6,223$	$-8,019$ $\pm 3,494$
(2) Negative Slope	$-9,986$ $\pm 6,097$	$+8,243$ $\pm 3,413$
(3) Zero Slope	$+1,718$ $\pm 6,020$	$-3,631$ $\pm 3,979$

where ω_2 and ω_3 are statistical weighting factors (inversely proportional to the square of the standard deviations).[11]

A fit to the magnetic scan of Fig. 4 gives

$$b \equiv b_2 = b_3 = 2,057,000 \pm 57,000 \text{ counts/1700 cycles} \quad . \tag{17}$$

Thus, according to Eqs. (14) and (15) we have for the measured AC phase shift

$$\Delta\Phi_{AC} = \frac{\overline{\Delta C}}{2b} = 2.19 \pm 0.52 \text{ milliradians (experiment)} \quad . \tag{18}$$

We note that the expected sign reversals of ΔC_2 and ΔC_3 for positive and negative slope operating points were observed. Furthermore, the expected difference counts ΔC_2 and ΔC_3 were zero to within statistical error for the zero slope operating point experiment, as expected. The experimental result given by Eq. (18) is to be compared with the theoretical prediction given by Eq. (8) resulting in the ratio

$$(\Delta\Phi_{AC})_{\text{exp}}/(\Delta\Phi_{AC})_{\text{theor}} = 1.46 \pm 0.35. \tag{19}$$

We note that $\Delta\Phi_{AC}$ is correct in sign, order of magnitude, and in reasonable statistical agreement with the theory.

At present, further statistics are being gathered at MURR to reduce the statistical errors. We note, that if the Aharonov-Casher experiment could be done by the interference of odd-electron neutral atoms, a phase-shift three orders of magnitude greater would be expected due to the larger magnetic dipole moment of the electron. However, the much larger electric polarizability of atoms may present a difficulty for such an experiment.

Acknowledgements

This work was supported by the Australian Research Council, and the Physics Division, National Science Foundation (NSF-PHY-8813253).

†Present address: National Institute of Standards and Technology, Gaithersburg, MD

References

1. A. Cimmino, G.I. Opat, A.G. Klein, H. Kaiser, S.A. Werner, M. Arif and R. Clothier, Phys. Rev. Lett. **63** (1989) 380.
2. Y. Aharonov and A. Casher, Phys. Rev. Lett. **53** (1984) 319.
3. Y. Aharonov and D. Bohm, Phys. Rev. **115** (1959) 485.
4. R.G. Chambers, Phys. Rev. Lett. **5** (1960) 3.
5. A. Tonomura, N. Osakabe, T. Matsuda, T. Kawasaki, J. Endo, S. Yano and H. Yamada, Phys. Rev. Lett. **56** (1986) 792.
6. J. Anandan, Phys. Rev. Letts. **48** (1982) 1660.
7. J. Anandan, Phys. Lett. **138A** (1989) 347.
8. A.G. Klein, Physica **137B** (1986) 230.
9. S.A. Werner and A.G. Klein, Methods of Experimental Physics **23A** (1986) 259 Academic Press.
10. R. Colella, A.W. Overhauser, and S.A. Werner, Phys. Rev. Lett. **34**, (1975) 1472; S.A. Werner, H. Kaiser, M. Arif and R. Clothier, Physica **B151** (1988) 22.
11. G.I. Opat, *Statistical Analysis of Interferometer Detection Systems*, Melbourne University internal report UM-P-89/31 submitted for publication.

NON LOCALITY OF THE AHARONOV CASHER - EFFECT*

B. REZNIK
School of Physics and Astronomy
Beverly and Raymond Sackler Faculty of Exact Sciences
Tel - Aviv University, Tel - Aviv 69978, ISRAEL.

ABSTRACT

The Aharonov-Casher effect is manifested in a (2 + 1)-dimensional model that screens the electromagnetic fields, in order to demonstrate that the effect is essentially non-local in its nature. The question of non-locality is discussed by means of a non-relativistic model for a superconductor. It is demonstrated that although the superconductor screens the electric field generated by an external charge it does not screen the modular electric field which is a constant of motion of the system. Consequently, a magnetic fluxon will accumulate the same phase as if the electric field were unscreened and the Aharonov-Casher effect will exist even in a force-free region.

1. Introduction

The work of Aharonov and Casher[1] reveals a new aspect of the Aharonov-Bohm effect[2]. Aharonov and Casher suggest that a neutral particle with a magnetic moment $\vec{\mu}$ may exhibit a topological force-free interference effect, as a result of an interaction with a charged wire[3]. Consequently, the Aharonov-Bohm effect admits a duality, meaning that we can reverse the roles of the charged particle and the solenoid in the Aharonov-Bohm effect and still get a force-free interference effect.

Let us consider the duality in (2 + 1)-dimensions more explicitly. Both effects are related to a two-particle system consisting of one particle with a charge q and location r, and a second particle with a magnetic moment μ and location R. In the non-relativistic limit when $\dot{r}/c \simeq \dot{R}/c \ll 1$ the corresponding two-particle Lagrangian is[1,4]

$$L = \frac{1}{2} m\dot{r}^2 + \frac{1}{2} M\dot{R}^2 + q/c\, \vec{A}(\vec{r} - \vec{R}) \cdot \dot{r} + \vec{\mu}/c \times \vec{E}(\vec{R} - \vec{r}) \cdot \dot{R} \qquad (1.1)$$

where A(r - R) is the vector potential generated by the magnetic moment at the location of the charged particle, and E(R - r) is the electric field generated by the charged particle at the location of the magnetic moment.

The duality is manifested by the numerical equality of the interaction terms

$$\oint q\vec{A}(\vec{r} - \vec{R}) = \oint \vec{E}(\vec{R} - \vec{r}) \times \vec{\mu}. \qquad (1.2)$$

As a result of the equality, two observers O′ and O″, situated at the rest frame of the

* This talk is based on the paper by B. Reznik and Y. Aharonov, Phys. Rev. D, 40 (1989) 4178.

magnetic moment and the rest frame of the charge respectively, will calculate the same phase accumulated as a result of relative motion of the particles, but may interpret the nature of the interaction in a completely different way. While observer O′ relates the phase $(Q'/\hbar c)\int A dl$ to the interaction of the charged particle with the vector potential generated by the magnetic moment, observer O″ will relate the phase $(1/\hbar c)\int E \times \mu dl$ to the interaction of the magnetic moment with a "vector potential" $E \times \mu$ generated by a charged stationary particle. Hence, according to observer O′, the system accumulates phase as a result of a force-free interaction of the Aharonov-Bohm type which will be interpreted as a non-local effect; while, according to observer O″, although the interaction $E \times \mu$ does not generate a dynamical force (there is however a non-vanishing hidden momentum[5]) on the magnetic moment, the accumulated phase can be interpreted as a result of a local interaction with a non-vanishing electric field.

According to the Galilean invariance of the interaction, both approaches are permitted and indeed give the same result. However, the resulting interpretations are not compatible. This suggests two possibilities: either the Aharonov-Bohm effect resulting from the interaction term $qA(r - R)$ is a result of a local interaction, or the interaction term $E \times \mu$, although it appears as a local interaction, also represents a non-local interaction and consequently, the Aharonov-Casher effect must be interpreted as a non-local force-free effect[5] (here and in the following the term force-free is related to the situation where all the classical field strengths vanish).

The purpose of this work is to show that the second possibility is the correct one. We demonstrate that in (2 + 1)-dimensions it is possible to arrange both, the charged particle and the magnetic moment, to be in force-free regions, and yet get the usual phase shift and interference as if the electric field were unscreened. Consequently, we conclude that the concept of duality does not contradict the non-locality of the Aharonov-Bohm effect and that its dual Aharonov-Casher effect is also essentially non-local in its nature[6].

The plan of this paper is as follows: In section 2 the effect is discussed in the context of a superconductor. We realize that in order to understand how the effect occurs in a force-free region we must examine the quantum behaviour of the superconductor. The effective Hamiltonian will be derived in section 3. In section 4, we discuss the correct limit for the model to be a faithful description of a superconductor and find in this limit the ground state of the system. Section 5 will present the modular variables, and discuss their importance. In section 6, the magnetic fluxons will be included and demonstrated to accumulate the correct phase.

2. The Effect in the Presence of a Superconductor

It is well known that a superconductor screens all the electromagnetic fields exponentially; hence it is suitable as a force-free medium. A type-2 superconductor may support quantized localized magnetic fluxons with a flux $\Phi = hc/q$ which pertains to the medium's unit charge $q = 2e$ (the Cooper pair). If we introduce into the superconductor an external particle with charge Q', that moves along a path enclosing a fluxon, it will accumulate the Aharonov-Bohm phase $(Q'/\hbar c)\int A dl$, where A is the vector potential generated by the fluxon

$$\frac{q}{\hbar c}\vec{A}(\vec{r}) = \frac{\hat{z} \times \vec{r}}{|\vec{r}|^2} + O(e^{-|\vec{r}|/\lambda}). \tag{2.1}$$

Here r is a two-dimensional vector, z is a unit vector perpendicular to the x-y plane, and λ is the penetration depth. The phase will be *non-trivial only for Q′ which* is unquantized with respect to q, $Q' \bmod q \neq 0$.

Is it possible to realize also the Aharonov-Casher effect by changing the roles of the charge and fluxon ? On first observation it appears that the answer should be negative. The electric field generated by the external charge is screened, hence the fluxon interacts locally with a field strength of $\exp(-r/\lambda)$, which cannot lead to the correct result. Nevertheless, there is the possibility that some part of the electric field still penetrates the superconductor, does not generate any dynamical force, but still is sufficient to cause the fluxon to accumulate the correct phase. Such a delicate effect cannot occur in a classical system, but is possible in a quantum system, as we shall see in the following.

In order to understand the mechanism that makes the effect possible, we must include the quantum behaviour of the shielder. This will be done in the following by means of a simple model for a superconductor in (2 + 1)-dimensions. Let us imagine a spherical superconductor with an external radius R_1 (and the corresponding boundary S_1). Insert the external charge Q' into a hole of radius R_2 (with a corresponding inner surface S_2) in the center of the superconductor. We assume that $R_2 << R_1$. In the following section we shall rely on the following assumptions and simplifications:

1. The electric field inside the superconductor for $R_2 << r << R_1$ depends on the induced charge (Cooper pairs) near the inner surface S_2, through the Gauss law. We shall neglect any dependence on local fluctuations inside the superconductor. This can be justified by noticing that time averaging over a short time (relative to the time scale of the interference effect) leads to a cancellation of the local fluctuations. Hence, it will be sufficient for our purpose to formulate the quantum behaviour of the induced charge.
2. The origin of the non-local behaviour is related to the fluctuation of the induced charge Q on S_2. A change in Q corresponds to a "jump" of a unit charge q from S_1 to S_2 and vice versa. This can be described as a collective radial fluctuation of the "macroscopic wave function" of the superconductor[7].

3. The Effective Hamiltonian

The first step is to establish the effective Hamiltonian and the corresponding Hilbert space. At first the contribution of the fluxon is neglected. To this end, let us examine the energy dependence of the superconductor on the induced surface charge, $E = E(Q)$. This function must have a minimum at the value $Q = -Q'$, where Q' is the external charge. Hence, it is most likely to be of the form

$$E = \frac{1}{2C}(Q + Q')^2 + \text{other contributions.} \tag{3.1}$$

The superconductor can be described as a spherical capacitor, where the inner and outer surfaces S_2 and S_1 play the role of the capacitor plates. The constant C plays the role of "capacitance" and is expected to depend on the geometry of the superconductor, $C = C(R_1,R_2)$.

Let q be the charge of a Cooper pair ($q = 2e$), and let n be the number of Cooper pairs near S_2 (n is negative for Cooper holes). Using these definitions, we can write the energy related to the induced charge Q in the form

$$E = \frac{(nq + Q')^2}{2C}. \tag{3.2}$$

Proceeding to a quantized theory we define first the corresponding Hilbert space. The most natural choice is to use the states $| n \rangle$, which correspond to n Cooper pairs on the surface S_2. By Equation (3.2) the matrix element of the Hamiltonian is given by

$$\langle n | H | n \rangle = \frac{(nq + Q')^2}{2C} \tag{3.3a}$$

and also

$$< n \mid H \mid n' > \; = \; 0 \tag{3.3b}$$

for $n \neq n'$. This is not satisfactory since we expect to get fluctuations in n and therefore a non-vanishing "tunneling" amplitude. Hence, define the non-diagonal matrix element as

$$< n' \mid H \mid n > = \frac{A}{2}\,\delta_{n,n'+1} + \frac{A}{2}\,\delta_{n+1,n'} \tag{3.4}$$

where we have included only the first-order "tunneling" effect, and neglected higher orders. The coefficient A is unknown, but as we shall see, knowledge of its numerical value is unnecessary for our needs.

The Schrödinger equation in the energy representation is

$$H \mid E > = E \mid E >. \tag{3.5}$$

Multiplying from the left by the vector $< n \mid$, and substituting to the right of H the unity operator $\Sigma \mid n >< n' \mid$, we reach

$$\sum_{n'} < n \mid H \mid n' >< n' \mid E > \; = \; E < n \mid E >. \tag{3.6}$$

Define now the wave function u_n as $u_n = < n \mid E >$, i.e. the amplitude of the superconductor to have energy E and n Cooper pairs on the inner surface. Then the Schrödinger equation for the wave function reads

$$i\hbar \frac{\partial u_n}{\partial t} = \frac{1}{2C}(nq + Q')^2 u_{n(t)} - \frac{A}{2}(u_{n+1} + u_{n-1}). \tag{3.7}$$

The last term on the right is of special significance since it represents the non-local jumps that were suggested in the previous section. Consequently, the stationary state of the superconductor will be an infinite superposition of the wave functions u_n and hence represent a state with an undefined induced charge.

In order to get a representation which diagonalizes the Hamiltonian, it is necessary to transform from the u_n to another basis. This can be done by the transformation

$$u_{n(t)} = \frac{1}{\sqrt{2\pi}} \int_{-\pi}^{+\pi} e^{-i(n+\alpha)\theta}\, U(\theta,t)d\theta \tag{3.8a}$$

$$U(\theta,t) = \frac{1}{\sqrt{2\pi}} \sum_{n} e^{i(n+\alpha)\theta}\, u_{n(t)} \tag{3.8b}$$

where α is an arbitrary constant, and θ takes the value $\theta \in (-\infty,+\infty)$. By definition $U(\theta)$ is a periodic function up to a phase

$$U(\theta + 2\pi) = e^{2\pi i\alpha}\, U(\theta). \tag{3.9}$$

If we choose the normalization of the wave function u_n to be

$$\sum_n |u_{n(t)}|^2 = 1, \tag{3.10a}$$

we get for the normalization of $U(\theta,t)$

$$\int_{-\pi}^{+\pi} U^*(\theta,t)U(\theta,t)d\theta = 1. \tag{3.10b}$$

Substituting Equation (3.8) into (3.7), and using the periodicity of the wave functions $U(\theta,t)$, gives the Schrödinger equation in the θ representation:

$$i\hbar \frac{\partial U(\theta,t)}{\partial t} = \frac{1}{2C}\left[\left(-iq\frac{\partial}{\partial\theta} - \alpha q + Q'\right)^2 - A\cos\theta\right]U(\theta,t). \tag{3.11}$$

Define charge and number operators:

$$\hat{N} = -i\frac{\partial}{\partial\theta} \tag{3.12a}$$

$$\hat{Q} = -iq\frac{\partial}{\partial\theta}. \tag{3.12b}$$

The operator $\hat{Q}$ is the canonical conjugate momentum of θ. The physical meaning of $\hat{Q}$ becomes clear if we calculate its expectation value

$$<\hat{Q}> = q\alpha + q\sum_n n\,|u_n|^2. \tag{3.13}$$

Clearly, $\hat{Q}$ represents the induced charge on S_2 plus a constant charge. If we choose the constant α to take the value $\alpha = Q'/q$, the operator represents the total charge $Q_t = Q_{induced} + Q_{external}$, and the Hamiltonian takes its simplest form

$$H_{eff} = \frac{1}{2C}\hat{Q}_t^2 - A\cos\theta. \tag{3.14}$$

The last specific representation ("gauge") of the Hamiltonian emphasizes the role of the external charge in the model. The external charge does not appear in the classical equations of motion as is evident from Equation (3.14) (or by constructing the effective Lagrangian and noticing that the effect of Q' for any α is to introduce a total time derivative $d\theta/dt$ which does not change the equations of motion). Its only effect is to change the structure of the state of the system. From equation (3.9) we see that in this "gauge" the external charge determines the phase of the wave function resulting from the transformation $\theta \rightarrow \theta + 2\pi$.

However, in the following we choose to use the value $\alpha = 0$ for which there is a clear distinction between the induced fluctuating superconductor's charge and the external charge. Hence, for the rest of the paper $Q_t = Q + Q'$.

Another consequence of the nature of the eigenfunction and of the symmetry of the Hamiltonian under $\theta \rightarrow \theta + 2\pi$ is that the energy must be a periodic fuction of Q':

$$E(Q') = E(Q' + q), \quad \text{or} \quad E = f\left\{\cos\left(\frac{2\pi Q'}{q}\right)\right\}. \tag{3.15}$$

This means that energetically the superconductor does not distinguish between Q' and Q' + q. Consequently, since the external electric field is screened, the most the fluxon can sense is some modular function $\cos(2\pi Q'/q)$. We shall see that indeed this is the case.

4. The Ground State

The operators $\hat{Q}$ and θ satisfy the usual canonical commutation relation

$$[\hat{Q}, \theta] = -iq \tag{4.1}$$

and therefore obey an uncertainty relation

$$\Delta\theta\, \Delta Q \geq q. \tag{4.2}$$

The last relation is analogous to the phase-charge uncertainty in the B.C.S theory. According to this theory, the wave function is the ground state of the system and obeys the relations $\Delta\theta \rightarrow 0$, and $\Delta Q >> q$. Hence, we shall require the ground state of the effective Hamiltonian to satisfy the same relations.

The ground state of the effective Hamiltonian can be constructed easily as a linear combination of localized wave functions at all the minima of the potential. First, construct a ground state wave function $u_0(\theta)$ in the well around $\theta = 0$. The Hamiltonian can be approximated as a harmonic oscillator

$$H_{osc} = \frac{1}{2C}(\hat{Q} + Q')^2 + \frac{A}{2}\theta^2 \tag{4.3}$$

and the solution is well known to be

$$u_0(\theta) = e^{-i\theta Q'/q}\, h_0(\theta) \tag{4.4a}$$

where

$$h_0(\theta) = (\sqrt{AC}/\pi q)^{1/4}\, e^{-\sqrt{AC}/2q\, \theta^2}. \tag{4.4b}$$

Second, construct a superposition of the unperturbed wave functions at the separate wells

$$U(\theta) = \sum_N u_0(\theta - 2\pi N) = \sum_N e^{-i(\theta - 2\pi N)Q'/q}\, h_0(\theta - 2\pi N). \tag{4.5}$$

The solution is periodic in θ, in agreement with equation (3.9).

Using this result we find for the uncertainties of the canonical variables

$$(\Delta\theta)^2 = \frac{1}{2}\frac{q}{\sqrt{AC}} \tag{4.6a}$$

$$(\Delta Q)^2 = \frac{1}{2} q\sqrt{AC}. \tag{4.6b}$$

Hence, taking the limit $\sqrt{AC} >> q$ leads to the required conditions $\Delta\theta \to 0$ and $\Delta Q >> q$.

5. Modular Variables

Having derived the wave function of the system, we can now examine the behaviour of the total charge $Q_t = Q_{induced} + Q_{external}$. For the expectation value of $(Q_t)^n$ we find

$$< (\hat{Q}_t)^n > = \int_{-\pi}^{+\pi} U(\theta)^* (\hat{Q} + Q')^n \, U(\theta)\, d\theta = \int_{-\pi}^{+\pi} h_0(\theta)^* \, \hat{Q}^n \, h_0(\theta)\, d\theta \,. \tag{5.1}$$

Notice that only the part that is localized at the well around $\theta = 0$ contributes to the integral. The contribution of the other parts can be neglected since according to (4.6) the spread $\Delta\theta \to 0$. Consequently, the expectation value of the total charge is independent of the external charge Q' (odd moments of Q_t vanish). Applying the Gauss law in two spatial dimensions we obtain the following operator equality for the radial electric field inside the superconductor

$$\hat{E} = \frac{1}{2\pi r}\hat{Q}_t = \frac{1}{2\pi r}(\hat{Q} + Q') \tag{5.2}$$

where r is a c - number. Since the last result for $< \hat{Q}_t >$ applies also for the expectation value of E, we conclude that in the superconductor (for $R_2 << r << R_1$) the field strength vanishes. Hence the interior of the superconductor can be regarded as a force-free region.

The expectation value of any moment of $\hat{Q}_t$ is by the right hand side of eq. (5.1) independent of Q'. Can we construct a function of $\hat{Q}_t$ that is sensitive to the external charge? The ground state, in the required limit of $\sqrt{AC} >> q$, is a superposition of orthogonal functions localized at $\theta = 2\pi N$. The dependence on the external charge Q' appears only in the relative phase of these functions. Hence, only functions with a nonvanishing matrix element between $u_0(\theta)$ and $u_0(\theta + 2\pi)$ can have an expectation value which depends on Q'. Any finite polynomial will not do. Therefore, we are led to consider a function of the form $\exp(2\pi i Q_t/q)$, which is a unitary shifting operator of $u(\theta)$ to $u(\theta+2\pi)$. For this function

$$e^{2\pi i \hat{Q}_t/q}\, U(\theta) = e^{2\pi i Q'/q}\, U(\theta)\,. \tag{5.3}$$

The operator $\exp(2\pi i \hat{Q}_t/q)$ will be referred to as the modular operator[8] of $\hat{Q}_t$. The corresponding expectation value will be Q'-dependent:

$$< e^{2\pi i \hat{Q}_t/q} > = e^{2\pi i Q'/q}\,. \tag{5.4}$$

The operator $\hat{Q}_t$ (or $\hat{E}$), or any moment of $\hat{Q}_t$, is not a constant of motion since it does not commute with the Hamiltonian. Nevertheless, the modular operator of $\hat{Q}_t$ satisfies[8]

$$\left[e^{2\pi i \hat{Q}_t/q} , e^{i\theta} \right] = 0 \tag{5.5}$$

and hence

$$\left[e^{2\pi i \hat{Q}_t/q} , H_{eff} \right] = 0 . \tag{5.6}$$

The modular operator $\exp(2\pi i \hat{Q}_t/q)$ is a constant of motion. Moreover, from equation (5.3) we have $Q \bmod q = 0$. This is not surprising since it means that the total induced charge is quantized with respect to the charge of a Cooper pair, and by equation (5.6) this quantization is constant in time. Consequently, although the induced charge screens the external charge ($< Q_t > = 0$) it cannot screen the modular charge ($Q' \bmod q$) and the result is a non-trivial phase for the expectation value of the modular operator in eq. (5.4).

Similarly, define a quantum of electric field (a c-number) and a modular electric modular field operator as

$$e_0 = \frac{q}{2\pi r} \quad \text{and} \quad e^{2\pi i \hat{E}/e_0} \tag{5.7}$$

respectively, where e_0 pertains to the field generated by a single particle of the medium. We obtain

$$< e^{2\pi i \hat{E}/e_0} > = < e^{2\pi i \hat{Q}_t/q} > = e^{i2\pi Q'/q} = e^{i/ch \int \vec{E} \times \vec{\mu} \, d\vec{l}} . \tag{5.8}$$

Notice that the expectation value of the modular electric field on the left-hand side is equal numerically to the phase shift in the unscreened Aharonov-Casher effect on the right-hand side. The modular electric field will not vanish if and only if Q' is unquantized, $Q' \bmod q \neq 0$, which is the same condition for which Q' displays the Aharonov-Bohm effect. This transparency of the superconductor to the modular electric field is exactly the property that enables the fluxon to accumulate the correct phase, since as we shall see in the following section the solution for the Shrödinger equation that includes the contribution of the fluxon involves a phase which is proportional to the modular electric field.

6. The Interaction of the Fluxon

Finally we would like to demonstrate that the fluxon accumulates the usual phase although its motion is entirely confined to a force-free region $R_2 << r << R_1$.

In order to include the contribution of the fluxon to the effective Hamiltonian we shall assume that the fluxon has an effective mass M and momentum $\vec{P}$. Then, the interaction with the electric field is given by the replacement[1]

$$\vec{P} \rightarrow \vec{P} + \frac{1}{c} \vec{E} \times \vec{\mu} . \tag{6.1}$$

Since the "vector potential" $\vec{E} \times \vec{\mu}$ has only a nonvanishing azimuthal component we can simplify the problem and regard only the azimuthal motion of the fluxon described by the angle $\phi \in (-\pi,\pi)$. Substituting the corresponding operator for E (eq. 5.2) and the explicit value for $\vec{\mu}$, ($|\vec{\mu}|$ = hc/q), we reach the following effective Hamiltonian:

$$H_{eff} = \frac{1}{2C}\hat{Q}_t^2 - A\cos(\theta) + \frac{1}{2M}\left\{P_\phi - \frac{h}{2\pi r}\frac{\hat{E}}{e_0}\right\}^2. \tag{6.2}$$

Let us substitute a solution of the form

$$\Psi(\theta,\phi) = e^{i\phi\,\hat{E}/e_0}\,\psi(\theta,\phi). \tag{6.3}$$

We find that the function $\psi(\theta,\phi)$ satisfies the "free" equation of motion

$$\left[\frac{1}{2C}\hat{Q}_t^2 - A\cos(\theta-\phi) + \frac{1}{2M}P_\phi^2\right]\psi(\theta,\phi) = E\,\psi(\theta,\phi) \tag{6.4}$$

There is a path dependence in the potential term, the motion of the fluxon moves the location of the wells.

An explicit solution for ψ can be found in the adiabatic approximation ($\hbar^2/2Mr^2 \gg q^2/2C$ is satisfied). The wave function takes the form

$$\psi(\theta,\phi) = \rho(\theta)\,\eta(\phi) \tag{6.5}$$

and the Schrödinger equation (6.4) separates into two equations

$$\left\{\frac{1}{2C}\hat{Q}_t^2 - A\cos(\theta-\phi)\right\}\rho_\phi(\theta) = E_{sc}(\phi)\,\rho_\phi(\theta) \tag{6.6a}$$

$$\frac{1}{2M}P_\phi^2\,\eta(\phi) = (E - E_{sc})\,\eta(\phi). \tag{6.6b}$$

In our approximation the motion of the fluxon is assumed to be adiabatic (in accordance with assumption 1. section 2.), its corresponding wave function $\eta(\phi)$ is well localized and hence the angle ϕ in eq. (6.6a) is regarded as a slowly varying parameter. For $\phi = 0$ the solution has already been computed in section 4. When the fluxon moves the solution (eq. 4.5) changes according to the shift of the wells and takes the form

$$\rho(\theta) = \sum_N u_0(\theta - \phi - 2\pi N) \tag{6.7}$$

Hence, the total wave function eq. (6.3) is given by

$$\Psi(\theta,\phi) = e^{i\phi\,\hat{E}/e_0}\sum_N u_0(\theta - \phi - 2\pi N)\,\eta(\phi). \tag{6.8}$$

This is in general not a permissible solution for the Schrödinger equation (6.2), since unless E mod $e_0 = 0$ (Q'modq = 0) it is a non-single-valued function of ϕ. However, for a simply connected region it is still possible to use this solution.

Consider a coherent beam of fluxons prepared at $\phi = \phi_0$. The beam is split into two localized wave packets $\Psi = \Psi_1 + \Psi_2$, where Ψ_1 represents the beam on one side of the external charge and Ψ_2 on the opposite side. Each of the beams travels in a simply connected region and hence can be described by the solution (6.8). Therefore we have

$$\Psi_1 = e^{i\Delta\phi_1\hat{E}/e_0}\,\psi_1(\theta - \Delta\phi_1,\Delta\phi_1) \tag{6.9a}$$

$$\Psi_2 = e^{i\Delta\phi_2\hat{E}/e_0}\,\psi_2(\theta - \Delta\phi_2,\Delta\phi_2) \tag{6.9b}$$

where $\Delta\phi_i = \phi_i - \phi_0$. The final wave fuction at the interfernce region ($\Delta\phi_1 = \pi$ for ψ_1, $\Delta\phi_2 = -\pi$ for ψ_2) is given by

$$\begin{aligned}\Psi_1 + \Psi_2 &= e^{i\pi\hat{E}/e_0}\,\psi_1(\theta-\pi,\pi) + e^{-i\pi\hat{E}/e_0}\,\psi_2(\theta+\pi,-\pi) \\ &= e^{2\pi i\hat{E}/e_0}\,\psi'_1 + \psi'_2\end{aligned} \tag{6.10}$$

where ψ' stands for $e^{-i\pi\hat{E}/e_0}\psi(\theta+\pi,\pi)$ and we have used the single-valuedness of ψ. Clearly, the interference between the two beams depends on the value of the modular electric field. According to the discussion in section 5, although the expectation value of the electric field vanishes the modular electric field is a constant of motion of the system and by eq. (5.3,5.8) takes the value $e^{2\pi i Q'/q}$. Hence, the interference between the two fluxon beams will depend of the phase difference $2\pi Q'/q$.

In conclusion, we have shown that because of the transparency of the superconductor to the modular electric field, the magnetic fluxon will accumulate the same phase as if the electric field were unscreened. Hence, in this case the Aharonov-Casher effect must also be interpreted as a force-free non-local effect. We also note that the non-triviality of the modular electric field (eq. 5.8) is a general property that will occur whenever Q' mod $q \neq 0$ (where q is a unit charge of the medium and Q' is an external charge), hence we expect other models that manifest the Aharonov-Bohm effect and screen the electromagnetic fields (e.g. Higgs model in (2 + 1)-dimensions) to manifest also the Aharonov-Casher effect through a similar mechanism.

7. References

1. Y. Aharonov and A. Casher, Phys. Rev. Lett. **53** (1984) 319 .
2. Y. Aharonov and D. Bohm, Phys. Rev. **115** (1959) 485.
3. It has been recently announced that the effect was observed by A. Cimmino, G.I. Opat, A.G. Klein, H. Kaiser, S.A. Werner, M. Arif and R.Clothier, Observation of the Topological Aharonov-Casher Phase Shift by Neotron Interferometry, Phys. Rev. Lett. **63** (1989) 380. This experiment was proposed from a different point of view, in reference 4.
4. J. Anandan, Phys. Rev. Lett. **48** (1982) 1660. See also J. Anandan, Phys. Lett. A, **138** (1989) 347.
5. It was claimed by T.H. Boyer, Phys. Rev. **A36** (1987) 5083, that the Aharonov-Casher effect results from a local classic force. However,it was demonstrated by Y. Aharonov, P. Pearle and L. Vaidman, Phys. Rev. **A37** (1988) 4052, that this force (hidden momentum) does not generate acceleration, its effect is to stabilize the mechanical structure of the solenoid. This problem does not exist in our case since the electric field is entirely screened and the fluxon is stabilized by thesuperconductor. On this issue see also: A.S. Goldhaber, Phys. Rev. Lett. **62** (1989) 482.
6. A related problem has been discussed in the article : A. S. Goldhaber, R. Mackenzie, and F. Wilczek, Mod. Phys. Lett. **A4**, (1989) 21.
7. R.D. Feynman, in Feynman Lectures in Physics. Vol. 3, Chapter 21.
8. Y. Aharonov, el. al. Int. J. of Theor. Phys. 2, No.3 (1969) 213.

The Aharonov-Bohm Effect in Small Resistive Devices

Richard A. Webb
IBM Thomas J. Watson Research Center
P.O. Box 218 Yorktown Heights, New York 10598

Abstract

The wave-like nature of the electron can now be seen directly in transport measurements on disordered solids. In spite of the presence of strong elastic scattering, quantum self-interference results in the observation of the Aharonov-Bohm effect in the measurement of the magnetoresistance of small wires and loops. Non-local effects are directly observable and contribute significantly to what is measured. By using the magnitude of these quantum interference effects as a measure of the degree of coherence in the sample, the role of sample size, temperature, and applied voltage in destroying this quantum coherence has been measured. In addition, we present evidence that when the number of electrons in a device becomes well defined, the phase of those electrons become less well defined and the Aharonov-Bohm effect begins to disappear.

Introduction

The transport properties of very small non-superconducting structures has recently gained increased attention with the experimental verification that samples can now be fabricated in the limit where the measured resistance exhibits new contributions arising from complete quantum mechanical phase coherence over the entire sample[1]. Perhaps the first surprising result is that this coherence is maintained in spite of strong elastic scattering[2] and is observable in samples containing 10^7 to 10^{10} electrons. In addition these systems exhibit some interesting universal properties but the details are extremely sensitive to the how the measurement is made. As will be shown, the non-local aspects associated with what is actually being measured is of central importance. Experimental work in this field has already provided a partial answer to one of the important questions in quantum mechanics "How does a quantum system evolve into the more familiar everyday classical system?". This transition can be studied by varying the temperature, or sample size, or by applying increasingly larger voltages across the phase coherent region.

I have organized the material in this paper to reflect the content of my lecture presented at the Fundamental Aspects of Quantum Theory. Although most of the following material has been published elsewhere, it is repeated in an effort to make the proceedings of this conference more complete. Most of the experimental results discussed here have been obtained over the last five years. They are, for the most part, unexpected because we have been trained in statistical physics to think about the properties of a large ensemble of particles and to to average over all possible configurations in order to obtain the bulk properties of the system. It is precisely this averaging that, in general, removes the contributions due to quantum interference effects to the final result. Although the quantum mechanical properties of the individual constituents are important over some length scale (typically on the order of a few lattice constants), they are usually uncorrelated when averaged over the whole sample. It turns out that the relevant length scale over which the wave-like properties of the electron are important is L_Φ, the distance the electron can travel in the sample without loosing phase coherence in its wavefunction. There are a growing number of examples where at, low temperatures, this effective length scale can be 100-10,000 times larger than the characteristic atomic scale. The wealth of novel quantum mechanical coherent phenomena that has been observed in this intermediate "mesoscopic" regime is the subject of this lecture.

The Aharonov-Bohm Effect

One of the primary experimental signatures that quantum interference effects are playing an important role in determining the transport properties of any electrical circuit is that the resistance fluctuates as either the magnetic field or an externally applied voltage is altered. In a short wire, these fluctuations are a random function of the externally changed variable but in a doubly connected structure, such as metallic loop of wire, the electrical resistance is found to fluctuate periodically as shown schematically in Fig. 1. All of these phenomena are directly related to the Aharonov-Bohm effect. They have no classical analogs and are a signature that the phase of the electron wavefunctions is being altered by the externally applied vector or scalar potentials and this in turn changes the measured electrical resistance.

As shown by Aharonov and Bohm in 1959[3], there exist simple ways to smoothly alter the phases of wavefunctions. Their two suggestions for experimental verification are depicted schematically in Fig.2a and 2b. In Fig.2a an electron beam source is split so that it encircles a region of space containing a magnetic flux. In the strictest sense, the magnetic flux Φ must be completely shielded from the electron paths. After recombining the beam, the intensity is measured at a point P on a screen. The equivalent experiment for the scalar potential is displayed in Fig.2b, where while the electrons are inside the Faraday cages (zero electric field E on the electrons), a different voltage is applied to each Faraday cage, and before they leave, the voltage is turned off. The prediction is that the phase φ of the wave-function is wound up or down by the potentials according to

$$\Delta\phi = \frac{e}{\hbar}\int(V dt - A\bullet ds) \ , \qquad (1)$$

where ds and dt are elements of the path and time respectively. Simply ramping the vector or scalar potential causes the current impinging at some point on the screen to oscillate periodically in $\Delta\Phi$ = h/e or $\int V dt$ = h/e. Classically, this cannot occur because the electron never "sees" the fields E and B, and so it cannot react to the classical force e(E + v×B), where E is the electric field and v is the velocity of the electron. These effects have been studied for electron beams in a vacuum[4] and in ballistic transport in solids[5], where the electrons suffer no collisions of any kind (no phase randomization), with the conclusion that when the magnetic field in the region where the electron travels is zero the magnetic vector potential indeed can alter the phase of the electron wave function.

The predictions by Aharonov and Bohm and the subsequent experimental verification demonstrate conclusively that the magnetic vector potential is much more than a mathematical convenience. Although the magnitude of the vector potential may depend upon the choice of gauge and is arbitrary up to a constant, the measured interference phenomena only depend upon differences between phases shifts accumulated along the two paths.

The Aharonov-Bohm Effect in a Disordered Solid

In the normal discussions of the Aharonov-Bohm effect it is assumed that the electrons experience no collisions or other phase altering interactions. Transport phenomena in an everyday solid, however, is quite a different environment. In general the motion of the electrons in a solid is diffusive. This means that the mean free path ℓ (typically $\simeq$ 100Å) is much smaller than the sample length. In the theory and in many of the experiments, ℓ is also smaller than the width w or thickness t of the wires which form the sample. The mean free path, in turn, is much larger than the Fermi wavelength λ_F ($\simeq$ 1Å in a good metal) of the electrons, so that treating the motion of the carriers semiclassically is a good approximation to the physics of the resistance. The electrons move through the sample as nearly plane waves, and collide elastically with impurities, potential fluctuations, grain boundaries, etc.. In addition, at finite temperature, electrons can scatter inelastically from phonons as well as other mechanisms. The elastic collisions with the impurities do not destroy the phase information contained in the electron's wave-function; as shown by Landauer, they merely shift the phase by some fixed amount. Another way of stating this is that elastic scattering does not change the quantum state of the electron or the scatterer. Inelastic collisions, those in which the quantum state of the electron or scatterer are changed, destroy the phase memory of the electrons irrevocably. Thus at low temperatures, where inelastic collisions due to phonons become rare, the electron can propagate large distances without loosing phase coherence. In typical metals, the average diffusion length between inelastic collisions, the phase coherence length L_Φ, can be several microns at temperatures $T \lesssim 1K$. This means, with the aid of modern circuit lithography, that samples can now be fabricated and studied on length scales where the details of the electrons wavefunction are important.

Much of the recent theoretical and experimental effort in this field has been focused on the calculation and the measurement of the conductance of small disordered metallic samples. The standard approach to to calculating electrical resistance has been to use the Kubo formula[6] which describes the response of the electrons to an electromagnetic potential. A conceptually different way of calculating the electrical resistance of a sample was found by Landauer in the middle fifties[2]. The starting point was to consider the resistance of an single scatterer in an otherwise perfect wire with ends connected to reservoirs. Scattering at the obstacle is elastic, and inelastic processes can only occur in the reservoirs. The reservoirs have the property that they absorb all carriers travelling toward them and re-emit carriers with a phase and energy which is unrelated to that of the absorbed carriers. A net current results if the two reservoirs have differing chemical potentials. The resistance is determined by studying the currents and the probability densities associated with the wave functions at the Fermi energy. In a one dimensional wire, the obstacle is characterized by the transmission probability T(E) for carriers to traverse the sample. For zero temperature Landauer finds a resistance,

$$R = \left(\frac{2\pi\hbar}{e^2}\right)\left(\frac{1 - T(E_F)}{T(E_F)}\right), \tag{2}$$

where E_F is the Fermi energy. Eq. (2) demonstrates that the resistance of the obstacle is zero if transmission through the sample becomes perfect, $T=1$. In a real sample there are many such scattering sites and Eq. (2) can be applied to calculate the resistance in this case as well. If the phase information in the electrons wavefunction is not lost (phase coherent transport), the transmission probability for carriers to traverse the sample is not in a simple way related to the transmission probabilities for carriers to scatter from one of the obstacles because of interference phenomena. In a long sample with an irregularly spaced sequence of barriers these interference effects give rise to an exponential increase of the resistance with increasing length as was predicted by Anderson[7] (Anderson Localization).

It should emphasize here that the transmission of electrons through a random potential where only elastic scattering occurs has a very peculiar property not generally associated with electrical resistance. The Joule heat due to the resistance in Eq. (2) is produced only in the reservoirs and not in the sample. While real systems may not be entirely described by this simple picture, they do at low temperatures exhibit a clear separation of elastic and inelastic scattering. Since Eq. (2) depends only on the elastic static scattering properties of the sample, one might conclude that elastic scattering leads to a resistance. This is not correct. The scattering process for electrons at the Fermi energy is described by four wave functions; the two wave functions describing waves of unit amplitude incident on the sample, from the left or the right, and the complex conjugate (time reversed) wave functions which describe waves of intensity T and R=1-T incident from both reservoirs. But these later two wave functions which require *coherent* incident streams from both reservoirs are not present in Landauer's treatment because the reservoirs cause phase randomization. It is the *elimination* of these two wave functions which gives rise to a positive resistance.

The simplest circuit analogue of the Aharonov-Bohm geometry is a loop of wire of small thickness and width fed by two leads. The possibility of observing magnetoresistance oscillations from such a loop was first discussed in detail by by Büttiker, Imry, and Landauer for loops without electrical leads attached and by Gefen, Imry, and Azbel where the leads were included[8]. Their results clearly indicated that in the presence of strong elastic scattering the electrical resistance would exhibit an Aharonov-Bohm type periodic oscillation as the flux threading the interior of the loop was increased. The flux period of the oscillation would h/e, the normal metal flux quantum. The observation of such an effect would be direct evidence of the interference of the electron wave-function in the measured resistance. Each electron is described by a wave-function which may be schematically written $Ae^{i\varphi}$, where φ develops in time according to the rules of the plane wave and is boosted or retarded deterministically by elastic collisions with impurities. For a ring, the electron has a finite probability of traveling diffusively in each branch. The phase shifts accumulated in the different paths are not the same because the microscopic details of the scattering are different in each branch. The superposition of the two partial wave components of the wave-function, which gives the resulting probability for finding the electron at a particular location in the sample (assume for the moment that the electron is restricted to only two paths through the conductor), is $|A_1 + A_2|^2 = A_1^2 + A_2^2 + A_1A_2\cos(\varphi_1-\varphi_2)$. In large samples, where measurements are made at length scale much larger than the phase coherence length, the last term contributes nothing because on average it is zero. In small samples at low temperatures, where $L \simeq L_\Phi$, the cross term has a specific value, which in general is not zero, but varies randomly throughout the sample. Ideally, a pair of voltage probes placed at particular points along a wire measure these interference probabilities. Another sample with the same dimensions and impurity concentration will have a resistance at zero magnetic field which varies at most by $\pm|A_1A_2|$. The application of a magnetic flux to the center of the ring alters the magnitude of the interference term because of the extra contribution to the phase by the vector potential. The averaged resistance from a large ensemble of such samples will, of course, produce the classical Drude value because the third term in the superposition will average to zero.

Figure 3 displays a photograph of a gold loop that is similar to the one which was used in 1985 to confirmed the above theoretical predictions. As shown in Fig. 4, the magnetic field dependence of the resistance of the ring contains oscillations that are periodic in magnetic field. As demonstrated by this and other experiments in several different laboratories[9], the period of oscillation is $\Delta\Phi = h/e$ through the area enclosed by the loop — just the period predicted by Aharonov and Bohm in their *in vacuo* idealization of the experiment. We see then, in this resistance measurement on a disordered metal loop, the direct signature of interference of the electron wave-function.

Universal Conductance Fluctuations

There is another quantum mechanical contribution to the resistance of this loop. The periodic oscillations are superimposed on a randomly fluctuating background resistance. This random variation of the conductance is shown in more detail in Fig. 5 where the magnetoresistance of the gold ring shown in Fig. 3 is displayed over an 8 Telsa field range. As shown by Stone[10], the origin of this random signal is also an Aharonov-Bohm effect. As the electron partial waves diffuse through the wires which form the ring, their trajectories randomly enclose the flux which pierces the metal. Since the trajectories are random walks, the enclosed fluxes are random functions of positions within the wire. The various partial waves interfere with each other at every point within the sample and contribute to Aharonov-Bohm interference effects. When averaged over the entire phase coherent region of the sample this interference appears in the magnetoresistance as random fluctuations in R(H). The characteristic magnetic field increment to induce a fluctuation is approximately one h/e flux quantum in the phase coherent area (for quasi one-dimensional wires this area is equal to the product of L_Φ and the sample width).

All the contributions to the electrical resistance, including the Drude resistance and classical magnetoresistance ($\propto B^2$) R_D, can be written as[11]

$$R(H) = R_D + R_0 + R_1 \cos\left\{\frac{\Phi}{h/e} + \alpha_1\right\} + R_2 \cos\left\{\frac{2\Phi}{h/e} + \alpha_2\right\} + \ldots \quad (3)$$

Empirically, one finds that R_n and α_n are random functions of H with field scales proportional to the area of the metal. It turns out that if one computes the RMS value of the conductance fluctuations, $\Delta G = \Delta R/R^2$, for a sample of length L_Φ and width $w<L_\Phi$, a universal value of $\simeq e^2/h$ is obtained independent of the material or average resistance (so long as the transport is diffusive and the temperature low enough)[1,12]. This universality has thoroughly verified in many experiments[13] but its origins are still under study[14]. In the above equation for the interference amplitude, the absolute phase change does not enter – only the relative phases of the two paths. This in turn implies that the interference is not necessarily destroyed as the magnetic field increases to rather strong values so that a large amount of flux pierces the metal in the device. The increasing of the magnetic field does not alter the character (average period, amplitude, ...) of the oscillations until the field reaches classically strong values where Landau orbits form and destroy the diffusive motion of the electrons. As seen in Fig. 5, the random conductance fluctuations persist to rather large magnetic fields without any attenuation what-so-ever – in fact they have been observed at $H > 15T$ without attenuation. (Landau orbits will not form in this metal until $B \simeq 200T$.) The same is true for the h/e oscillations as is shown by the insets in Fig. 5. The peak-to-peak amplitude of the h/e oscillations can be as large as 2 e^2/h but because of the random contributions to both the amplitude and phase (R_n and α_n in Eq. (3)) the RMS value, averaged over the full 8 T range, is only $\simeq 0.3\ e^2/h$.

The interference effects discussed above are "sample specific": the pattern of resistance changes as a function of the magnetic field is periodic with period $\Phi_0 = h/e$ for the ring, but the oscillation "phase" can differ from sample to sample. Similarly the aperiodic conductance fluctuations are a "fingerprint" of the impurity arrangement in the sample. There are also effects that survive upon averaging over many samples. In the first experiments by Sharvin and Sharvin[15]. in 1981 on long metal magnesium cylinders (5 mm long and 1μm in diameter), periodic oscillations in the electrical resistance as a function of the applied magnetic field (parallel to the axis of the cylinder) were observed. The Aharonov-Bohm effect discussed earlier predicts a periodicity in magnetic flux, Φ, equal to h/e. Their experiment showed a clear period of h/2e or half the normal period with the amplitude of the oscillations decreasing rapidly with magnetic field. Both these observations confirmed the earlier theoretical predictions of Al'tshuler, Aronov, and Spivak[16] that were based upon the standard ensemble averaging techniques used successfully in the calculation of the weak localization quantum corrections to the electrical conductivity of metals[17]. They are concerned with a special interference mechanism which is not a sample specific one, it survives ensemble averaging. In a metal wire these "localization" corrections arise from self interference effects occurring on a length scale equal to L_Φ and contribute to a slight decrease of the conductivity of the wire (on the order of e^2/h). For the cylinder experiments, the electron can be thought of as a pair of partial waves; one which travels clockwise around the circumference and the other traveling counter clockwise but following the time reversed trajectory. The two waves interfere back at the origin and the phase of each wave can be changed by the application of a magnetic field. But at zero field, the phases are equal and hence the interference is constructive (ie. increases the probability of finding the electron back where it started and thus lowering the conductivity in the simplest case). Each partial wave travels once around the circumference of the ring before interfering and the phase difference of each wave is given by the Aharonov-Bohm effect. Thus the total phase change is actually twice the normal Aharonov-Bohm phase change yielding an interference which is periodic in flux with a period of h/2e. The primary condition for the observation of these oscillations is that the phase coherence length L_Φ be longer that the perimeter of the cylinder. It has been shown that the phase of these h/2e oscillations is the same in each coherent segment of the sample (L_Φ long) and the conductance is a minimum at zero field so that these quantum corrections do not average to zero with increasing length of the cylinder. The attenuation of the amplitude of the oscillations with increasing magnetic field arises from the extra phase shifts induced by the presence of flux in the walls of the metal cylinder which destroy the equality between each partial wave and its time reversed state.

Electrostatic Aharonov-Bohm effects

The electrical circuit analogue of Fig.2b has also been studied in order to test the effects of a scalar potential on the quantum interference in disordered systems[18]. The sample we used is shown schematically in Fig. 6a and consists of loop with capacitor probes along the arms. A static voltage applied to the capacitor probes can contribute to the phase accumulated by the electrons in the arms of the loop according to the time integral $(e/\hbar)\int V dt$. As shown in Fig. 6b, the phase of the h/e oscillations from this loop can be shifted by $\simeq\pi$ when the voltage applied to the capacitor plates is changed by 0.75 V, but the amplitude of the oscillations is unaffected. Thus at any constant magnetic field, the output voltage from the ring can be reproducibly changed by the application of a potential to either of the capacitor probes. In fact the voltage can be used to smoothly shift the position of the oscillation along the magnetic field axis. This is demonstrated in Fig. 6c which contains several magnetoresistance traces recorded with different voltages on the probes. The voltage between the probes increments by 0.2V for each subsequent trace. At first, the oscillations shift toward negative field, and then after the fourth trace shift back toward positive field. This is precisely the effect one expects if a global phase change is added to all of the electrons in an arm of the loop. We expect that the time integral which accumulates phase from the electric potential is cut off by the mean time between inelastic collisions. Using this time scale and accounting naively for the screening of the potential by the metal in the loop, we find that the expected voltage to shift the oscillations by π is $\simeq$ 1-2V, which is in reasonable agreement with the observed value. In a similar fashion, a voltage applied to the capacitor plates also affects the random fluctuations. The net result of phase changes caused by the capacitor voltage is to make R_n and α_n from Eq. 3 random functions of this voltage as well as of the magnetic field. In Fig. 6c we see that the background resistance changes shape slightly each time voltage changes. The characteristic voltage scale to cause fluctuations is about twice that to shift the oscillations. This factor of two in scale is reminiscent of the corresponding magnetic field scales which also differ by a factor of about two.

The experiments discussed are not measuring the Aharonov-Bohm effect in the ideal environment because we are unable to completely shield the electric or magnetic field away from the current. This means that there are also classical forces at work on the electrons in the loop. The electric field tends to push the electrons toward or away from the edge of the wire in the loop. (These same electrons screen the field.) Moving the paths in the impurity potential is equivalent to changing the potential[19], and we expect that this too will cause the interference pattern (and hence the magnetoresistance pattern) to change. Just how gradually it changes with the capacitor voltage is an open question at this time. We remark here that displacing so much as a single impurity by more than a Fermi wavelength will substantially change the conductance: it will cause a change of order e^2/h.

The Approach to the Classical Domain

One question frequently arises: "How does a quantum system evolve into the more familiar classical system?" Experiments on these small phase coherent systems have already provided a partial answer to some aspects of this question. It turns out that one can consider the wires and loops we have discussed thus far as a *Quantum Interference Detector* and use the magnitude of the measured interference phenomena to gain a better understanding of how the long range phase coherence is destroyed as we change some important parameter of the system. In general temperature, sample size, diffusion coefficient, and details of how the measurement is made play a crucial role in determining whether or not quantum interference effects will be a significant contribution to the electrical resistance in a disordered system. Of these, phase randomization due to elastic scattering and finite temperature effects has historically been thought to be the most devastating. As demonstrated in the Sharvin and Sharvin cylinder experiments and our single loop experiments, elastic scattering is not destructive. The question of thermal averaging arises because, in a system as large as the gold ring shown in Fig. 3, containing 10^8 atoms, the average separation between electron energy levels is $\Delta E \simeq 10^{-8}$eV, which corresponds to a temperature of 10^{-4} K. When the thermal energy exceeds some characteristic energy of the system, the delicate quantum effects should begin to disappear. The naive analogy to Shubnikov-de Haas oscillations leads one to expect that the oscillations will decay according to a Dingle factor $\Delta R \propto \exp(-\Delta E/k_B T)$. The surprising discovery from the resistance fluctuations is that the spacing between energy levels is not the characteristic energy scale and that the decay is much slower. As shown by Imry, Stone and others[1,10,11] the energy scale that governs the physics in these systems is related to the Thouless energy[20].

$$E_C = \frac{\hbar D}{L_\Phi^{\,2}} \quad . \tag{4}$$

As shown in Fig. 7a, once k_BT exceeds E_C, the magnitude of the quantum corrections to the resistance, ΔR, decays slowly as the square-root of the number of uncorrelated energy "bands",

$$\Delta R(T) \sim \sqrt{\frac{E_C}{k_BT}} \ . \tag{5}$$

For the gold ring displayed in Fig. 3, $E_C/k_B \simeq 0.03$K, but because of the weak algebraic averaging, the quantum interference effects are readily observable above 1 K. In semiconductors where E_C can be ~10K, quantum interference effects have been reported above 100 K.

This weak algebraic averaging of the quantum interference also appears in the size dependence with the characteristic length scale being L_Φ. When the length of the sample is increased beyond $L = L_\Phi$, the resistance fluctuations grow as[21],

$$\Delta R(L/L_\Phi) \sim (L/L_\Phi)^{1/2} \ , \tag{6}$$

but the average resistance grows linearly with L, thus the *relative* magnitude ($\Delta R/R$) of the quantum interference contributions in a long wire decay as the square-root of the number of uncorrelated length segments. The same is true for N phase coherent loops in series, with each loop separated by L_Φ. The magnitude of the h/e oscillations grow as shown in Eq. 6. The conductance change, ΔG, however decays as $(L/L_\Phi)^{-3/2}$ because for ΔR small compared to R, $\Delta G = \Delta R/R^2$. Both the behavior of ΔV and ΔG as a function of the number N of uncorrelated regions is shown in Fig. 7b. For a single ring having half the perimeter larger than L_Φ, the loss of phase coherence reduces the oscillations severely: the oscillations decrease exponentially[22] as the path length L increases. The reason is simply that the probability of an electron reaching the terminus of the loop retaining phase coherence depends exponentially upon L/L_Φ so that

$$\Delta R(L/L_\Phi) \sim e^{-L/L_\Phi} \ . \tag{7}$$

Since all of these averaging mechanisms depend upon L_Φ, the temperature dependence of L_Φ determines the observability of quantum interference effects. Generally as the temperature increases, the probability of an electron absorbing and re-admitting a phonon increases. This inelastic process changes the quantum state of the electron and destroys its the phase coherence. The temperature dependence of the inelastic scattering rate depends upon the dimensionality of the phonons and on the degree of disorder. For narrow wires, the temperature dependence of the phase coherence length is usually $L_\Phi \propto 1/T^p$, where p is between 0.33 and 1.0. Thus $L_\Phi(T)$ is also a weak power law in T and not an exponential.

There are other ways to destroy the phase coherence of electron wave functions, and the dominant one at low temperatures is paramagnetic scattering. Since the phase shift acquired by scattering from an impurity with a fluctuating spin depends on the instantaneous direction of the spin, the electron will in general acquire a *time dependent* phase shift, and this will destroy the interference effects. In experiments where a low concentration of paramagnetic impurities was adsorbed onto the surface of a gold ring, the h/e oscillations disappeared at low fields, but as the field increased beyond some critical value, the full amplitude of the resistance fluctuations was recovered[23]. The reason for this striking recovery is that when the magnetic field energy $g\mu_B H$ (where μ_B is the Bohr magneton and g is the Landé g factor), becomes larger than the thermal energy the paramagnetic spins align with the field — classically they cease to fluctuate. Above this field, the magnetized impurity does not scatter inelastically; every electron acquires the *same* phase shift from this impurity, which now acts like a stationary elastic scatterer.

One obvious way to destroy quantum interference effects in a disordered conductor is to pass too much current through it. Joule heating effects are familiar from classical electronics, and an excessive amount of current will certainly raise the temperature of the sample. Perhaps not so apparent is the observation that currents too small to cause Joule heating are important. If the voltage drop across a phase coherent segment exceeds E_C/e, then averaging over incoherent energy regions will take place again, and the magnitude of the quantum interference effects will be reduced by $(E_C/eV)^{1/2}$. An example of this effect is shown in Fig. 7c where the current dependence at 40 mK of the h/e oscillations is displayed for a Au loop similar to that shown in Fig. 3. At very low temperatures, there are subtle effects of V on the magnitude of the quantum interference at voltages lower than E_C. It has been theoretically predicted with a hint of experimental confirmation, that when $E_C > eV > k_B T$ another type of averaging takes place. This averaging mechanism means that in order to preserve quantum interference effects the voltage across a phase coherent region of the sample must be kept below $\sqrt{\hbar D/k_B T}$.

Nonlocal Quantum Interference

There have been several unexpected experimental consequences of long range phase coherence in disordered systems. One of the most surprising is that the RMS amplitude of the voltage fluctuations measured as a function of magnetic field in a four terminal sample become independent of length[24]. when the separation L between the voltage probes is less than L_Φ. In a typical four wire sample configuration, the current is injected via lead 1 and removed by lead 4 and the voltage difference is measured between leads 2 and 3. Classically, the average resistance $R_{14,23} = V_{23}/I_{14}$ increases linearly with L. The reason is that classically there is a local linear relationship between the current density and the electric field and any point in the sample. The average voltage difference between any two points along the sample is given by $\Delta V_{23} = E\,L$. The quantum interference contributions to the voltage fluctuations (measured by changing the magnetic field), however, behave much differently. Figure 8 displays the length dependence of the voltage fluctuations measured in two samples: a long Sb wire 0.08 nm thick, and 0.1 nm wide; and a long gold wire 0.03 nm thick and 0.3 nm wide. The samples have eight leads attached with average spacings that vary from 0.2 μm to 4.0 μm. The symmetric and antisymmetric parts (see below) of the voltage fluctuations were determined from ± 3 T field sweeps for each of the 15 different combinations of voltage probe spacings. The phase coherence length L_Φ was estimated from the weak localization behavior observed in one of the segments of the sb line and from the half width of the autocorrelation function of the conductance fluctuations. We point out that not all segments actually exhibit a clear weak localization magnetoresistance because the conductance fluctuations frequently distort the shape of the curve. Using our values of 1.10 μm for L_Φ in the Sb line and 1.6 μm for L_Φ in the Au line, and assuming that the conductance fluctuation we would measure if our voltage probes were spaced L_Φ apart is e^2/h, the rms value of the voltage fluctuation for this phase coherent segment, ΔV_φ, is computed and used to normalize all our measured voltage fluctuation data. The open circles are the symmetric part of the voltage fluctuation and for $L/L_\Phi > 1$ the fluctuations grow in magnitude as $(L/L_\Phi)^{1/2}$ just as expected from Eq. 4. Translated into conductance fluctuations, this behavior indicates that the rms value decays as $(L/L_\Phi)^{-3/2}$ in agreement with the predictions of the conductance fluctuation theory. However for $L/L_\Phi < 1$, the rms value of the voltage fluctuations become independent of the distance L between voltage probes. The RMS values of the voltage fluctuations are plotted as a function of L. The length scale has been normalized to L_Φ, and the voltage scale has been normalized to $V_\varphi \simeq IR_\varphi{}^2e^2/h$, where R_φ is the resistance when the voltage probes are exactly L_Φ apart. The striking feature of these data is that as $L \rightarrow 0$ the RMS amplitude of the voltage fluctuations does *not* go to zero. The quantum interference contributions to the measured voltages give the same result when the probes are separated by 1 μm or 0 μm (zero average resistance) !

The explanation of this length independence is that the voltage probes do not define the length scale for the measurement. The effective length of the sample is really determined by the properties of electrons themselves. The electron is a wave whose phase coherence extends over a region L_Φ long. This means that the electron waves exist in classically forbidden regions of the sample and in particular the waves can propagate coherently into the current and voltage probes a distance on the order of L_Φ. The application of a magnetic field alters the interference over the entire phase coherent region. The microscopic details of the scattering impurities are different in each section of the sample as well as in the voltage leads. Thus a voltage difference is measured between leads 2 and 3, even when the two leads are attached to the same point on the sample, because of the propagation of electrons into the probes.

This apparent nonlocal contribution to the measured voltage fluctuations is best understood by the following experiment[25]. Take a long line with many voltage probes connected to it. Put current in the first probe and remove the current through the next closest probe. The voltage is measured by using any two probes away from the classical current path. The average voltage will be nearly zero. However when a magnetic field is applied to the sample the voltage will fluctuate randomly due to phase coherent carriers propagating into the classically forbidden regions of the sample. The results of such an experiment performed on the above sample are also shown in Fig. 8 for the total voltage fluctuations. The distance L is now the distance between the end of the classical current path and the closest voltage probe. Both the symmetric and antisymmetric contributions to the non-local voltage fluctuation decay exponentially with increasing length with an approximate dependence on length given by $\Delta V/\Delta V_\varphi \simeq \exp(-aL/L_\Phi)$ with $a = 1.2 \pm 0.3$. This exponential decay in not surprising, but that the scale for the decay is L_Φ is surprising (in classical physics, the width sets the scale). This also demonstrates that carriers will propagate into all leads of the sample and that adding probes to a phase coherent sample actually changes the sample. Recent theoretical analysis[26,27] is consistent with the general shape of the curve in Fig. 12 but the analytic theories seem to be nearly a factor of 2.5 different than the experimental data while the numerical studies predict exactly the results obtained in the experiments. The length independent part of the data clearly demonstrates that there is a nonlocal relationship between E and j at any point in the sample due to quantum interference throughout a region L_Φ long.

Another consequence of these nonlocal effects is that the magnetoresistance R(H) measured in any four terminal arrangement on a small wire is not the same when the magnetic field is reversed; $R(H) \neq R(-H)$. Classically, voltage probes on the same side of a long homogenous sample, measure R(H) which is symmetric about zero magnetic field. The problem is due to a lack of understanding of what the voltage probes actually measure, and how they change the phase coherent properties of the sample. If the quantum-mechanical wave properties of the electron play an important role in the measured transport properties of a multi-terminal sample, the relationship between measured voltages and internal electric fields in the sample has a nonlocal contribution, as discussed earlier. The electron waves are sensitive to interference effects that can occur within a phase coherent section of the sample and any leads that are connected to the sample are obviously part of the sample. Thus interference effects that come from electron excursions into the voltage probes contribute to the measured properties of the sample. Classically, these effects are extremely small and would appear to violate our understanding of microscopic time-reversal symmetry as well as the Onsager-Casimir symmetry relations.

An example of the importance of measurement probes in contributing to the asymmetry in the magnetoconductance[31] of a 0.8 μm average diameter gold ring is shown in Fig. 9. Three magnetoconductance traces on the **same** sample with three different ways of measuring voltage and applying current are displayed. For clarity of display the h/e oscillations have been digitally filtered out and only the aperiodic conductance fluctuations are shown. The same qualitative features, however, apply to the h/e oscillations as well. The top trace was measured by applying current to lead 1 and removing it via lead 4. The voltage difference was measured between lead 2 and lead 3. The conductance $G_{14,23}$ is clearly asymmetric about zero field. The other two traces result from permutations of current and voltage leads as indicated below each curve. Each trace exhibits a different asymmetric pattern of fluctuations about zero field. The dashed curve on the top trace illustrates that after four days of lead switching experiments the data is extremely reproducible. These asymmetries have been observed in every experiment on conductance fluctuations in rings and lines in our laboratory over the last four years. The above observations serve to illustrate another important aspect to quantum transport in disordered systems: not only are the interference phenomena sample dependent but the results of any one measurement depend upon how the measurement is made. Although these words are somewhat familiar to every student of quantum mechanics, experimentalists in this field are now forced to deal with these effects on a daily basis.

A particularly clear explanation for these asymmetries has been given by Büttiker[29]. He has pointed out that Onsager-Casimir symmetry relations of the local conductivity tensor do not take into account how the sample is measured; they only relate the properties of the local conductivity tensor, $\sigma_{\alpha\beta}(H)=\sigma_{\beta\alpha}(-H)$, where α and β index the spatial coordinates. The proper way to state the symmetry properties of electrical conductors is through the reciprocity theorem, $R_{12,34}(H)=R_{34,12}(-H)$, where now the indices refer to current and voltage leads. This symmetry property is present in Fig. 9 if one compares the positive field data in the top trace with the negative field data in the bottom trace. In fact it is now clear that by doing lead reversal measurements both the symmetric and antisymmetric parts of the conductance fluctuation can be determined by the following formulas,

$$G_S = \frac{2}{[R_{14,23}(H) + R_{23,14}(H)]} , \tag{8}$$

$$G_A = \frac{2}{[R_{14,23}(H) - R_{23,14}(H)]} . \tag{9}$$

We point out that although it is tempting to think of G_S as equivalent to G_{xx} and G_A as a Hall coefficient G_{xy}, this identification is not correct when there is a nonlocal contribution to the measured resistance. In a four probe measurement there are six different combinations of current and voltage measurements, and each of the three antisymmetric pairs of lead reversal measurements will, in general, give a different G_A but for a sample like our rings all three G_S will be the same.

The top part of figure 10 displays the actual conductance obtained by adding and subtracting the top and bottom traces of Fig. 9 over a ± 8 T field range. The residual asymmetry in G_S, $G_S(H)$-$G_S(-H)$, is also shown and is very nearly equal to the noise in the measurement. The data also show that the antisymmetric part of the conductance fluctuation is also a random function of magnetic field. The rms value of the symmetric part of the fluctuations is 0.56 e^2/h and the antisymmetric part is 0.44 e^2/h. Both have the same characteristic magnetic field scale, B_C, for the fluctuations. Thus the symmetries predicted by Büttiker are indeed the ones found in the experiments and the asymmetries in a magnetoresistance measurement should no longer be mysterious.

Perhaps the clearest experimental evidence of nonlocal resistance is shown in Fig. 11. Two identical four probe wires were fabricated, but on the second sample a small ring was attached 0.2 μm from the classical current path[30]. Both samples exhibit random conductance fluctuations with the same characteristic field scale, but in addition, the second sample has more high frequency "noise". Fourier transforms of both sets of data clearly show that this high frequency "noise" is in fact an h/e oscillation with the characteristic period equal to what is expected based on the diameter of the ring. In order to observe this effect, some large fraction of the electrons had to encircle the ring coherently. Thus the quantum interference measured in any small sample will have nonlocal contributions, and the distance between voltage probes will not define the sample length.

The Role of $\Delta N \Delta \Phi > \hbar$

Recently we have been trying to extend the observation of the Aharonov-Bohm effect to very high resistance samples. That is, samples in which the electron transport is by thermally assisted quantum tunneling through localized states. In such samples the electrical resistance generally increases exponentially with decreasing temperature in a form given by the Mott variable length hopping description, $R = A\ \exp(T_0/T)^{1/2}$, valid in one dimension. The random potential disorder localizes the electrons over a characteristic distance given by the localization length ξ which can vary from 0.005 μm to many tenths of microns. Once the localization length becomes much larger than the sample length, the transport reverts back to the standard diffusion description. These localized states are distributed randomly both in position and energy throughout the sample. The tunneling of electrons from one end of the sample to the other can only be through states of energy with-in a few k_BT of the Fermi energy defined by the reservoirs at either end of the sample. The tunneling probability is greatly enhanced by the absorption or emission of phonons. Since the characteristic distance the electron tunnels is given by a balance between the energy difference of the initial and final states and the separation between these states, the most probable tunneling distance is given by $L_M = (\xi/4)(T_0)/T)^{1/2}$. A typical distance at 0.05 K can be on the order of 0.2μm which is much larger than the typical localization length of 0.03 μm.

The real question is "can we see an Aharonov-Bohm effect in these systems?". At first thought the answer is absolutely NO. The reason is that every time the electron absorbs or emits a phonon the quantum state of the electron changes and this is surely a phase randomizing process. However, if one asks what the motion of the electron is *after* it has absorbed or emitted the phonon, one realizes that the ensuing motion is purely elastic. Our picture is that we have an electron trapped in a localized state at a position X_0 with energy E_0 and after the interaction with the phonon bath it is in the same localized state with a new energy E_1. Now the electron can tunnel to the final state located at position X_1 with energy E_1. But just as in the case described for diffusive transport systems, the electron has many many ways it can get to the final state. So which path does the electron choose to get to the final state? We believe that the electron chooses all paths to get to the final state, just as in all the cases in quantum mechanics. There is a big difference here though. The probabilities for the electron to elastically visit the large number of possible states in the sample must have a much stronger exponential energy and distance dependance than in the usual cases we are familiar with. None-the-less, there are many partial waves involved in this process and the phase of the wave function for these partial waves can be altered by the Aharonov-Bohm effect. Thus the probability of an electron for tunneling from it's initial site to it's final site should show a periodic h/e oscillation in a doubly connected ring geometry as the vector potential on the sample is smoothly changed. At low temperatures, the number of available sites for transport can become very small (the number of states in an energy width k_BT goes to zero as T goes to zero), the total electrical resistance can be dominated by one or a few states and large h/e oscillations might be expected.

We have been searching[31] for the existence of h/e oscillations in single rings and arrays of rings in samples where the electrical resistance varies from 10^4 to 10^9 Ohms. The rings are 0.3 μm in diameter, 0.02 μm thick and have line width of 0.03 to 0.05μm. The experiments are exactly the same as described in the earlier sections of this paper with the exception that the currents we are measuring are typically in the range of 10^{-10} to 10^{-15} amps when the voltage across the phase coherent segment of the sample is kept well below k_BT/e. Thus far we have only seen clear h/e oscillations in samples where the localization length is comparable to the ring size. For samples where the localization length is much smaller than the ring size but the most probable tunneling length, L_M, is comparable to the ring diameter, the h/e oscillations are almost non-existent. Occasionally we will see a few oscillations which appear to be Aharonov-Bohm related but they disappear very fast as the magnetic field is changed.

One possible explanation for the reduction in the h/e amplitude in high resistance samples is that the two arms of the ring are not balanced in the sense that each arm may have an exponentially different transmission probability thus reducing the size of the interference effects. Our next idea was to place a gate electrode near one arm of the ring (just as in the electrostatic Aharonov-Bohm experiments) with the hope that by applying voltage to the gate the energy of the localized states in one arm could be changed more than those in the opposite arm and thus increase the probability of obtaining a more balanced ring. As shown in figure 12, much to our surprise something totally unexpected happened when we measured the electrical conductance of a ring as a function of the voltage on a nearby gate electrode! The electrical conductance was found to oscillate periodically with increasing gate voltage[32]. The period of the oscillation ΔV_G for this data is about 0.02 volts on the gate. If we calculate the capacitance of the gate electrode to the sample we find that $C_G = 1\text{x}10^{-17}$ farads and if we assume that $Q=C_GV_G$ where Q is the charge induced on the sample, we find that $Q=1.2$ e where e is the charge of the electron. Thus to within the errors we have in our knowledge of the gate capacitance, the oscillations shown in Fig. 12 correspond to a change in the transmission coefficient which is periodic with the addition of a single electron.

Figure 13 shows the dc I-V curve of such a device. At low voltages the current through the sample is very small and remains small until the voltage across the sample exceeds a threshold value V_C. Above V_C the current through the sample increases linearly with increasing voltage. This I-V curve is exactly what one would expect for a Coulomb blockade effect. The basic idea here is that in order to transfer a single electron from one end to the sample to the other the electron-electron interaction causes a change in energy of $E_C=e^2/2C_s$, where C_s is the capacitance of the sample (not the gate). If the thermal energy is much smaller than this charging energy, the electron is essentially blocked from moving through the sample. If it moves, it must pay the Coulomb energy and at low voltages and low temperatures there is no way the electron can get enough energy to pay this price. However, when the voltage across the sample exceeds E_C/e, this Coulomb barrier can be easily overcome, the energy comes from the now large electric field. The gate electrode plays a very interesting role. Although the electron's charge is quantized in units of e (forget about quarks for the moment) the voltage on the gate electrode essentially plays the role of changing the potential difference V across the sample capacitor. If the potential difference across the sample corresponds to an integral multiple of e, $Q=Ne=C_sV$, the full Coulomb energy must be paid before an electron can move. If, however, the charge difference across the sample capacitor corresponds to -e/2, the transfer of an electron corresponds to changing the charge difference to +e/2. These two states have the same Coulomb energy and thus the electron can move freely through the sample without paying the energy E_C. The periodic oscillations result because the gate voltage is *continuously* changing the "effective potential difference across the sample capacitor. This Coulomb blockade effect has another interesting consequence. Only one electron at a time can pass through the sample. Indeed, if there were two electrons in the sample the cost in Coulomb energy would be twice as great and this is exponentially less likely.

We now find ourselves in an unusual situation (at least for condensed matter physicists). We have manufactured a ring shaped sample and we know that only one electron at a time can travel through the sample. If we apply the uncertainty relationship $\Delta N \Delta \Phi > \hbar$, we arrive at the suggestion that the phase of the electron in our device is not well determined. If that is the case, it would appear that the Aharonov-Bohm effect will disappear. Although, this may appear to be speculation at the moment, our data would appear to support this conclusion. Coulomb blockade effects in our samples disappear when the resistance of the ring is less than about 60,000 Ohms and thus far only clean h/e oscillations in the electrical resistance have been observed in samples where the resistance is less than 30,000 Ohms.

The second trace shown in figure 13 was obtained by placing an ac voltage on the gate electrode while the dc voltage on the sample was slowly changed. The amplitude of the ac gate voltage corresponded to a $\pm e/4$ charge change in the device. The resulting I-V curve demonstrates that even above the Coulomb blockade energy the conductance of the device is very sensitive to the fractional charge state. In fact, it turns out that this modulated I-V curve is the exact electrostatic analog of the DC SQUID[33]; charge plays the role of flux and E_C plays the role of the critical current. Instead of a super-current we have a "super-voltage" and the single electron tunneling frequency, I/e. plays the role of the Josephson frequency.

Concluding Remarks

I have tried to review some of the more exciting developments in the field of quantum effects in disordered solids that have been obtained over the last few years. Similar results have been obtained in the ballistic regime of transport and it now appears that even insulators exhibit some of the quantum mechanical interference properties discussed here. It therefore would seem that *small condensed matter samples will* continue to play a fundamental role in increasing our understanding of the quantum properties of many-body systems.

References

1). See the reviews by A.G. Aronov and Yu.V. Sharvin, Rev. Mod. Phys. **59**, 755 (1987); Y. Imry, in *Directions in Condensed Matter Physics*, eds. G. Grinstein and E. Mazenko, (World Scientific, Singapore, 1986); S. Washburn and R.A. Webb, Adv. Phys. **35**, 375 (1986).

2). R. Landauer, IBM J. Res. Dev. **1**, 223 (1957); R. Landauer, Philo. Mag. **21**, 863 (1970).

3). Y. Aharonov and D. Bohm, Phys. Rev. **115**, 485 (1959).

4). A. Tonomura, N. Osakabe, T. Matsuda, T. Kawasaki, J. Endo, S. Yano, and H. Yamada, Phys. Rev. Lett. **56**, 792 (1986).

5). N.B. Brandt, D.B. Gitsu, V.A. Dolma, and Ya.G. Ponomarev, JETP, **65**, 515 (1987); and references cited therein; G. Timp, A.M. Chang, J.E. Cunningham, T.Y. Chang, P. Mankiewich, R. Behringer, and R.E. Howard, Phys. Rev. Lett. **58**, 2814 (1987).

6).R. Kubo, J. Phys. Soc. of Japan **12**,570 (1957); D.A. Greenwood, Proc. Phys. Soc. **71**, 585 (1956).

7). P.W. Anderson, Phys. Rev. **109**, 1492 (1958).

8). M. Büttiker, Y. Imry, and R. Landauer, Phys. Lett. **96A**, 365 (1983); Y. Gefen, Y. Imry, and M.Ya. Azbel, Phys. Rev. Lett. **52**, 129 (1984).

9). R.A. Webb, S. Washburn, C.P. Umbach, and R.B. Laibowitz, Phys. Rev. Lett. **54**, 2692 (1985); V. Chandrasekhar, M.J. Rooks, S. Wind, and D.E. Prober, *ibid*. **55**, 1610 (1985); S. Datta, M.R. Melloch, S. Bandyopadyay, R. Noren, M. Varizi, M. Miller, and R. Reifenberger, *ibid* **55**, 2344 (1985).

10). A.D. Stone, Phys. Rev. Lett. **54**, 2692 (1985).

11). A.D. Stone and Y. Imry, Phys. Rev. Lett. **56** 189 (1986).

12). B.L. Al'tshuler, JETP Lett. **41**, 641 (1985); B.L. Al'tshuler and B.I. Shklovskii, JETP, **64**, 127 (1986); P.A. Lee and A.D. Stone, Phys. Rev. Lett. **55**, 1622 (1985); P.A. Lee, A.D. Stone, and H. Fukuyama Phys. Rev. B **35**, 1039 (1987).

13). J.C. Licini, D.J. Bishop, M.A. Kastner, and J. Melngailis, Phys. Rev. Lett. **55**, 2987 (1985); S. Washburn, C.P. Umbach, R.B. Laibowitz, and R.A. Webb, Phys. Rev. B **32** 4789 (1985).

14). Y. Imry, Europhys. Lett. **1**, 249 (1986); P.A. Lee, Physica A **140**, 169 (1986).

15). D. Yu. Sharvin and Yu. V. Sharvin, Pis'ma Zh. Eksp. Teor. Fiz. **34**, 285 (1981) (JETP Lett. **34**, 272 (1981)).

16). B.L. Al'tshuler and B.Z. Spivak, JETP Lett. **42**, 447 (1985).

17). P.A. Lee and T.B. Ramakrishnan, Rev. Mod. Phys. **57**, 287 (1985).

18). S. Washburn, H. Schmid, D.P. Kern, and R.A. Webb, Phys. Rev. Lett. **59**, 1791 (1987).

19). S. Feng, P.A. Lee, and A.D. Stone, Phys. Rev. Lett. **56**, 1960 (1986).

20). D.J. Thouless, Phys. Rev. Lett. **39**, 1167 (1977).

21). C.P. Umbach, C. van Haesendonck, R.B. Laibowitz, S. Washburn, and R.A. Webb, Phys. Rev. Lett. **56**, 386 (1986); W.J. Skocpol, P.M. Mankiewich, R.E. Howard, L.D. Jackel, D.M. Tennant, and A.D. Stone, *ibid.*, **56**, 2865 (1986).

22). F.P. Milliken, S.Washburn, C.P. Umbach, R.B. Laibowitz, and R.A. Webb, Phys. Rev. B **36**, 4465 (1987).

23). A.D. Benoit, S. Washburn, C.P. Umbach, R.A. Webb, D. Mailly, and L. Dumoulin, in *Anderson Localization* ed H. Fukuyama (Springer, Heidelberg, 1988).

24). A. Benoit, C.P. Umbach, R.B. Laibowitz, and R.A. Webb, Phys. Rev. Lett. **58**, 2343 (1987); W.J. Skocpol, P.M. Mankiewich, R.E. Howard, L.D. Jackel, D.M. Tennant, and A.D. Stone, *ibid*, 2347 (1987).

25). R.A. Webb, S. Washburn, H.J. Haucke, A.D. Benoit, C.P. Umbach, and F.P. Milliken, in *Physics and Technology of Sub-micron Structures* (Springer, Heidelberg, 1988).

26). Y. Isawa, H. Ebisawa, and S. Maekawa, in *Anderson Localization*, ed. H. Fukuyama (Springer, Heidelberg, 1988); J. Phys. Soc. Jpn. **56**, 25 (1987).

27). H.U. Baranger, A.D. Stone, and D. diVincenzo, Phys. Rev. B **37**, 6521 (1988); C.L Kane, P.A. Lee, and D. diVincenzo, *ibid.* **38**, 2995 (1988); S. Hershfield and V. Ambegaokar, **38**, 7909 (1988).

28). A.D. Benoit, S. Washburn, C.P. Umbach, R.B. Laibowitz, and R.A. Webb, *ibid.*, 1765 (1987).

29).. M. Büttiker, IBM J. Res. Dev. **32**, 317 (1988); M. Büttiker, Phys. Rev. Lett. **56**, 1761 (1987); M. Büttiker, Phys. Rev. B **35**, 4123 (1987).

30). C.P. Umbach, P. Santhanam, C. van Haesendonck, and R.A. Webb, Appl. Phys. Lett. **50**, 1289 (1987).

31). Z. Ovadyahu, F.P. Milliken, and R.A. Webb, to be published.

32). R.A. Webb and Zvi Ovadyahu, submitted to Phys Rev. Lett.

33). V. Chandrasekhar, Z. Ovadyahu, and R.A. Webb, to be published.

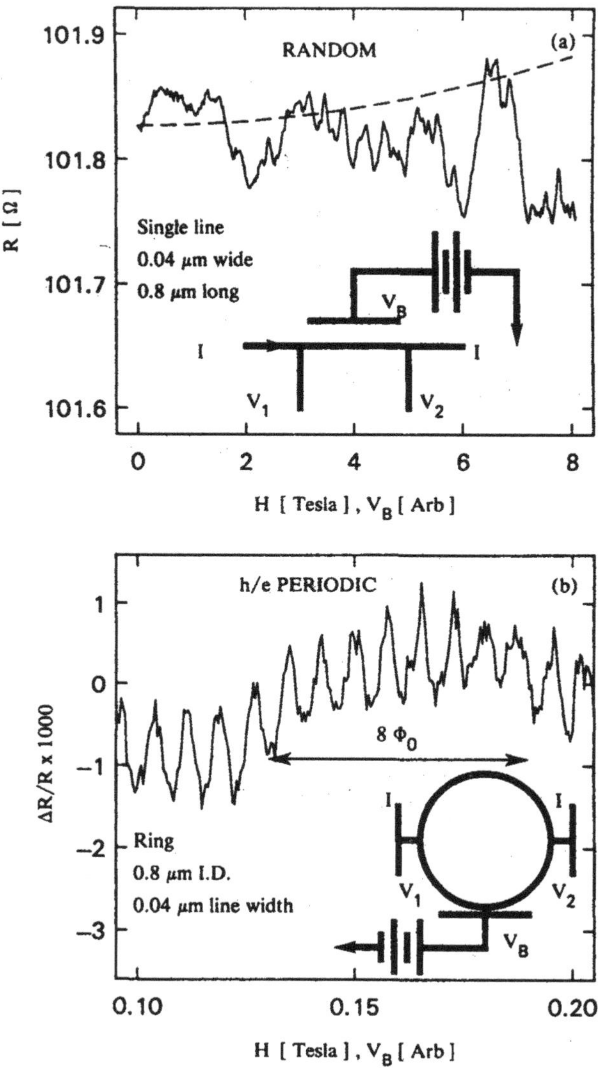

Fig. 1 A schematic representation of the resistance of a 100 Ω small wire (a) or 30 Ω small loop (b) as a function of an externally applied magnetic field H or an external voltage V_B. The resistance is measured by applying a current to the I leads of the sample and measuring the resulting voltage that develops across the sample with the V leads. The dashed line (a) is representative of the classical magnetoresistance expected in this wire.

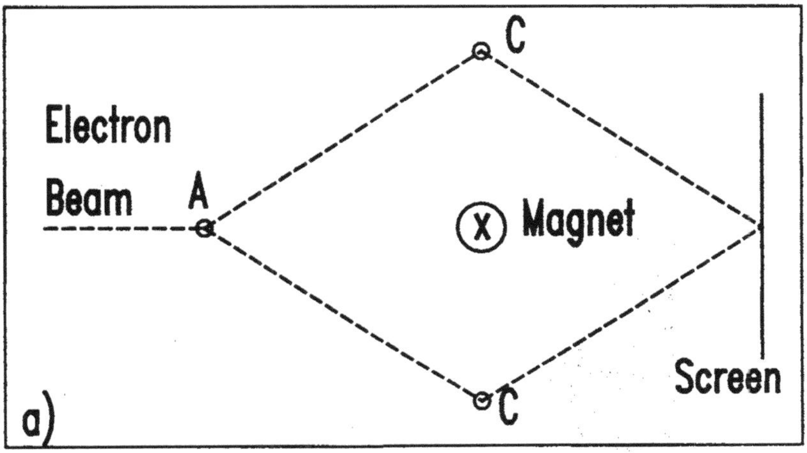

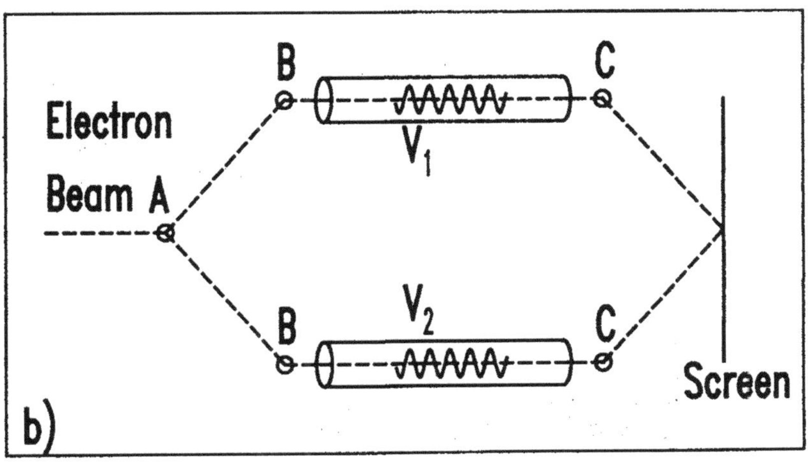

Fig. 2 (a) The Aharonov-Bohm mechanism for adjusting the electron phase with a vector potential. An electron beam is split by A and diverted by C. A magnetic field produced by an infinite solenoid is in the region between the two paths, and the field is identically zero at the paths. Interference from the recombined beam is observed at a screen. (b) The Aharonov-Bohm effect for electrostatic potentials. The shutters B chop the beam into wave packets and divert the beam into Faraday shields. After the wave packets enter the cylinders, scalar potentials V_1 and V_2 are turned on for a time t and turned off before the wave packets exit the tubes.

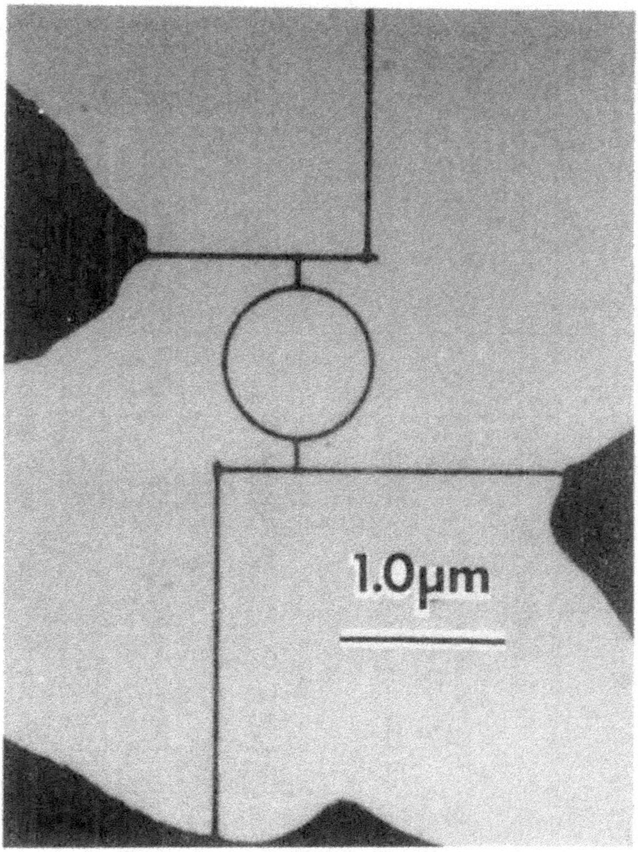

Fig. 3 Transmission electron photograph of a gold loop fabricated from a 30 nm thick polycrystalline film. The average diameter is 0.82 μm and the width of the wires forming the device are 0.03 μm.

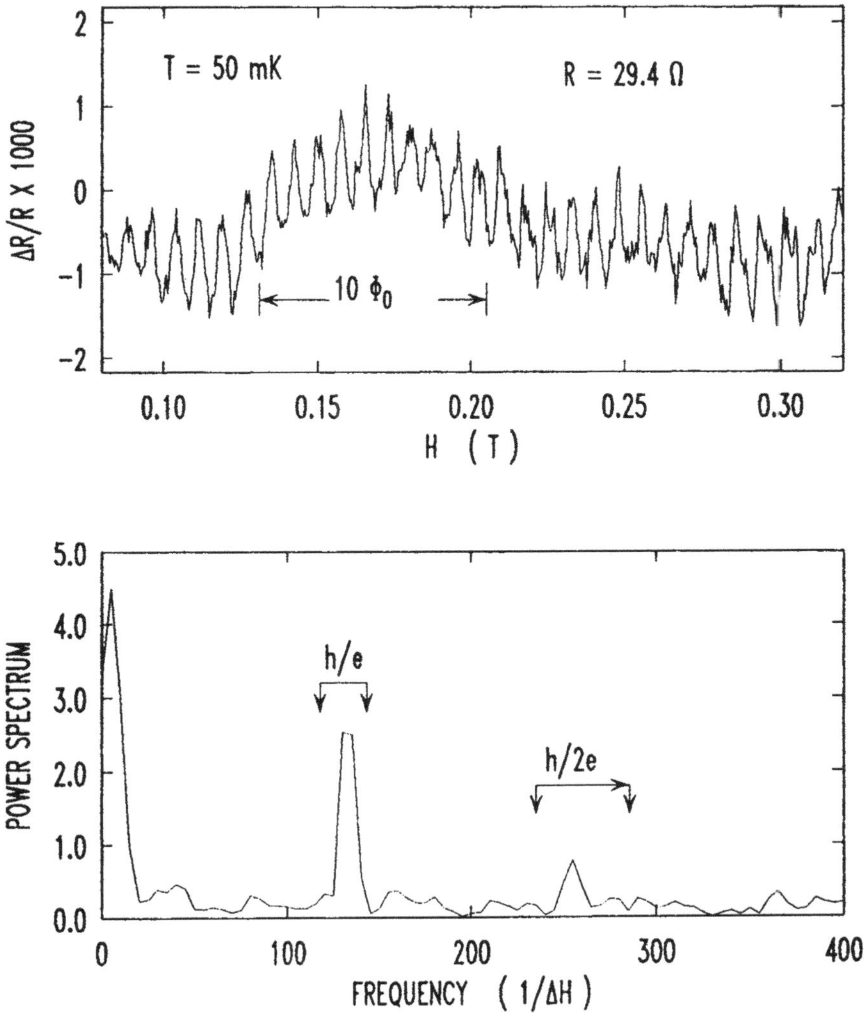

Fig. 4 The oscillations (top) in the magnetoresistance from a sample similar to that shown in Fig. 3 over a small field range. The bottom figure is the Fourier transform of the data and shows both an h/e and its harmonic, h/2e, are contained in the signal. The arrows indicate the expected region where these signals should appear based upon the measured inside and outside diameters of the ring.

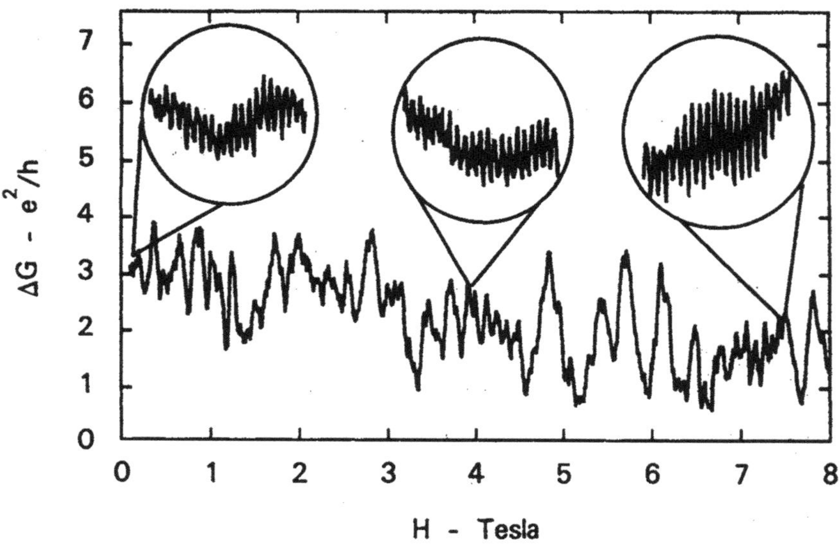

Fig. 5 The magnetoresistance of the Au loop shown in Fig. 3 over an 8 T field range demonstrating the random aperiodic variations due to magnetic field piercing the wires forming the ring. The insets show the h/e oscillations at selected magnetic fields that are also contained in the data.

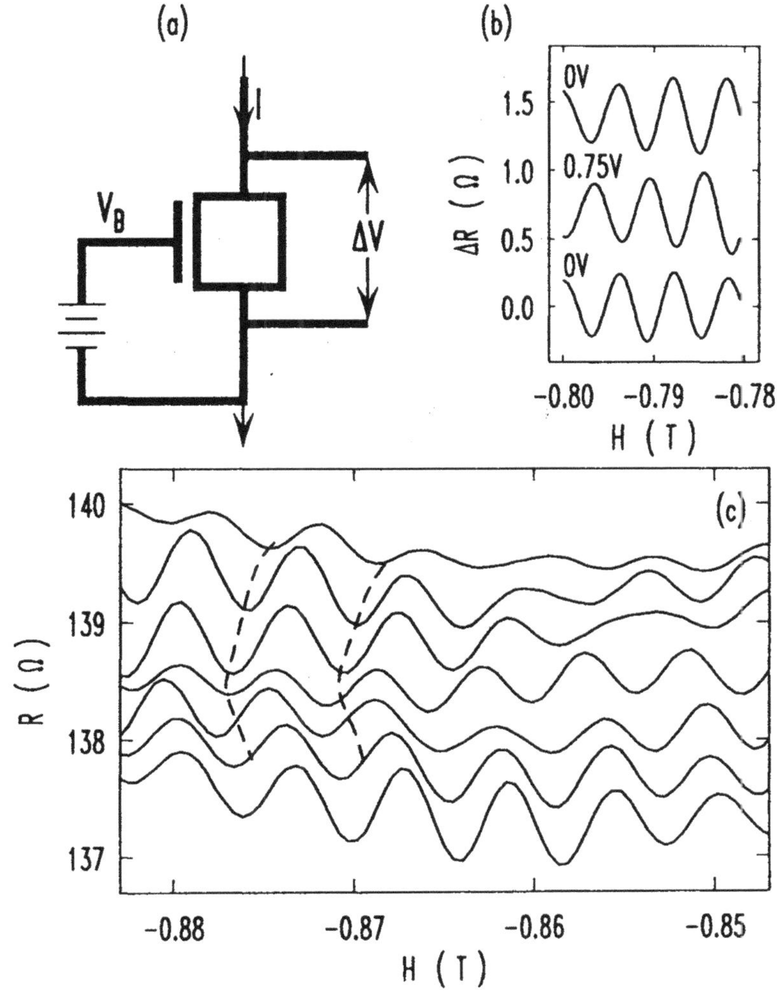

Fig. 6 (a) Schematic representation of the Sb loop with a capacitor "gate" along one arm used in the experiments. The actual sample had w= 0.08μm, t=0.08 μm, and the ring is 0.8 μm on a side. The capacitor probes are 0.16 μm away from the arms of the ring. (b) Demonstration that the "gate voltage" can shift the phase of the oscillations reproducibly. In the top and bottom traces (before and after), no voltage is applied to the capacitor probes. In the middle trace, 0.75V has been applied between the probes. (c) Wider field sweeps showing the gradual change of the magnetoresistance pattern as a function of the capacitor voltage.

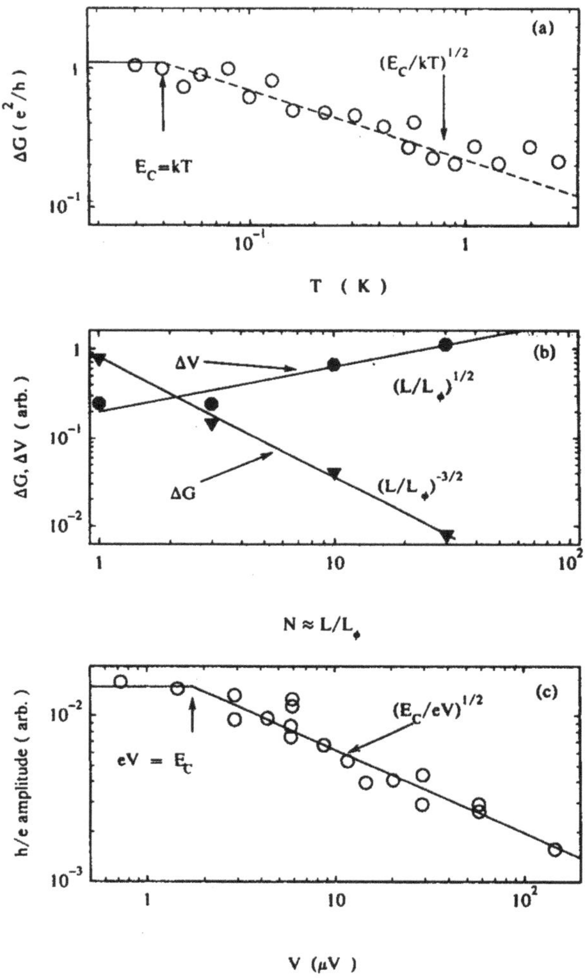

Fig. 7 Three ways to destroy quantum interference effects in a disordered system by averaging away the signal. (a) The role of energy averaging where, due to the increase of the number of uncorrelated energy channels with increasing temperature, the quantum interference signal decays. (b) The role of sample length in changing the rms value of the voltage and conductance fluctuations for a thin wire or series array of phase coherent rings. (c) This figure illustrates the effect of increasing the voltage across a phase coherent region of the sample on the amplitude of the h/e oscillations. This is not a heating effect but another form of energy averaging.

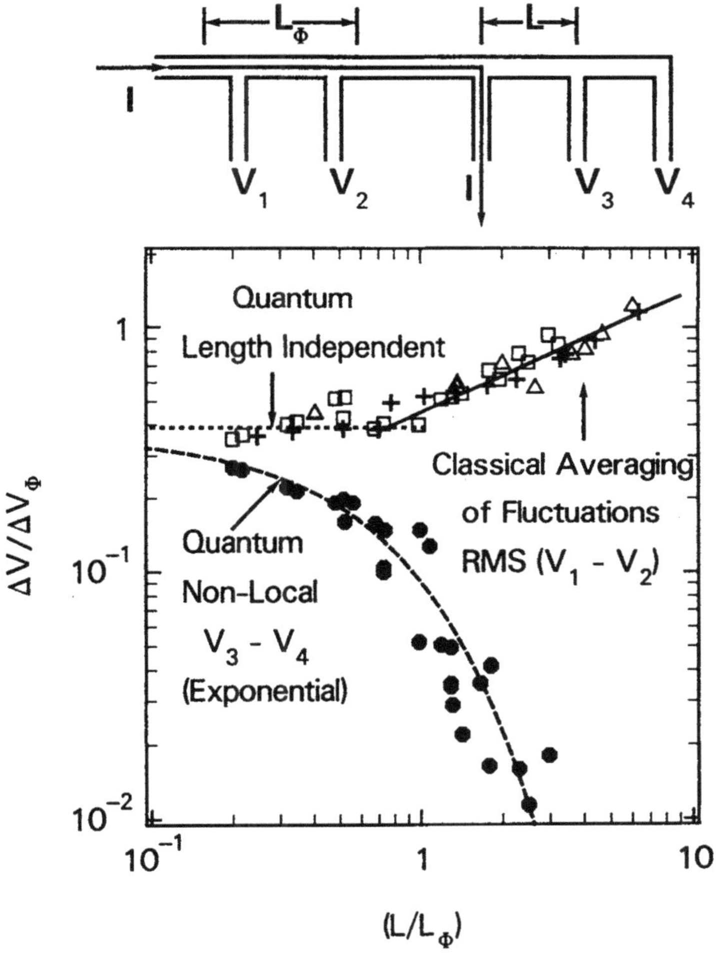

Fig. 8 Length dependence of the average voltage fluctuation amplitude in a long line for local (V_1-V_2) and non-local (V_3-V_4) measurements. The length scale has been normalized to the phase coherence length L_Φ and the voltage scale has been normalized to the voltage that would be measured if the separation between leads is L_Φ.

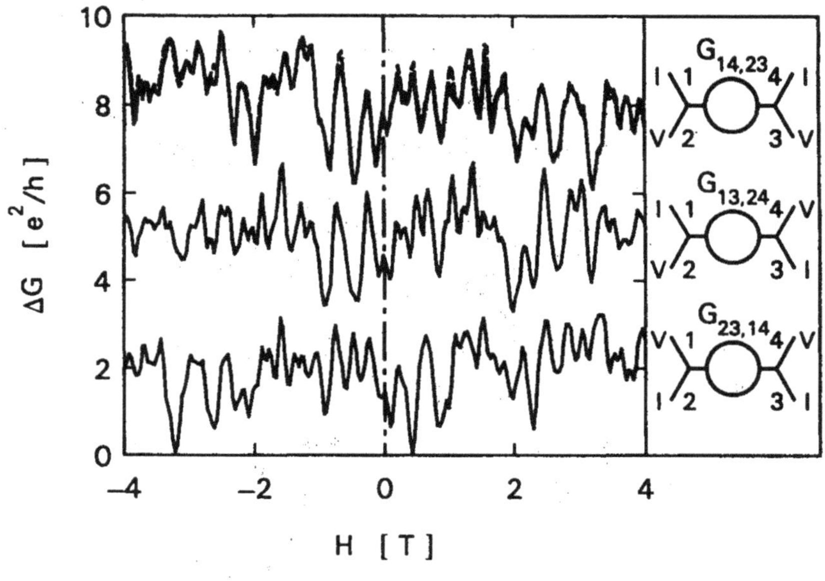

Fig. 9 The variation of the conductance fluctuation pattern with different lead configurations. The h/e oscillations have been filtered from the data for clarity of display. The inset illustrates the measurement configuration for each of the traces.

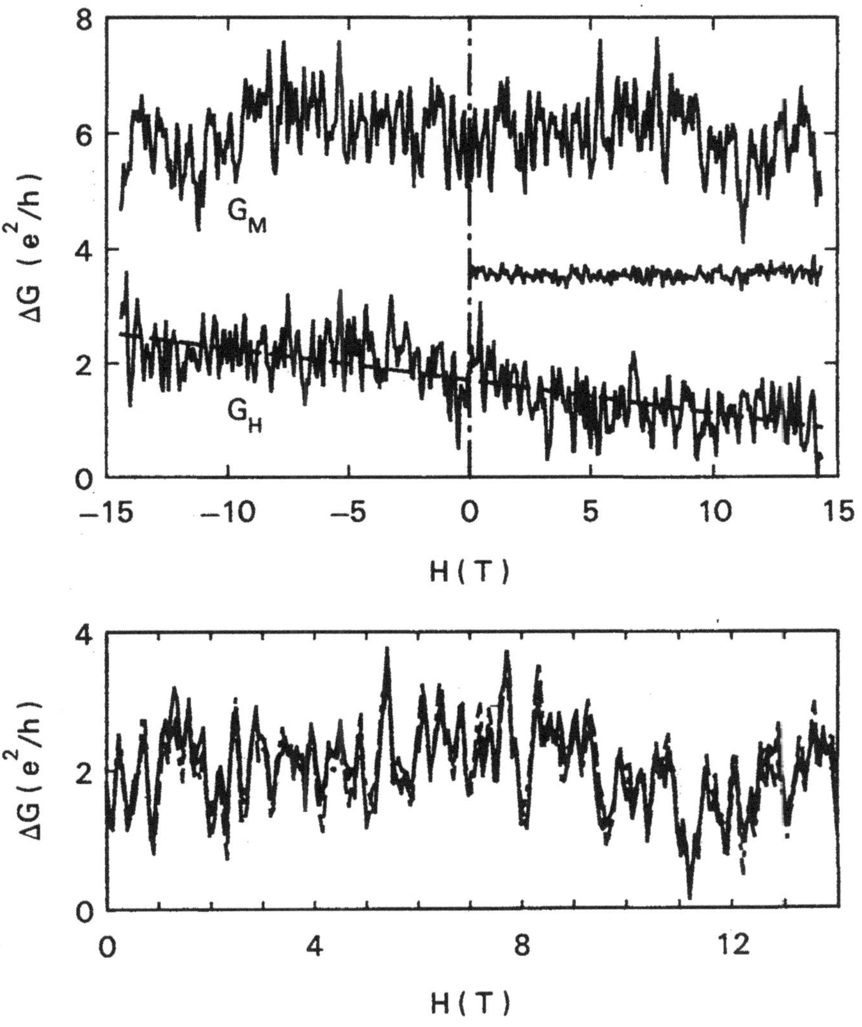

Fig. 10 (Top) The magnetic field dependence of the symmetric (G_M) and antisymmetric (G_H) components of the conductance fluctuations shown in Fig. 9 constructed from G_S(Eq. 8) and G_A (Eq.9). The residual asymmetry in G_S is also shown. (bottom) The symmetric part of the conductance fluctuations for each of the traces shown in Fig. 9 showing that they correlate to better than 90%.

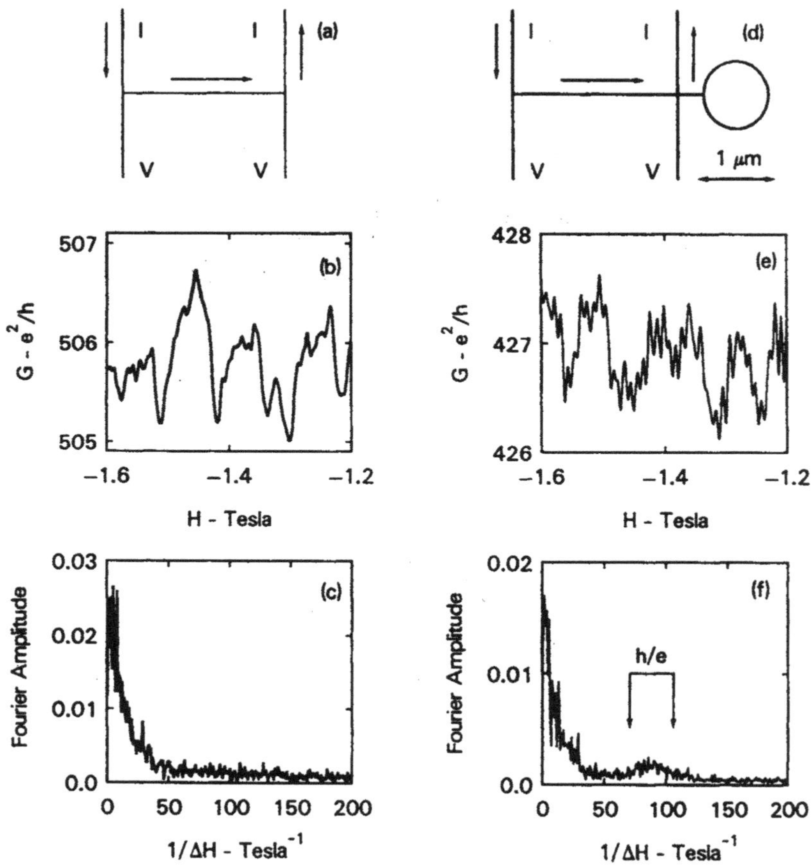

Fig. 11 The magnetoresistance and Fourier transforms of two similar wires, one of which has a ring dangling outside of the classical current path.

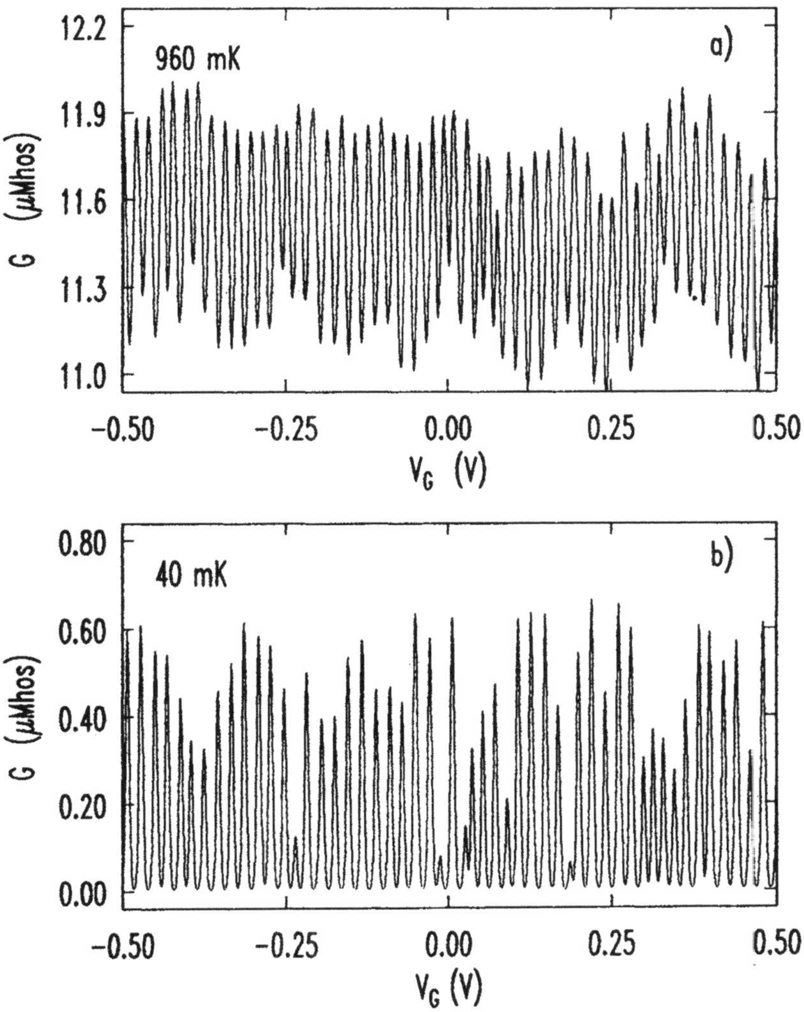

Fig. 12 The electrical conductance of a high resistance ring sample 0.02 μm thick, 0.05μm linewidth, and 0.3μm average diameter as a function of the voltage placed on a gate electrode separated by 0.1μm from one arm of the ring. Two different temperatures as displayed: (a) 960 mK; and (b) 40 mK. The sample is shown schematically in the inset to Fig. 13.

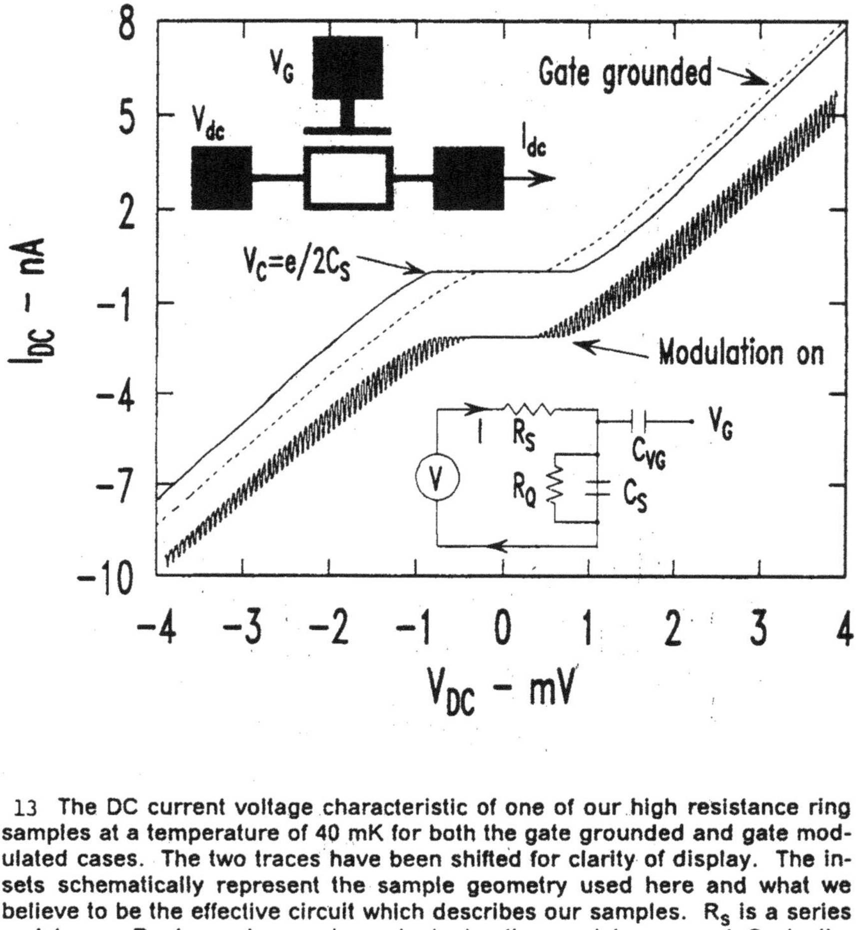

Fig. 13 The DC current voltage characteristic of one of our high resistance ring samples at a temperature of 40 mK for both the gate grounded and gate modulated cases. The two traces have been shifted for clarity of display. The insets schematically represent the sample geometry used here and what we believe to be the effective circuit which describes our samples. R_S is a series resistance, R_Q is a charge dependent shunting resistance, and C_S is the Coulomb blockade capacitance of our sample.

Dephasing of Interference by a Back Reacting Environment

Ady Stern[(1)] Yakir Aharonov[(1),(2)] and Yoseph Imry [(3),(4)]

[(1)] Raymond and Beverly Sackler Faculty of Exact Sciences
School of Physics and Astronomy, Tel Aviv University,69978, Tel Aviv, Israel
[(2)] Department of Physics, University of South Carolina, Columbia, SC 29208
[(3)] Physics Department , Weizmann Institute of Sciences , Rehovot , Israel
[(4)] IBM Thomas J. Watson Research Center, Yorktown Heights, NY 10598

Abstract

We discuss the dephasing of quantum interference due to the interaction of the interfering particle with its environment. We interpret this dephasing using two descriptions. One examines the way the interfering waves change their environment, and the other examines the quantum uncertainty in the phase shift induced by the environment. We demonstrate the equivalence of the two descriptions, and focus on the case in which the potential exerted by the environment on the interfering particle depends on the history of that particle.

1. Introduction

The two slit interference experiment is often used as a starting point in the description of quantum mechanics [(1)]. When the path of the interfering particle is not measured, says the moral of that experiment, an interference pattern is seen on the screen. In principle, such an experiment can use interfering objects of all scales, from microscopic electrons to macroscopic billiard balls. Yet, in practice, the first is experimentally observed, while the latter is not. The disappereance of quantum interference in the macroscopic scale is believed to be due to two reasons. First, the de–Broglie wave length of macroscopic objects is very small, and second, the interaction of those objects with their environment supresses interference effects [(2)(3)]. In this work, and in earlier works on the subject [(4)], we investigate the second reason. We do not discuss the first reason here, but only comment that recent experiments in mesoscopic systems have allowed the observation of interference effects even when the relevant wavelengths are very short [(5)].

In our previous works, we have used two descriptions of the way the interaction of an interfering particle with its environment might supress quantum interference. The first regards the environment as measuring the path of the interfering particle. When the environment has the information on that path, no interference is seen. The second description answers the question naturally raised by the first: How does the interfering particle "know", when it reaches the screen, that the environment has identified its path? We have proven that the two descriptions are equivalent. In the present work, we will review those two descriptions, and dwell on some subtleties arising when the interaction of the particle with its environment depends on its history, i.e., it is non-local in time.

As a guiding example, we consider an Aharonov–Bohm (A–B) interference experiment on a ring (see Fig. 1). The A–B effect has been proven to be a convenient way to observe interference patterns in mesoscopic samples, because it provides an experimentally easy way of shifting the interference pattern. This experiment starts by a construction of two electron wave packets, $l(x)$ and $r(x)$ (l, r stand for left, right), crossing the ring along two opposite sides. We assume that the two wave packets follow well defined classical paths, $x_l(t), x_r(t)$. The interference is examined after each of the two wave packets traverses half of the ring's circumference. Therefore, the initial wave function of the electron (whose coordinate is x) and the environment (whose set of coordinates is denoted by η) is:

$$\psi(t=0) = \left[l(x) + r(x)\right] \otimes \chi_0(\eta) \tag{1.1}$$

At time τ_0, when the interference is examined, the wave function is, in general,

$$\psi(\tau_0) = l(x, \tau_0) \otimes \chi_l(\eta) \; + \; r(x, \tau_0) \otimes \chi_r(\eta) \tag{1.2}$$

and the interference term is,

$$2 \; Re \left[l^*(x, \tau_0) r(x, \tau_0) \int d\eta \chi_l^*(\eta) \chi_r(\eta)\right] \tag{1.3}$$

Had there been no environment present in the experiment, the interference term would have been just $2Re[l^*(x, \tau_0) r(x, \tau_0)]$. So the effect of the interaction is to multiply the interference term by $\int d\eta \chi_l^*(\eta) \chi_r(\eta)$. The first way to understand this effect is seen directly from this expression, which is the scalar product of the two environment's states coupled to the two partial waves. At $t = 0$ these two states are identical. During the time of the experiment, each partial wave has its own interaction with the environment, and therefore the two states become different.

Since the environment is not observed in the interference experiment, its coordinate is integrated upon, i.e. the scalar product of the two states is taken. When the two states do not overlap at all, the final state of the environment identifies the path the electron took. Quantum interference, which is the result of an uncertainty in this path, is then lost. Thus, interference is lost when the two interfering partial waves shift the environment into states orthogonal to each other, i.e., when the environment has the information on the path the electron takes.

The second explanation for the loss of quantum interference regards it from the point of view of how the environment affects the partial waves, rather than how the waves affect the environment. It is well known that when a static potential $V(x)$ is exerted on one of the partial waves, this wave accumulates a phase (A system of units where $\hbar = 1$ is applied.):

$$\phi = -\int V(x(t))dt \tag{1.4}$$

and the interference term is multiplied by $e^{i\phi}$. "A static potential" here, and throughout this paper, is a potential which is a function of the particle's coordinate and momentum only, and does not involve any other degrees of freedom. For a given particle's path, the value of a static potential is well defined. When V is not static, but created by environment degree(s) of freedom, its value is not well defined any more. The uncertainty in its value results from the quantum uncertainty in the state of the environment. Therefore, ϕ is not definite, too. In fact, ϕ becomes a statistical variable, described by a distribution function $P(\phi)$. (For the details of this description see ref. [4].) The effect of the environment on the interference is then to multiply the interference term by the average value of $e^{i\phi}$, i.e.,

$$\langle e^{i\phi}\rangle = \int P(\phi)e^{i\phi}d\phi \tag{1.5}$$

The averaging is done on the interference screen that shows a sum of many interference patterns, corresponding to different environment states. Since $e^{i\phi}$ is periodic in ϕ, $\langle e^{i\phi}\rangle$ tends to zero when $P(\phi)$ is slowly varying over a region much larger than one period, i.e., 2π. Hence, when the uncertainty in the phase becomes of the order of the interference pattern, interference is lost. In the Feynman–Vernon terminology, $\langle e^{i\phi}\rangle$ is the influence functional of the two paths taken by the two partial waves. This is, then, the second explanation for the loss of quantum interference.

Our statement of equivalence between the two explanations is then put into an equation,

$$\langle e^{i\phi}\rangle = \int d\eta \chi_l^*(\eta)\chi_r(\eta) \tag{1.6}$$

When the environment measures the path taken by the particle, it induces a phase shift whose uncertainty is of the order of 2π.

2. Two examples of dephasing

We will now demonstrate the equivalence of the two descriptions on two simple examples, before we proceed to the more complicated case of an history–dependent interaction with the environment. First, we consider an environment composed of one heavy free particle. Initially, this particle is localized somewhere along the right arm of the ring, and its average momentum is zero. The coordinate and momentum of this particle are denoted by η and p_η, respectively (see Fig. 2). The particle's mass, M, is large enough so that $\frac{\delta p_\eta}{M}\tau_0 \ll \delta\eta$, where τ_0 is the duration of the experiment. Consequently, the particle's wave function does not change appreciably during the experiment, and its kinetic energy plays no role in its time evolution. So, our environment is composed now of one degree of freedom, with no significant dynamics of its own, but with an infinite dimensional Hilbert space. The interaction between the interfering electron and the environment is assumed to have the form

$$H_{int} = V(x - \eta) \tag{2.1}$$

where V is non zero within some short range of interaction. x here is the position of the interfering electron on the ring, while η is, e.g., the position of the environment particle on the plane of the ring (see Fig. 2) .

As before, the experiment starts with the initial state

$$\Big(l(x) + r(x)\Big) \otimes \chi_0(\eta) \tag{2.2}$$

The left partial wave does not interact with the environment, so that its propagation does not affect the latter. The right partial wave interacts with the environment, and this interaction results in a momentum exchange between the right wave packet and the environment particle. When this momentum transfer is larger than the momentum uncertainty of the environment particle, this particle measures the path the electron traverses, and quantum interference can no longer be seen. On the other hand, the uncertainty in the environment particle position, $\delta\eta$, makes the potential $V(x - \eta)$ uncertain, and then the shift in the right partial wave phase becomes uncertain, too. As the following paragraphs show, when the phase of the right partial wave becomes completely uncertain, and the interference pattern becomes a sum of mutually canceling terms, the environment particle measures the electron's path.

The wave function at time τ_0 is,

$$\psi(\tau_0) = |l(\tau_0)\rangle \otimes \chi_0(\eta) \; + \; |r(\tau_0)\rangle \otimes exp\Big(-i\int_0^{\tau_0} V(x_r(t),\eta)dt\Big)\chi_0(\eta) \tag{2.3}$$

The interference term then becomes

$$2Re\left[l^*(x,\tau_0)r(x,\tau_0)\;\; \langle\chi_0(\eta)|e^{-i\int_0^{\tau_0} V(x_r(t),\eta)dt}|\chi_0(\eta)\rangle\right] \tag{2.4}$$

Hence, the whole effect of the environment on the interference term shows up in the expression,

$$\langle\chi_0(\eta)|e^{-i\int_0^{\tau_0} V(x_r(t),\eta)dt}|\chi_0(\eta)\rangle \tag{2.5}$$

As before, this expression can be interpreted in terms of the effect of the electron on the environment, or in terms of the effect of the environment on the electron. The first follows from looking at eq. (2.5) as a scalar product of two environment states, one coupled to the left partial wave, and the other coupled to the right partial wave. The second is a bit more subtle. For a given η, $\phi(\eta) \equiv \int_0^{\tau_0} V(x_r(t),\eta)dt$ is the phase shift the electron would have experienced, had the environment particle been in a position η. Eq. (2.5) is then just

$$\begin{aligned} &\int \chi_0^*(\eta)e^{i\phi(\eta)}\chi_0(\eta)\; d\eta \\ &= \int \left|\chi_0(\eta(\phi))\right|^2 e^{i\phi}\frac{d\eta}{d\phi}\; d\phi \end{aligned} \tag{2.6}$$

The limits of the last integral depend on the function $\phi(\eta)$, and are not relevant for our present discussion. (In cases where several values of η lead to the same ϕ, the right hand side of this equation has to include a summation over these values of η.) The last integral leads us to the second interpretation: this integral is the average value of $e^{i\phi}$, averaged over a probability distribution for the phase shift ϕ. This distribution function, as seen in (2.6) , is

$$P(\phi) \equiv |\chi_0(\eta(\phi))|^2\frac{d\eta}{d\phi}. \tag{2.7}$$

The effect of the environment on the electron, which would have been a well determined phase shift for a "classical" environment, becomes the average value of all possible phase shifts when the quantum nature of the environment is taken into account.

When does the environment destroy the interference pattern ? Again, there are two ways to answer this question, corresponding to the two interpretations discussed above. The interference term reduces to zero when the environment's state coupled to the right partial wave is orthogonal to the one coupled to the left partial wave, or, alternatively, the interference term reduces to zero when its average value, averaged over all possible states of the environment, is zero. By further inspection of (2.6) , the second condition can be rephrased in terms of the phase distribution function $P(\phi)$. The environment multiplies the interference term by the factor given by (2.6) , and this factor is,

$$\left\langle e^{i\phi}\right\rangle = \int P(\phi)e^{i\phi}d\phi \;=\; e^{i\langle\phi\rangle}\int P(\phi)e^{i(\phi-\langle\phi\rangle)}d\phi \tag{2.8}$$

where $\langle\phi\rangle \equiv \int \phi P(\phi)\, d\phi$ is the average phase shift. If $P(\phi)$ is a narrow distribution function, that is, if $\langle\delta\phi^2\rangle \ll 1$, equation (2.8) can be approximated,

$$\langle e^{i\phi}\rangle \approx e^{i\langle\phi\rangle} = e^{i\int_0^{T_0}\langle V(x_r(t))\rangle dt} \tag{2.9}$$

The environment induces then a phase shift proportional to the expectation value of the potential. Hence, as far as the interfering particle is concerned the environment potential can, in this limit, be replaced by a static potential $V(x)$.

On the other hand, when $P(\phi)$ is very broad, i.e. $\langle\delta\phi^2\rangle \gg 1$, there is an appreciable probability for various values of phase shifts, extending over a range much larger than 2π. Then, excluding pathological cases of extremely non–smooth distribution functions, the intensity of the interference pattern tends to zero. In this limit, the environment's potential cannot be approximated by a static potential, and the analysis of the experiment has to be done in terms of the system + environment. The experiment involves then a transition of the interfering electron from a pure state to a mixed state.

The equivalence between the two descriptions discussed above can also be understood in terms of the uncertainty principle. Suppose that the free particle's state is initially a minimal wave packet, i.e., $\delta\eta\delta p_\eta \sim 1$. Suppose also that the interaction between the free particle and the interfering particle is smooth enough so that it can be expanded as,

$$V(x_r(t),\eta) \approx V(x_r(t),\eta=0) \;+\; \nabla_\eta V(x_r(t))\cdot\eta \tag{2.10}$$

where $\eta = 0$ is chosen to be the center of the minimal wave packet. The zeroth order contribution to the expansion is just a static potential, which induces a well defined phase shift. By (2.3) , the state of the environment (i.e., the free particle) after the interaction is

$$e^{-i\int_0^{\tau_0} \nabla_\eta V(x_r(t))dt\cdot\eta}\chi_0(\eta) \tag{2.11}$$

so that the interaction shifts the free particle's momentum by $\int_0^{\tau_0} \nabla_\eta V(x_r(t))dt$. When this momentum shift is larger than the uncertainty in the free particle's momentum, δp_η, the environment is shifted into a state orthogonal to its original one and the interference is destroyed. The condition for this destruction is, therefore,

$$\int_0^{\tau_0} \nabla_\eta V(x_r(t))dt > \delta p_\eta \tag{2.12}$$

On the other hand, by looking at this loss of interference from the point of view of phase uncertainty, we find that the condition for this loss is,

$$\langle\delta\phi^2\rangle^{\frac{1}{2}} = \int_0^{\tau_0} \nabla_\eta V(x_r(t))dt\ \delta\eta > 1 \tag{2.13}$$

The assertion $\delta\eta\delta p_\eta \sim 1$ yields the agreement of the two conditions. When the uncertainty in momentum is large, it takes a large momentum shift, i.e., a long interaction time, in order to shift the environment to a state orthogonal to its initial one. On the other hand, large momentum uncertainty yields small uncertainty in position. This, in turn, yields a well defined potenial, and again, a long interaction time is needed for the phase to become uncertain. Thus, the equivalence between the two descriptions is a consequence of the momentum - position uncertainty in the minimal wave packet.

The above example of an environment lacks an important feature, commonly found in real situations, namely, the independent dynamics of the environment. These dynamics exist in the following example. The interference of two electron's trajectories $x_r(t)$ and $x_l(t)$ is examined in a cavity at zero temperature. The electron interacts with the electromagneic vector potential. It is well known that such an interaction causes a relative phase shift of $\frac{e}{\hbar c}\int(A(x_r(t),t)\cdot\dot{x}_r(t) - A(x_l(t),t)\cdot\dot{x}_l(t))\,dt$. The average value of the vector potential in vacuum is zero, therefore the mean phase shift is also zero. However, there are quantum fluctuations of the vector potential in the vacuum, and those will cause an uncertainty in the phase shift. The uncertainty will be,

$$\begin{aligned}\langle\delta\phi^2\rangle = (\frac{e}{\hbar c})^2 \int dt \int dt' &\left[\vec{A}(x_r(t),t)\cdot\dot{\vec{x}}_r(t) - \vec{A}(x_l(t),t)\cdot\dot{\vec{x}}_l(t)\right] \\ &\times \left[\vec{A}(x_r(t'),t')\cdot\dot{\vec{x}}_r(t') - \vec{A}(x_l(t'),t')\cdot\dot{\vec{x}}_l(t')\right]\end{aligned} \tag{2.14}$$

Now, let us look at the simple case in which the two partial waves follow a $1D$ harmonic oscillator trajectory in two opposite directions [6], i.e.,

$$x_r(t) = -x_l(t) = R\sin\omega_0 t \tag{2.15}$$

Then, the phase uncertainty is,

$$\begin{aligned}\langle\delta\phi^2\rangle = (\frac{e}{\hbar c})^2 \int dt\omega_0 R\cos\omega_0 t(A_x(R\sin\omega_0 t, t) + A_x(R\sin\omega_0 t, t)) \\ \times \int dt'\omega_0 R\cos\omega_0 t'(A_x(R\sin\omega_0 t', t') + A_x(R\sin\omega_0 t', t'))\ .\end{aligned} \tag{2.16}$$

The correlation function $\langle A_x(x,t)A_x(x't')\rangle$ in vacuum is

$$\langle A_x(\vec{x},t)A_x(\vec{x}',t')\rangle = \sum_\alpha \frac{\hbar c^2}{\Omega\omega_\alpha} e^{(ik_\alpha\cdot(\vec{x}-\vec{x}')-i\omega_\alpha(t-t'))}(1-\frac{(k_{\alpha x})^2}{|k|^2}), \tag{2.17}$$

where $\omega_\alpha = ck_\alpha$, and Ω is the cavity's volume.

Substituting this expression into the expression for $\langle\delta\phi^2\rangle$, equation (2.14) we obtain,

$$\langle\delta\phi^2\rangle = 4\omega_0^2 R^2 \int d\vec{k}\frac{\hbar c}{|k|} \left|\int_0^T dt\cos[\frac{1}{2}k_x(x_1(t)-x_2(t))]\cos\omega_0 t e^{ic|k|t}\right|^2 (1-\frac{k_x^2}{|k|^2}) \tag{2.18}$$

Most of the contribution to this integral comes from $|k| \approx \frac{\omega_0}{c} \pm \frac{1}{cT}$. Since $\omega_0 R << c$, the dipole approximation is valid, i.e., $k_x\big(x_r(t) - x_l(t)\big) << 1$. The expression for $\langle\delta\phi^2\rangle$ is then exactly the probability of a photon field at T=0 to absorb a photon from an oscillating dipole of frequency ω_0. Since the photon field is at T=0, the most effective way to leave a trace on it is to emit a photon. The time it takes for the phase to be completely uncertain is exactly the time it takes for the electron to emit a photon. Since the accelerations of the two partial waves are exactly opposite, the radiation emitted from each of them is in a phase shift of π relative to the other. The phase of the radiation contains the information about the path taken by the electron. Therefore, although the two partial waves emit the same amount of radiation (and they necessarily do, since in the dipole approximation they both feel the same potential fluctuations), the trace they leave on the environment must be different. This is seen in terms of the environment's wave function. The environment's wave function is, up to first order in the potential,

$$|\text{vac}\rangle\ \pm\ i\sum_k |a_k|\ |1 \text{ photon in } k \text{ mode}\rangle\ , \tag{2.19}$$

where the $\pm$ sign refers to the environment state coupled to the first and second partial waves. The difference in sign reflects the phase difference in the radiation.

Had the two partial waves propagated with equal velocities in part of their trajectory, then for that part the dipole radiation would have caused uncertainty in each of the partial waves' phase, but not in their relative phase. The relative phase uncertainty would then have been accumulated by the quadropole interaction (next order in $k \cdot r$). Then, the $\cos[\frac{1}{2}k_x(x_1(t) - x_2(t))]$ in (2.14) should be replaced by $[k_x(x_1(t) - x_2(t))]^2$, yielding

$$\langle \delta\phi^2 \rangle = 4\omega_0^2 R^2 \int d\vec{k} \frac{\hbar c}{|k|} \left| \int_0^T dt [k_x(x_1(t) - x_2(t))] \cos \omega_0 t e^{ic|k|t} \right|^2 (1 - \frac{k_x^2}{|k|^2}) . \quad (2.20)$$

In both cases, we conclude that the time it takes for the phase to get uncertain is just the time it takes the partial waves to emit photons that will identify them. By suitably placed reflecting mirrors one can isolate the photons of each path from that of the other path, thus creating a situation where any photon identifies its path.

3. Dephasing by a back reacting environment

As demonstrated above, the potential the environment exerts on the interfering partial waves might induce a shift and/or a reduction of the interference pattern. However, we haven't discussed yet what determines this potential. In particular, is the potential exerted by the environment on the interfeing particle affected by the presence of that particle? Classically, the answer is negative for a "test particle", while it might be positive for a real particle. The formal generalization of the examples we have considered, given below, opens a way to discuss this question quantum mechanically. Consider a dynamical environment interacting with an interfering particle. The hamiltonian of the environment will be denoted by $H_{env}(\eta, p_\eta)$, while the interaction term is $V(x_r(t), \eta)$. For simplicity, we restrict ourselves to the case where only the right wave interacts with its environment.

Starting with the initial wave function, eq. (1.1) , the wave function at time τ_0 is,

$$\psi(\tau_0) = l(\tau_0) e^{-iH_{env}\tau_0} \chi_0(\eta) \; + \; r(\tau_0) \hat{T} e^{-i\int_0^{\tau_0} (H_{env} + V) dt} \chi_0(\eta) \qquad (3.1)$$

where $\hat{T}$ is the time ordering operator. It is useful at this point to write $\psi(\tau_0)$ in terms of $\hat{V}_I(t) \equiv e^{iH_{env}t} V(x_r(t), \eta) e^{-iH_{env}t}$, i.e., the potential V in the interaction picture. Note that $\hat{V}_I(x, t)$ is an operator in the *environment* Hilbert space, with parameters (x,t). Its time evolution is not affected by the presence of the particle. Therefore, $\hat{V}_I(x,t)$ repersents the local potential exerted by the environment at (x,t), in the absence of the interfering particle. Using $\hat{V}_I$, $\psi(\tau_0)$ can be written as

$$\psi(\tau_0) = l(\tau_0) \otimes e^{-iH_{env}\tau_0} \chi_0(\eta) \; + \; r(\tau_0) \otimes e^{-iH_{env}\tau_0} \hat{T} e^{-i\int_0^{\tau_0} V_I(x_r(t),t)dt} \chi_0(\eta) \quad (3.2)$$

Hence, the interference term is multiplied by

$$\begin{aligned} &\langle \chi_0 | e^{iH_{env}\tau_0} \hat{T} e^{-i\int_0^{\tau_0}(H_{env}+V)dt} | \chi_0 \rangle \\ =&\langle \chi_0 | \hat{T} e^{-i\int_0^{\tau_0} V_I(x_r(t),t)dt} | \chi_0 \rangle \end{aligned} \quad (3.3)$$

Obviously, this is the scalar product of the two environment state coupled to the two partial waves. The description in terms of the statistical properties of the phase stems from the observation that equation (3.3) is the expectation value of the time evolution operator, which is a unitary transformation. Being a unitary transformation, the time evolution operator can be represnted as $e^{i\hat{\phi}}$, where $\hat{\phi}$, naturally defined as the phase operator, is hermitian.

Hence, the expression in equation (3.3) becomes,

$$\langle \chi_0 | e^{i\hat{\phi}} | \chi_0 \rangle \quad (3.4)$$

and $\hat{\phi}$ is implicitly defined by,

$$e^{i\hat{\phi}} = \hat{T} e^{-i\int_0^{\tau_0} dt \; \hat{V}_I(x_r(t),t)} \quad (3.5)$$

The time ordering operator in the implicit definition suggests an investigation of the commutation relation

$$[\hat{V}_I(x_r(t),t) \; , \; \hat{V}_I(x_r(t'),t')] \quad (3.6)$$

An immediate mathematical observation is that when this commutation relation is zero for all $0 < t,t' < \tau_0$, i.e., during the interference experiment, the time ordering operator can be ignored, and equation (3.5) is easily translated into an explicit expression for the phase. This expression is,

$$\hat{\phi} = -\int dt \; \hat{V}_I(x(t),t) \quad (3.7)$$

i.e., the instantaneous rate at which a partial wave accumulates its phase is given by the local potential the environment exerts on it at that instant. When the commutation relation in equation (3.6) is non–zero for parts of the right path, the translation of equation (3.5) into an explicit definition of the phase is not trivial. In particular,

$$\dot{\phi}(t) \neq V_I(x_r(t), t) \tag{3.8}$$

Physically, we expect equation (3.8) to be true when the potential the environment exerts on the interfering particle (say, an electron) differs from the local potential V_I, and the difference arises from to a back reaction of the environment to the presence of the interfering electron. The local potential V_I, being local in time, does not allow present electron – environment interaction to modify the potential the electron will feel in the future. Hence, it does not allow a back reaction of the environment to the presence of the interfering electron. We will therefore demonstrate now the equivalence between the condition

$$[\hat{V}_I(x_r(t), t) \, , \, \hat{V}_I(x_r(t'), t')] = 0 \text{ for all } 0 < t, t' < \tau_0 \tag{3.9}$$

and the statement that the potential exerted on the electron is not affected by the presence of the electron. We will demonstrate this equivalence for an electromagnetic–type electron–environment interactions, such as interactions with photons, phonons and other electrons.

Consider an interfering electron interacting with the (not necessarily free) electromagnetic field, A(x,t). The potential in the interaction picture is then

$$-\frac{e}{c}\dot{x}_r(t) \cdot A_I(x_r(t), t) \tag{3.10}$$

where the time evolution of $A_I(x_r(t), t)$ is determined by the independent dynamics of the field (independent of the presence of the interfering particle). The commutation relation in equation (3.9) is then proportional to the commutation relation

$$[A_I(x, t) \, , \, A_I(x', t')] \tag{3.11}$$

at $x = x_r(t), x' = x_r(t')$. This commutation relation satisfies the same equation of motion as does $A_I(x, t)$, i.e., a wave equation, with the initial conditions,

$$[A_I(x, t) \, , \, A_I(x', t)] = 0 \text{ for all } x, x' \tag{3.12}$$

$$[A_I(x, t) \, , \, \dot{A}_I(x', t)] \propto i\delta(x - x') \tag{3.13}$$

Therefore, the value of the commutation relation in equation (3.11) is just the amplitude of an electromagnetic excitation created at x', t' to be present at x, t.

When this amplitude is zero for all points $(x_r(t), t)$ and $(x_r(t'), t')$ the local potential expression for the phase (equation (3.7)) is exact. But, when that amplitude is zero there is also no way the potential exerted on the electron at $(x_r(t), t)$ can be affected by the electromagnetic interaction the electron has experienced before, at previous points along its path. Therefore, there can be no back reaction. Such is, e.g., the case of a *free* electromagnetic field. For a free electromagnetic field, the commutation relation in equation (3.11) is non-zero only when (x, t) is on the light cone of (x', t'), while the motion of the partial wave is, of course, inside the light cone. Hence, the phase accumulated by the right partial wave is given by the integrated local potential (equation (3.7)). And, indeed, the potential exerted by a free electromagnetic field on the interfering partial wave is not affected by the partial wave's history. The interfering electron might emit photons along its path, thus changing the state of the field, but these photons travel faster than the partial wave, so that it can never re-encounter them. Hence, the potential it feels along its path is independent of the existence of *those* photons. On the other hand, if the field is not free, e.g., in the presence of mirrors or electric charges, photons emitted by the electron might be reflected, re–encounter the electron, and thus change the potential it feels.

Our conclusion at this point is then that if the potential exerted on a partial wave along its path is independent of its history, then the phase accumulated by that partial wave is given by the integrated local potential (equation (3.7)). This is an important conclusion since it allows the use of equation (3.7) in many circumstances. In particular, when the interfering electron is coupled to a thermal bath, the short memory time of the bath usually justifies the no–back–reaction assumption. The back reaction might be important when the interfering electron shifts the bath significantlly out of equilibrium for a long enough time scale.

What happens when there is a significant back reaction of the environment? In particular, do the fluctuations in the *local* potential, $\hat{V}_I$, have any relevance to the dephasing of interference? We first give a positive, somewhat surprising, answer to the second question, and then discuss the first. Regarding the dephasing of interference in a back–reacting environment, we show that the fluctuations in the local potential can be used for a determination of a lower limit to the intensity of the interference term, i.e., a lower limit to

$$\left|\langle\chi_0|\hat{T}e^{-i\int_0^{\tau_0} V_I(x_r(t),t)dt}|\chi_0\rangle\right| \tag{3.14}$$

A sufficient, though not necessary, condition for interference to be seen, is that this lower limit is not too small, compared with unity.

To obtain the lower limit, we write the environment state coupled to the right wave as a superposition of the initial state $|\chi_0\rangle$ and a state $|\xi\rangle$ orthogonal to $|\chi_0\rangle$, with amplitudes $A_\parallel, A_\perp$, respectively:

$$\hat{T}e^{-i\int_0^{\tau_0} V_I(x_r(t),t)dt}|\chi_0\rangle = A_\parallel|\chi_0\rangle \;+\; A_\perp|\xi\rangle \tag{3.15}$$

where $|A_\parallel|^2 + |A_\perp|^2 = 1$ and $|\chi_0\rangle, |\xi\rangle$ are both normalized. Obviously, $A_\parallel = \langle\chi_0|\hat{T}e^{-i\int_0^{\tau_0} V_I(x_r(t),t)dt}|\chi_0\rangle = \langle\chi_0|e^{i\hat{\phi}}|\chi_0\rangle$. $|A_\parallel|$ is a measure for the intensity of the interference term, so that as long as $|A_\perp| \ll 1$ interference is observed.

On the other hand, the time ordered product can be broken into an ordered product of infinitely many exponents,

$$\hat{T}e^{-i\int_0^{\tau_0} V_I(x_r(t),t)dt} = \prod_{k=1}^{N} e^{-\frac{i\tau_0}{N}V_I(x_r(t_k),t_k)} \equiv \prod_{k=1}^{N} e^{-iV_I^k\delta t} \tag{3.16}$$

where $t_k \equiv \frac{k}{N}\tau_0$, and the limit of $N \to \infty$ is taken. For each instantaneous operator $e^{-iV_I^k\delta t}$ we define instantaneous amplitudes $a_\parallel^k, a_\perp^k$ such that,

$$e^{-iV_I^k\delta t}|\chi_0\rangle = a_\parallel^k|\chi_0\rangle \;+\; a_\perp^k|\xi^k\rangle \tag{3.17}$$

where again $|\xi^k\rangle$ is a normalized state, orthogonal to $|\chi_0\rangle$. Now, since $|a_\parallel^k| \le 1$, and $|\langle\xi^k|\xi^{k'}\rangle| \le 1$, equations (3.15) - (3.17) imply that

$$|A_\perp| \;\le\; \sum_{k=1}^{N}|a_\perp^k| \tag{3.18}$$

In the limit of $N \to \infty$, $a_\parallel^k$ and $a_\perp^k$ can be expressed in terms of moments of V_I^k. Substituting the expression for $a_\perp^k$ in equation (3.18) , we get,

$$|A_\perp| \le \sum_{k=1}^{N} \langle\delta(V_I^k)^2\rangle^{\frac{1}{2}}\,\delta t \;=\; \int_0^{\tau_0} dt\,\langle\delta V_I(x_r(t),t)^2\rangle^{\frac{1}{2}} \tag{3.19}$$

Therefore, a sufficient condition for the observation of interference is that the fluctuations in the *local* potential along the partial wave's path do not sum up to unity. If this condition is fulfilled, no additional fluctuations due to a back reaction potential can dephase the interference.

An interesting application of this conclusion is given in the following problem. A particle is located at the top of a symmetric potential "hill" (more quantitavely:

The particle's initial wave function $\psi(x)$ is a minimal wave packet of width δx, centered around $x = 0$, and symmetric, i.e., $\psi(x) = \psi(-x)$. The potential is, say, $V(x) = V_0 e^{-\left(\frac{x}{a}\right)^2}$, with $a >> \delta x$. See Fig. 3) An external force whose magnitude is F, but its direction is unknown, is applied on the particle for a duration τ, in which the particle's wave function does not significantlly spread. For $t > \tau$, is there any measurement that can identify the direction of F ? Classically, the answer is, of course, positive, and it doesn't matter if F is infinitesimal. The particle falls down to the right or to the left, according to the direction of F. Quantum mechanically, however, a measurement can distinguish between two states only if they are orthogonal. And, as we have seen above, if the phase fluctuations induced by the uncertainty in the applied potential are much smaller than unity, the particle on the "hill" (corresponding to the "environment" in the discussion above) is not shifted to a state orthogonal to its initial state. This is true,even though the unstable nature of the initial state tends to "amplify" the effect of the external force. The applied potential is here,

$$V = -Fx \tag{3.20}$$

so that the phase uncertainty it induces is,

$$\delta\phi = \delta V\tau = F\delta x\tau \tag{3.21}$$

Thus, as long as $F\tau \ll \frac{1}{\delta x}$, no measurement, not even at $t \to \infty$ will identify the direction of F.

We conclude our discussion of the back reaction of the environment with a simple, exactly solvable, problem, in which the phase shift due to the back reaction can be pointed out. We go back to the one dimensional interference experiment, in which two electron partial waves follow two opposite harmonic oscillator's trajectories (Eq. (2.15)). This oscillator's frequency is ω_0. This time we turn off the coupling to the electromagnetic potential, and consider an environment composed of one phonon mode, whose frequency is $\Omega \gg \omega_0$, and its coupling to the electron is linear. The hamiltonian of the environment is the free hamiltonian of a phonon mode, i.e.,

$$\hat{H}_{env} = \frac{p^2}{2} + \frac{1}{2}\Omega^2 q^2 \tag{3.22}$$

and the interaction hamiltonian is

$$V = Cqx \tag{3.23}$$

In the above p, q are the phonon's momentum and position, respectively, and x is the interfering particle's position. The interaction can be represented in the interaction picture as,

$$\hat{V}_I(x_r(t),t) = Cx_r(t)\left(q\cos\Omega t + \frac{p}{\Omega}\sin\Omega t\right) \equiv Cx_r(t)q_I(t) \tag{3.24}$$

Since the phonon mode is in the ground state, $\langle q\rangle = 0$, and $\delta q = \langle q^2\rangle^{\frac{1}{2}} = \frac{1}{\sqrt{\Omega}}$. The motion of the electron partial wave, being slow relative to the phonon mode's frequency, shifts adiabatically the equilibrium point of the phonon mode, without exciting it. Thus, the mode's wave function at time t is the ground state of an oscillator centered at $-\frac{Cx_r(t)}{\Omega^2}$. In the beginning and the end of the partial wave's path, when $\frac{C|x_r(t)|}{\Omega^2} \ll \frac{1}{\sqrt{\Omega}}$, i.e., $|x_r(t)| \ll \frac{\Omega^{\frac{3}{2}}}{C}$, the mode's state overlaps with its initial state. In the intermediate stage, the phonon mode's state is nearly orthogonal to its initial state. It has the information on the electron's path, and this information results in a back reaction potential. Therefore, we expect the interference pattern, examined after the electron (and, essentially, also the phonon mode) has returned to its initial point, to show an extra phase, accumulated during that intermediate period. Using the commutation relation of potentials at two different points:

$$[\hat{V}_I(x(t),t), \hat{V}_I(x(t'),t')] = i\frac{C^2}{\Omega}x(t)x(t')\sin\Omega(t-t') \tag{3.25}$$

we can derive an explicit expression for the phase $\hat{\phi}$, out of the implicit definition (equation (3.5)). This expression is:

$$\phi = -\int_0^{\tau_0} dt Cx_r(t)q_I(t) - \frac{1}{2}\int_0^{\tau_0} dt \int_0^t dt' \frac{C^2}{\Omega}x_r(t)x_r(t')\sin\Omega(t-t')$$
$$\equiv \phi_0 + \phi_{br} \tag{3.26}$$

The first term in equation (3.26) is the phase accumulated due to the local potential. Its rate of accumulation is $\dot{\phi}_0 = Cx_r(t)q_I(t)$, the local potential. Being proportional to $q_I(t)$, this rate has a zero average value, and an uncertainty $\delta\dot{\phi}_0 = Cx_r(t)\delta q_I(t) = Cx_r(t)\frac{1}{\sqrt{\Omega}}$. The second term is the back reaction contribution. As generally expected for a back reaction, this contribution is of second order in the coupling constant, and it is non-local in time. In this special example, the back reaction contribution is a C-number, so that it does not contribute to the phase uncertainty, i.e., to dephasing. Substituting the right path, $x_r(t) = R\sin\omega_0 t$

(equation (2.15)), and taking the adiabatic limit, $\frac{\omega_0}{\Omega} \to 0$, we find that the back reaction contribution to the phase is,

$$\phi_{br} = -\frac{C^2 R^2}{2\Omega^2} \int_0^{\tau_0} dt \sin^2 \omega_0 t \tag{3.27}$$

and the rate at which it is accumulated is,

$$\dot{\phi}_{br} = -\frac{C^2 R^2}{2\Omega^2} \sin^2 \omega_0 t = -\frac{C^2}{2\Omega^2} x_r^2(t) \tag{3.28}$$

This rate becomes larger as $x_r^2(t)$ increases, i.e., as the phonon mode is shifted far from its equilibrium point. In particular, for $x_r(t) \gg \frac{\Omega^{\frac{3}{2}}}{C}$, this rate is higher than the uncertainty in the local potential, which is $Cx_r(t)\langle q^2\rangle^{\frac{1}{2}}$. As long as the environment state significantly overlaps its initial state, $\left|\dot{\phi}_{br}\right| \ll \delta\dot{\phi}_0$. Once the environment state is orthogonal to its initial one, most of the phase accumulation rate is due to the back reaction.

A generalization of the above picture to a general back reacting environmnet should be done in a careful way, since some of the above conclusions originate in the simple nature of the example. In general, back reaction of the environment might contribute also to the phase uncertainty. It is only in our simple example that ϕ_{br} is a C-number. However, while $\delta\phi_0$ is of second order in the interaction potential, $\delta\phi_{br}$ is of higher order, since ϕ_{br} by itself is of second order.

Acknowledgments

Research at USC was partially supported by NSF grant no. PHY-8807812. Research at TAU and WIS was partially supported by the fund for basic research administered by the Israel Academy of Sciences and Humanities. Research at TAU was partially supported by the German – Israel foundation for scientific research and development. One of us (A.S.) acknowledges the support of the David Ben–Gurion foundation administered by the Israel Labor Federation. We thank K. Mullen, D. Loss, A.J. Leggett, E. Ben-Jacob, and S. Popescu for instructive discussions.

REFERENCES

1. R.P. Feynman, A.R. Hibbs, *Quantum mechanics and path integrals*, Mc. Graw–Hill, 1965.
2. Feynman R.P. and Vernon F.L. *Ann. Phys.* **24**,118 (1963).
3. Caldeira A.O. and Leggett A.J. *Ann.Phys.* **149**,374(1983).
4. A. Stern, Y. Aharonov and Y. Imry, to appear in *Phys. Rev. A.*
5. Imry Y., in *directions in condensed matter physics*,ed. G. Grinstein and G. Mazenko, Singapore, world scientific (1986).
6. Caldeira A.O. and Leggett A.J. Phys. Rev. A **31**, 1059 (1985) .

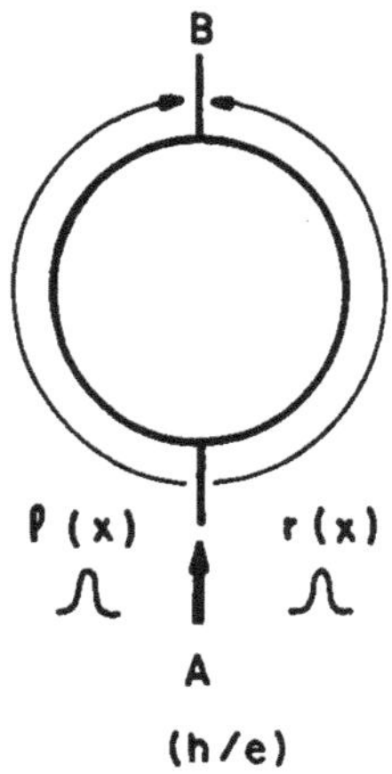

Fig.1: Schematics of interference experiments in A-B rings: Each partial wave traverses half the ring, and the interference is examined at the point B. This kind of interference gives rise to $\frac{h}{e}$ oscillations of the conductance.

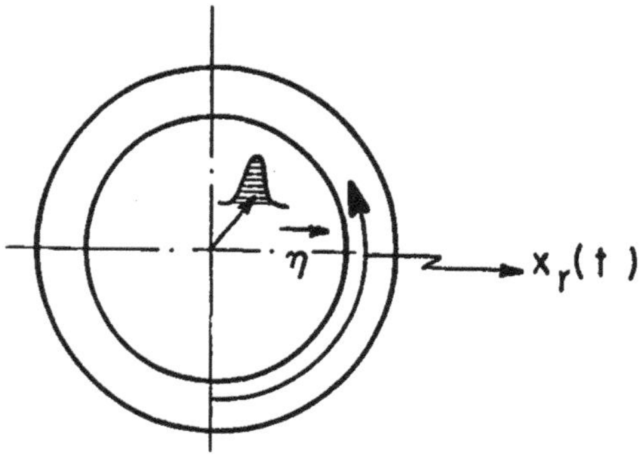

Fig.2: The right path $x_r(t)$ interacts with an environment composed of one heavy free particle, whose coordinate is $\vec{\eta}$.

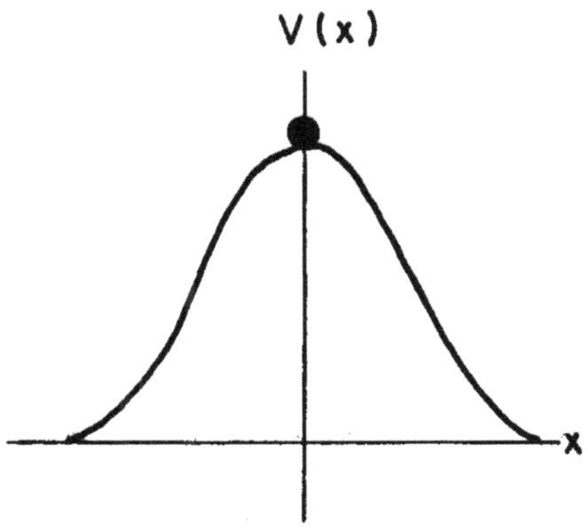

Fig.3: A particle is initially localized on the top of a potential hill.

MACROSCOPIC QUANTUM TUNNELLING OF THE TOTAL MAGNETIZATION IN SMALL FERROMAGNETIC GRAINS

ANUPAM GARG

Dept. of Physics and Astronomy, Northwestern University, Evanston, Illinois 60208, U.S.A.

ABSTRACT

A novel system in which to look for the phenomena of macroscopic quantum tunneling and coherence is a small ($\sim 100\,\mathring{A}$), single domain ferromagnetic particle. The total magnetization, M, in such a particle can tunnel between various easy directions as determined by the magnetocrystalline anisotropy and the external magnetic field, if any. The barrier to tunneling is provided by the magnetocrystalline anisotropy, and the tunnelling is provided by the magnetocrystalline anisotropy, and the tunnelling rate (calculable by a WKB approach[1] can achieve values of the order of 10^8 sec^{-1}. This is true for all types of anisotropy.

Since the observation of such tunnelling would provide dramatic evidence that superpositions of macroscopically distinct states can exist, we have studied the effect of including dissipation (in the sense defined by Calderia and Leggett[2], since this is known to suppress quantum effects severely. In the present problem, dissipation is provided mainly by the magnetoelastic coupling to nonmagnetic solid medium in in which the magnetic particle may be embedded, since the waves in an isolated particle are at too high a frequency to couple effectively to M. We find[3], surprisingly, that this mechanism is very benign; the WKB exponent is only increased by a factor of relative size $10^{-4} - 10^{-6}$!

We have also considered the inclusion of other effects such as stray fields, shape anisotropy, and interparticle interactions when there are several particles.

---

[1]E.M. Chudnovsky and L. Gunther, *Phys.Rev.Lett.* **60**, 661 (1988).

[2]A.O. Calderia and A.J. Leggett, *Ann.Phys.* (NY) **149**, 374 (1988).

[3]Anupam Garg and Gwang-Hee Kim, *Phys.Rev.Lett.* **63**, 2512 (1989).

TOWARDS A TWO VECTOR FORMULATION OF QUANTUM MECHANICS

Y. AHARONOV
Department of Physics & Astronomy, The University of South Carolina
Columbia, South Carolina 29208
USA
and Physics Department, Tel Aviv University
Tel Aviv
Israel 69978

and

D. Rohrlich
Physics Department
Jerusalem University
Jerusalem Israel

Abstract

A two-vector formulation of Quantum Mechanics is proposed. The new formalism permits a covariant description of the "collapse" of the wave function. A number of surprising predictions of the new formalism are discussed.

Quantum systems are generally described via wave-functions, or states. The state is supposed to represent all our knowledge about a quantum system; form the state we predict its behavior. However, this standard description via states has serious shortcomings. Indeed, the standard formulation fails to accommodated basic symmetries of spacetime: Lorentz covariance and time-reversal symmetry at the microscopic level. These flaws are usually glossed over, but efforts to deal with them are gradually yielding clues how quantum mechanics might better be formulated. We find that a description of quantum systems in terms of two state vectors is needed. One state evolves forward from initial conditions, as usually, while the second evolves backwards in time from future boundary conditions. Such a formulation generates all the results of standard quantum mechanics, and in addition leads to surprising new results that are hard to obtain using the conventional description. We will present the two vector formulation here, and also summarize the research that has inspired it.

1. The standard formulation of quantum mechanics is not time-symmetric even at the microscopic level. The breakdown of time symmetry reflects our habit of predicting future events on the basis of initial conditions; the latter have occurred and are given data, the former have not occurred and fully depend on the initial conditions. We inherit this habit of thinking from classical mechanics. There, it entails no violation of time symmetry, because classical physics is deterministic. But quantum physics is not deterministic, and indeed the usual dichotomy of initial conditions vs. equations of motion is not particularly natural there, for quantum mechanics sets strict limits to what we can know about initial conditions. We have the additional problem of inferring, from given data, what values dynamical variables took at earlier times. When such retrodictions are considered, along with predictions, time symmetry is restored, as the following example shows.

Consider a sequence of measurements, at successive time t_1, t_2, and t_3, on a quantum system. At time t_1 we measure the operator A so the system is prepared in an eigenstate which we will denote $|a>$. At time t_3 we measure the operator B and find the system in the state $|b>$. (A and B need not commute.) In between, at time t_2, the operator C is measured; we are informed of this measurement but not told the result. Nevertheless, we can make a good prediction about the outcome of the measurement. We know that the probability of an outcome $|c_i>$ for the measurement C, given the initial state $|a>$, is

$$|< c_i|U(t_2,t_1)|a>|^2$$

where $U(t_2,t_1)$ is the unitary operator for time evolution from t_1 to t_2. Likewise the probability of the outcome $|b>$ given an initial state $|c_i>$ is

$$|< b|U(t_3,t_2)|c_i>|^2$$

These are the only quantum mechanics results we need to know. It now follows[1] from classical laws of probability that the probability that the state $|c_i>$ was found at t_2 is

$$P_i = \frac{|< b|U(t_3,t_2)|c_i>|^2 |< c_i|U(t_2,t_1)|a>|^2}{\sum_j |< b|U(t_3,t_2)|c_j>|^2 |< c_j|U(t_2,t_1)|a>|^2} \tag{1}$$

Why? Suppose we begin with an ensemble of N identical systems. Each system begins in the state $|a>$, but we do not always end up in the state $|b>$; that only happens in

$N \cdot \sum p(b|c_j)p(c_j|a)$ cases, on the average. (We use the notation p (f|g) to denote the probability of f given g.) In this smaller sample that remains, the outcome $|c_i>$ is found Np ($b|c_i$) p ($c_i|a$) times on average. Hence the probability p_i is given by Eq. (1).

With a slight rewriting, Eq. (1) becomes manifestly time-symmetric:

$$P_i = \frac{|<c_i|U(t_2,t_3)|b>|^2 |<c_i|U(t_2,t_1)|a>|^2}{\sum_j |<c_j|U(t_2,t_3)|b>|^2 |<c_j|U(t_2,t_1)|a>|^2} \qquad (1')$$

where $U(t_i, t_j) = U^{-1}(t_j,t_i) = U^\dagger(t_j, t_i)$. Eq. (1′) suggests the following description: the outcome of a measurement of C at t_2 can be calculated in terms of two vectors $|a>$ and $|b>$, which represent initial and final boundary conditions, respectively. Evolving $|a>$ forwards in time from t_1 to t_2, and $|b>$ backwards in time from t_3 to t_2, we project $|c_i>$ onto these two vectors to find the amplitude for the outcome $|c_i>$ (up to a normalization). Such an interpretation [2] is just as tenable as the conventional one, at least in the present example, and has the added virtue of being time -symmetric.

Eq. (1) suggests something more. (For simplicity let the operators $U(t_i, t_j)$ equal 1 here.) Note that if the operator C had been chosen to be A, the outcome $|a>$ would have been certain at t_2; whereas if the operator B had been chosen for C, the outcome $|b>$ would have been certain. Even if A and B are non-commuting observables, we seem to be able to assign a definite value to both these operators at the intermediate times. Could there be some sense in which both have simultaneous reality? That question will be explored below. However, we are so far dealing with the special case where a measurement was definitely made between t_1 and t_3, only we do not know what the measurement was. Eqs. (1) and (1′) do not entitle us to talk about values of an operator C at t_2 if no measurement was made then. As long as only particular operator was measured, we cannot say much about values of other operators, for the measurement of C represents a disturbance of the system. But can the time-symmetric description be extended to a broader class of interactions that do not disturb the system between initial and final states? We will see that an extension is indeed possible. But we defer the question and focus on another shortcoming of the conventional description.

2. The problem of relativistic covariance in quantum mechanics[3] arises as soon as we consider as simple a system as two spin-1/2 particles prepared in a spin-0 state:

$$x \equiv \frac{1}{\sqrt{2}}\left[|\uparrow>_r|\downarrow>_s - |\downarrow>_r|\uparrow>_s\right]$$

for particles located to points r and s, spacelike separated. Since the total angular momentum is zero, a measurement of any component of spin on either of them represents a measurement of that component on both of them. Therefore the conventional (Copenhagen) interpretation of quantum mechanics asserts that a measurement on either particle instantly reduces their combined state X to a direct product of one-particle states, regardless of how far apart the particles may be. We are led directly into a paradox, for what is simultaneous in one Lorentz frame is not simultaneous in another; while a measurement of the spin of either particle acts locally and so is covariant. Thus it seems each Lorentz frame implies a different version of when the two particles decouple, and we are left without any covariant state definition.

More formally, the paradox is this: on the one hand, we have different assertions about what is the actual quantum state for these two particles, by different observers; on the other hand these different assertions correspond to physically different worlds. What makes these worlds physically *different*? It is the fact that, in principle, we have ways to measure non-local properties.[3] For example, the state X is characterized by $J = 0$, where J is the spin quantum number for the spin-correlated two -particle system. This property of X is non-local: it belongs to neither particle alone. A verification that J = 0 constitutes a non-local measurement. And it can be accomplished by means of two local devices, on at r and the other at s. Consider an interaction Hamiltonian

$$H_{\text{int}} = -g(t) \sum_{i=x,y,z} (\sigma_i^r \; q_i^r + \sigma_i^s \; q_i^s).$$

Here g(t) is a function with support only in a short interval (t_1, t_2); its integral over the interval is one. The matrices σ_i^r represent the i-th component of spin for a particle located at r, etc. The variables q_i^r belong to the measuring device located at r, etc.; their conjugate variables, π_i^r correspond to actual pointer or indicator positions on the device. Now choose the initial state of the two devices to be as follows:

$$q_i^r - q_i^s = 0, \qquad \pi_i^r + \pi_i^s = 0. \tag{2}$$

This choice is always available since $[q_i^r - q_i^s , \pi_i^r + \pi_i^s] = 0$. After completion of the measurement, we have

$$(\text{change in } \pi_i^r) \;=\; i\int dt\left[H_{\text{int}}, \pi_i^r\right] = \sigma_i^r, \tag{3}$$

and similarly for the change in π_i^s. So imagine that immediate after t_2, each local device records the final value of π_i^r and π_i^s for posterity. Even without any further lapse of time (necessary for comparing the results at r and s) a measurement that J = 0 is essentially complete. From Eqs. (2-3) we deduce that

$$(\text{change in } \pi_i^r + \pi_i^s) = \left[\pi_i^r + \pi_i^s\right]_{t_2} = \sigma_i^r + \sigma_i^s = 0, \quad \text{if } J \;=\; 0 \tag{4}$$

Conversely, if Eq. (4) is found true when the measurements at r and s are compared, we have verified (in retrospect) that the two particles were in the combined state J = 0. The interval (t_1,t_2) can in principle be made arbitrarily short, so the verification of J=0 can be practically instantaneous. Furthermore, if in the initial state J = 0 then the measurement leaves the state unchanged: $q_i^r + q_i^s = 0$ remains true throughout the measurements and so is true at all times as well, for all components i.

When one observer, say in the lab frame, measures one particle of a pair, he concludes that the state of both reduces instantly, The reduced state must be $|\uparrow\rangle_r|\downarrow\rangle_s$ or $|\downarrow\rangle_r|\uparrow\rangle_s$ after a measurement of either σ_z^r or σ_z^s. Furthermore, he could have determined, an instant earlier, that the pair was in the non-local state X. For an observer in a rocket frame, neither of these conclusions is acceptable. The first conclusion fails because state reduction of the two systems cannot be simultaneous to both observers. The second conclusion fails because in the rocket frame, the non-local measurement of the state X is not instantaneous: the devices at r and s act during different time intervals. These actions thus disturb the state of the system from the point of view of the rocket frame. Either observer alone can verify the state of an extended system in his own Lorentz frame; but if both want to verify their independent descriptions, they will interfere with each other. There is no covariant description of the state that can be confirmed by experiment on both frames. the state is not something covariant: two observers cannot both verify their independent versions of the state history.

The two-vector formulation offers a way out of the paradox. We have two vectors, one propagating forwards in time from an experiment in the past, and a second propagating backward in time from an experiment in the future. The only reduction takes place locally, when an experiment is made. Therefore, in the example of the two correlated spins, we can always appeal to future measurement of the second particle which will complete the definition of the future boundary conditions. If the initial state was X and experimenters later find the particle at r polarized along the positive z-axis and the particle at s polarized along the positive x-axis, then the final boundary condition for the system is

$$\frac{1}{\sqrt{2}}|\uparrow>_r\left[|\uparrow>_s+|\downarrow>_s\right].$$

But even without knowing the result of the measurement a s, we know that any measurement of the z-component of spin at s will find it polarized along the negative z-axis. The boundary condition at r enforces this correlation, without our needing to specify a moment when it becomes true as a result of non-local state reduction. It is easy to check, from Eq. (1′), that between these boundary values the state $|\downarrow>_s$ has probability 1. And since the past and future boundary values the state $|\downarrow>_s$ has probability 1. And since the past and future vectors used in this description are determined by events (local experiments) in spacetime, the two-vector formulation is automatically covariant: the two vectors transform covariantly along with the events that define them.

Here we begin to see how the language of two vectors may actually be more consistent that the conventional language. In non-relativistic quantum mechanics, the state is a useful tool for calculating all possible outcomes--not only of actual experiments, but also of experiments that could be or could have been performed. It transforms covariantly under the appropriate group, the group of Galilean transformation. When we consider Lorentz transformations, we find that the state does not exist at all in a covariant way. The two-vector formulation restores covariance and at the same time answers the same need: it allows us to predict what will result, or what would result, or what would have resulted, from an experiment in between the initial and final boundary conditions. These predictions will in general be uncertain, and they will agree in every respect with predictions of the conventional formulation. The advantage is simply that the two-vector description will itself be covariant, not just its predictions.

The reader may object: predictions for actual or potential experiments can be made using the two vectors, but if these experiments are performed, they will in general change the evolution of the system. Since the future boundary condition depends on this evolution, the analogy with the conventional formulation does not work! Let us consider this objection in relation to the two-spin system described above. There, we saw that once the spin of one of the particles had been measured, the two-vector formalism predicts the other to have the opposite spin (if the same component is measured); this result holds no matter what the future boundary condition on the second particle may be. True, if the *measurement is actually performed, it disturbs the two vectors that previously described the* system, by inserting new intermediate conditions. But these intermediate conditions do not destroy the original two vectors; they only limit their propagation. Thus, if we are given boundary conditions at t_1 and t_3 and insert an experiment at t_2, the state at t_3 still propagates forwards and backwards, but it only propagates backwards as far as t_2, not to t_1. Similarly the state at t_1 propagates to t_2 and not to t_3. The experiment at t_2 now propagates forwards and backwards. However, the post selection of the ensemble at t_3 may look different now, because the future boundary condition is being applied to a disturbed ensemble. In contrast with conventional quantum mechanics, the ensemble depends on the experiment chosen. A departure from the conventional notion of an ensemble is apparent, but the ensemble of the two-vector formulation is in many cases the natural one, e.g. for any experiment that is described by an S-matrix.

3. Another objection may be raised: Perhaps the two-vector formulation provides a more flexible language in cases where the conventional formulation has problems. But does it justify introduction of a new formalism? Is it not just a matter of taste, how a theory is best stated? This point is well taken. The introduction of the two vectors in place of one is a big step, and it is not obvious that it leads to a useful formalism. In particular, we have the following question. Conventional quantum mechanics defines expectation values of operators in states via

$$<A> \equiv \frac{\langle\psi|A\psi\rangle}{\langle\psi|\psi\rangle}$$

where $<A>$ is the expected value of the operator A in the state ψ. In a two-vector formalism, this expression would naturally generalize to [4]

$$<A>_w \equiv \frac{\langle\psi_2|A\psi_1\rangle}{\langle\psi_2|\psi_1\rangle}$$

in a interval between states ψ_1 and ψ_2. But conventional quantum mechanics does not assign any particular meaning to this expression.

Remarkable, a physical meaning can indeed be given to the expression in Eq. (5). This kind of expectation value, measured between two different states, will be referred to as the weak value of an operator A. (Of course, when $\psi_2 = \psi_1$ the weak value becomes strong.) It turns out that by 'weakening' the measurement of A, we can strengthen the claim that the two vectors actually describe a system during the intervening period. To 'weaken' in this context means to diminish interference with the system, at the cost of deceased accuracy. This notion of a 'weak measurement' complements the two-vector formalism by introducing a broad class of measurements that can be made without significantly disturbing boundary conditions. It sounds paradoxical, and leads to apparently paradoxical behavior, but in fact accords with conventional quantum mechanics. Let us take an example:

We want to measure the orientation of a ferromagnet made of N spins.[5] To do this we use a measuring device which interacts with the ferromagnet according to a Hamiltonian of the von Neumann type:

$$H_{int} = -g(t)gA.$$

A is an operator which we wish to measure; in this case, a spin component. As before, q is canonically conjugate to π, the pointer setting, and g(t) is a normalized function with compact support. But this time we choose the initial state of the device to be a gaussian in q, with spread equal to $1/\sqrt{N/2}$. In the π-representation the initial state will also be a gaussian, with spread equal to $\sqrt{N/2}$. We select for an ensemble in which all the spins were initially polarized along the positive x-axis, and finally polarized along the positive y-axis. Naturally, for large N this is an unlikely set of boundary conditions, but there is no limit to how many tries we can make until we get the ensemble, so let us assume we have it. Because we have chosen a diffuse initial state for the measuring device, a measurement of L_x induces a change in L_y of magnitude

$$\Delta L_y = -i\int dt\, g(t) q\left[L_{x_t}, L_y\right] = qL_z \leq \frac{1}{\sqrt{N/2}}\cdot\frac{N}{2}$$

which is of order $\sqrt{N/2}$, or less. Similarly, a measurement of L_y entails a disturbance in the magnitude of L_x of this order. Given the initial and final boundary conditions, we are justified in taking the value of both L_x and L_y to be N/2 (up to corrections of order $\sqrt{N/2}$) throughout the entire period.

However, we can draw from these results another one which looks quite paradoxical: Let us measure the component of spin along the line bisecting the two axes, i. e. choose the interaction Hamiltonian to be

$$H_{int} = -g(t)q(L_x + L_y)/\sqrt{2};$$

then the state of the system after the measurement is

$$\exp\left\{iq(L_x + L_y)/\sqrt{2}\right\}\left|L_x = \tfrac{N}{2}\right\rangle\left|\pi \approx 0\right\rangle,$$

where $|\pi \approx 0\rangle$ represents the gaussian in π centered about zero. Including the post-selection of states with $L_y = N/2$, the state of the apparatus after the measurement is

$$\left\langle L_y = \tfrac{N}{2}\right| e^{iq(L_x+L_y)/\sqrt{2}} \left|L_x = \tfrac{N}{2}\right\rangle \left|\pi \approx 0\right\rangle$$
$$\approx e^{iqN/\sqrt{2}}\left|\pi \approx 0\right\rangle$$

up to a term of order 1 in the exponent, which is otherwise of order $\sqrt{N}$. (The exponential of $L_x + L_y$ has been written as a product of exponential, using the Baker-Campbell-Hausdorf relation and estimates of the size of q.) The result, then, is that the gaussian in p has been shifted by approximately $N/\sqrt{2}$, exactly as if L_x and L_y added classically to yield a vector $\sqrt{2}$ times larger than possible!

Thus, by 'weakening' the disturbance induced by successive measurements of L_x and L_y, we have shown that the weak value

$$\left\langle \frac{L_x + L_y}{\sqrt{2}} \right\rangle_w = \frac{\left\langle L_y = \frac{N}{2} \middle| \frac{L_x + L_y}{\sqrt{2}} \middle| L_x = N/2 \right\rangle}{\left\langle L_y = N/2 \middle| L_x = N/2 \right\rangle} = \frac{N}{\sqrt{2}}$$

corresponds to a genuine (and surprising) physical process, albeit an infrequent one. Clearly, we are just getting started; this is just one of many possible phenomena. To exhibit a similar result involving a single spin, we consider a weak measurement of a z-component of spin, σ_z, between an initial state

$$|\xi\rangle \equiv \cos\tfrac{\alpha}{2}\left|\sigma_x = \tfrac{1}{2}\right\rangle + \sin\tfrac{\alpha}{2}\left|\sigma_x = -\tfrac{1}{2}\right\rangle$$

and a final state polarized along the positive x-axis. The expected weak value of σ_z is then arbitrarily large:

$$\langle\sigma_z\rangle_w = \frac{\left\langle \sigma_x = \frac{1}{2} \middle| \sigma_z \middle| \xi \right\rangle}{\left\langle \sigma_x = \frac{1}{2} \middle| \xi \right\rangle} = \tan\tfrac{\alpha}{2}.$$

Such a result looks far removed from anything physical. Nevertheless, an experiment designed to measure σ_z under these conditions will indeed yield this 'impossible' value! The result can be derived rigorously,[4] but it may be more enlightening to follow what happens intuitively: A beam of spin-1/2 particles passes through a stern-Gerlach device which polarizes them alone the z-axis, splitting the up and down components which then impinge on a screen. However, the polarization is so weak that the width of the beam is much larger than the separation induced by the polarizer. The wave function near the screen, for the variable z, would look like two very broad gaussian peaks slightly separated one from the other. But the effect of the boundary conditions is to give those two peaks a relative phase that makes them cancel almost every where, except for a tiny peak far from either of the original peaks, at a deflection corresponding to the predicted absurd value of $\tan\frac{\alpha}{2}$. Of course, to get these results requires a large beam population, since many of the trials will not pass the boundary conditions. However, the experiment can proceed as slowly as desired, so that at any moment only one particle is present in the apparatus, and thus this is really an effect belonging to the particles in the deflected peak as individuals.

This effect may be regarded as a quantum stunt, a straightforward if unanticipated consequence of standard quantum mechanics, or it may be that we are seeing a qualitatively new aspect of quantum behavior that was never understood before. In either case, the weak measurement supplied by the two-vector description proves itself physically meaningful, and the tow vector formulation makes it easier to discover such effects. A number of other examples provide further reason to accept these effects as physically meaningful and not just quantum stunts. As a final example, let us consider a particle trapped in a finite potential well. As is well known, such a particle can 'tunnel' to a classically forbidden region where its total energy is necessarily less than its potential energy V. We ask in what sense we can observe that such tunneling takes place. If we measure the position well enough to localize the particle strictly outside the well, it will be in a superposition of energy eigenstates, some with energies greater than V. On the other hand, if the particle is in a definite eigenstate of energy with E < V, and a measurement of position is made, the particle may turn up either in the well or in the forbidden region. A genuine tunneling situation has the particle in the forbidden region with energy E < V, but quantum mechanics apparently forbids us from observing such tunneling.

Now let us consider an ensemble of particles pre-selected to an energy eigenstate and post-selected by a measurement of position. In the intermediate time we make a measurement of the kinetic energy. The measurement proceeds via an interaction Hamiltonian

$$H_{int} = -g(t)q\frac{p^2}{2m}$$

where the canonical conjugate of q is π, a pointer setting on the device. In an ideal measurement the initial and final settings would be free of error, but any realistic experiment contains errors. Thus we can take the initial state of the pointer to be

$$\phi_{in} = \exp\left(-\pi^2 / \delta^2\right)$$

introducing errors of order δ in the kinetic energy. The pre-selected state of the particle will be, by assumption, $|E_0\rangle$. The post-selected state will be a position eigenstate $|x_0\rangle$ deep inside the classically forbidden region., Realistically, this will also be a gaussian centered around x_0 with dispersion Δ. The final state of the apparatus will thus be (taking into account the interaction Hamiltonian):

$$\phi_{fin} = \left\langle x = x_o \middle| e^{i\alpha p^2/2m} \middle| E = E_o \right\rangle \phi_{in}.$$

Now comes the logic of the weak measurement. We would like to choose and x_0 so far from the potential well that the particle must have been outside the potential well before the measurement of kinetic energy. An explicit calculation shows that the probability of finding particles inside the potential well becomes negligible when

$$x_0 >> \frac{1}{n\delta\,\Delta}.$$

When the precision of the kinetic energy measurement is increased ($\delta \to 0$), the post-selected position must be ever farther out ($x_0 \to \infty$). We now ask what is the final state of the pointer p for particles satisfying this condition. Another explicit calculation shows that the final state is again a gaussian with the same spread, but with the peak at the value E - V, a negative value, Thus precisely in the case of particles found so far from the well that they could not have been there before the kinetic energy measurement we find the 'impossible' negative values for the kinetic energy.

From the point of view of standard quantum mechanics all this is a game of errors of measurement. An ideal measurement of kinetic energy can yield only positive values, since the eigenvalues of $p^2/2m$ are positive. Realistic measurements, however, have errors and thus negative values are possible. They also disturb particle positions, but the disturbance is bounded along with the accuracy of the measurement. Hence it is possible to localize a tunneling particle far enough from the well while getting negative readings for the kinetic energy. But this fantastic conspiracy of cancellations seems to be telling us, in straight forward language, that genuine tunneling can be observed if we design subtle experiments, as prescribed by a two-vector formalism involving weak measurements.

To conclude, this paper has pointed out some shortcomings of the conventional formulation of quantum mechanics, relating to Lorentz covariance of states and time symmetry of predictions and measurements.

A new approach, based on pairs of vectors evolving forwards and backwards in time, has been offered to remove these shortcomings. But the two-vector description offers much more than a reformulation of familiar results. As a formalism of recalculations, it introduces a host of unexpected and peculiar phenomena. We have seen examples of spins vectors adding classically, 'impossible' values of spin projections and observable negative kinetic energy. The new formalism paints an entirely new picture of quantum reality for us. This new picture of quantum reality is not yet fully clear to us, but we may hope it will involve a new understanding of time, since in this formalism quantum states propagate in both time directions.

REFERENCES

1. Y. Aharonov, P. G. Bergmann, and J. L. Lebowitz, Phys. Rev. 134, B1410 (1964).
2. D. Z. Albert, Y. Aharonov, and S. D'Amato, Phys. Rev. Lett. 54, 5 (1985); Y. Aharonov, D. Z. Albert, and S. D'Amato, Phys. Rev. D34, 1975(1985).
3. Y. Aharonov, D. Z. Albert, Phys. Rev. D21, 3316 (1980); D24, 359 (1981); D29, 228 (1984); Y. Aharonov, D. Z. Albert, and L. Vaidman, Phys. Rev. D34, 1805 (1986).
4. Y. Aharonov, D. Z. Albert, and L. Vaidman, Phys. Rev. Lett. 60 1351 (1988); Y. Aharonov and L. Vaidman, in Bell's Theorem, Quantum Theory, and Conceptions of the Universe (M. Kafatos, ed.), Kluwere Academic Pub., 1989, p. 17; Y. Aharonov and L. Vaidman, Phys. Rev. A41, 11 (1990).
5. Y. Aharonov, D. Z. Albert, A. Casher, and L. Vaidman, Ann. N. Y. Acad. Sci. 480, 620 (1986); Phys. Lett. A124, 199 (1987).
6. Y. Aharonov and S. Popescu, Tel-Aviv preprint (1990).

A Suggested New Translation Gauge Invariance for Space-time

David Bohm
Physics Department, Birkbeck College, University of London,
Malet Street, London, WC1E 7HX, England.

Abstract

In this paper a new concept of translation gauge invariance is developed which is significant at short distances, while the usual Lorentz invariance emerges at longer distances. This is illustrated in terms of two examples. The first of which involves a field of frames, while the second brings in spinors and the twistors of Penrose.

1. Introduction

In this paper I want to make some suggestions aimed at giving geometric insight into the meaning of gauge theories. This will lead to the possibility of extensions to new kinds of translational gauge invariance. In doing this, I shall discuss reasons why Lorentz invariance should be limited, failing at very short distances. The main discussion in this paper will be classical, since its purpose is to explore and clarify the geometric concepts involved rather than to be directly applied, for example, to particle theory.

2. The geometric picture of the meaning of gauge invariance

I shall start by giving a geometric picture of gauge invariance. To do this I shall begin with rotations. Gauge invariance to rotations imply that neighbouring regions of space can be rotated relative to each other in an arbitrary way. Geometrically this means that objects are able to rotate independently rather than, for example, having to be aligned in fixed orientations relative to each other. Clearly this possibility is necessary for space to have its commonly accepted meaning. Dynamical theories which imply this possibility will therefore have to be invariant to local rotations (extended, of course, in a relativistic theory to local Lorentz transformations).

What this means can be seen in more detail by introducng an arbitrary set of frame vectors $e^i{}_\mu(x)$. In general relativity, parallel displacement of these frames is expressed by

$$\frac{\partial e^i_\mu}{\partial x^\lambda} + \Gamma^\nu_{\mu\lambda} e^i_\nu = R^\nu_{\mu\lambda} e^i_\nu \tag{1}$$

where $R^\nu{}_{\mu\lambda}$ are the Ricci rotation coefficients, which describe infinitessimal Lorentz transformations of the vectors $e^i{}_\mu$, when one makes the requirement that $R^\nu{}_{\mu\lambda}$ correspond to a Lorentz transformation and that $\Gamma^\nu{}_{\mu\lambda}$ are torsion free. We can then solve for the $R^\nu{}_{\mu\lambda}$ and find, as is well known, that they can be expressed as a certain linear combination of the curls

$$e^i_{[\mu\nu]} = \frac{\partial e^i_\mu}{\partial x^\mu} - \frac{\partial e^i_\mu}{\partial x^\lambda} \tag{2}$$

The result is given by

$$R_{\mu\nu\lambda} = \frac{1}{2}\sum_i \eta_i \left[h^i_\lambda h^i_{[\mu\nu]} + h^i_\nu h^i_{[\mu\lambda]} + h^i_\mu h^i_{[\lambda\nu]} \right] \tag{3}$$

where η_i is the usual Lorentz pesudo-Euclidian metric.

It has already been shown that the concepts of Reimanian geometry can be applied to a crystal (Bilby et al.[1]). We can give an easily visualisable image of the meaning of geometric concepts by turning this approach around and approximating geometry by describing it through a set of micro-crystals. Later we will, in fact, take such micro-crystals as our basic model of space-time, with the aim of ultimately questioning the usual ideas of the continuity of the latter. But for the present we shall begin by developing our picture of geometry in terms of microcrystals.

To see how to do this we begin by noting that the frame vectors determine a set of four planes through the relationships

$$e^i{}_\mu dx^\mu = n^i \tag{4}$$

where n^i is an integer. Dislocations are implied whenever the $\oint e^i{}_\mu dx^\mu \neq 0$. In order for neighbouring micro-crystals to have different orientations, it is necessary to have a distribution of dislocations on the boundary between them. For example if $e^i_\mu = \lambda^i(x)\,\partial\alpha^i/\partial x^\mu$ where α^i is any function of x, then on any surface on which $\lambda^i(x)$ changes discontinuously, there will be a distribution of dislocations.

General relativity implies that neighbouring frames can be arbitrarily Lorentz transformed relatively to each other. Such a transformation clearly implies a corresponding change in the distribution of dislocations. In general, this will introduce torsion in the affine connection. Such torsion is cancelled out, however, by suitable Lorentz transformations of neighbouring frames. Equation 3 implies that it is always possible to choose Ricci rotation coefficients, $R^\nu{}_{\mu\lambda}$, corresponding to Lorentz transformations which bring about an affine connectiuon $\Gamma^\nu{}_{\mu\lambda}$ that is free of torsion. What happens is that the opening of a parallelogram due to dislocations within it is cancelled by rotating one pair of parallel lines relative to the other, so that the figure closes.

If we regard e^i_μ as the potential of a set of gauge fields, this suggests a model in which freedom of neighbouring regions to move relative to each other is brought about by dislocations. These are determined by $e^i_{[\mu\nu]}$, which are fields corresponding to the potentials. To put it pictorially, the gauge field is not only the "glue" that holds the microcrystals together, but also the "lubricant" that allows them to "slip" relative to each other, thus permitting free movement (i.e. rotation and acceleration) of objects despite the rigidity in the individual micro-structures.

If these ideas were to be regarded as a description of the actual space-time, one would be led to ask what is the size of these micro-crystals. It seems reasonable to suggest that this might be of the order of the Planck length ($\sim 10^{-33}$cms). For dimensions thus far investigated in physical experiments, the shortest distances are of the order of 10^{-16} cms. Therefore there is room for an isotropic and almost Lorentz invariant distribution of (4-dimensional) micro-crystals. We can compare this to a glass which behaves essentially isotropically for longer wave lengths, while it deviates from this for sufficiently short wave lenghts.

3. Reasons for possible limits on Lorentz Invariants

I wish now to raise an important question: is Lorentz invariance a fundamental concept for space-time? I shall give some reasons suggesting that, in fact, it is not.

If the basic gauge transformation in general relativity were a Lorentz transformation, the potentials ought to be the $\Gamma^{\nu}_{\mu\lambda,}$ (or alternatively $R^{\nu}_{\mu\lambda}$.) In this case the field would be the curvature tensor $R^{\nu}_{\mu[\lambda\alpha]}$. In analogy to other gauge theories, the field equations would be of the general form

$$R^{\nu}{}_{\mu}{}^{[\lambda\alpha]}{}_{;\,\alpha} = S^{\nu\lambda}{}_{\mu}. \tag{5}$$

The $S^{\nu\lambda}{}_{\mu}$, which would clearly be the source of the curvature field, would have to be the angular momentum tensor. But actually according to Einstein's theory, the source is the energy momentum tensor $T^{\mu\nu}$ and the field equations are given by

$$R_{\mu\nu} - {}^{1}\!/_{2}Rg_{\mu\nu} = T_{\mu\nu}. \tag{6}$$

Here it is basically a function of the <u>field</u> rather than of its exterior derivitives that is equated to the source. And even then, only a part of the field tensor enters into these equations.

All of this is very different from what happens in other gauge theories. It suggests that the fundamental gauge invariance underlying general relativity is not to the Lorentz transformation, but rather that it has to be something else. As explained in detail by P. Bergmann et al .[2] this role is played by the general transformation of coordinates

$$\delta x^{\mu} = \varepsilon^{\mu}(x) \tag{7}$$

when the above is applied to make an infinitesimal transformation of the Lagrangian, this generates the energy-momentum tensor $T^{\mu\nu}$. This is well known for the special case of

$\varepsilon^{\mu}(x) = C^{\mu}$, where C^{μ} are constants. However Bergmann showed that it still holds (though with a generalised form for $T^{\mu\nu}$) where $\varepsilon^{\mu}(x)$ are not constants). It is therefore clear that if gravity is to be included in the framework of gauge theories in general, the basic gauge transformation should be given by equation 4 and not by a Lorentz transformation. In essence this gauge transformation simply represents the possibility of shifting neighbouring objects (e.g. micro-crystals) by allowing them to be displaced relative to each other in space and time. Evidently this requirement is fundamental for space and time to have their ordinary meanings. And what we have said here suggests that this is even more fundamental than Lorentz invariance.

Hehl et al.[3] have, in fact, proposed theories in which the angular momentum tensor is part of the source of the field and have thus been considering questions that are related to those described above. However they expressly limit themselves to maintaining the concept of metric and, therefore, are not addressing the question raised here of whether the Lorentz group is fundamental. What I am doing is to suggest that it may be worth while to explore the implications of a new kind of gauge invariance, i.e. invariance to translations. This would imply that metric is not universally preserved in such a gauge theory. Indeed Bergmann already suggested that metric need not play a fundamental role in space-time geometry. He seems to have had the original Weyl theory in mind which suggested the original idea of invariance to the "gauge" or scale of length - the first notion of gauge invariance from which this concept obtained its name.

It should be noted however that general relativity already implies that Lorentz invariance is not basic on the large scale where curvature is significant. Indeed it has to be restricted to infinitesimal regions which are treated mathematically in terms of tangent spaces. However with quantum fluctuations, $g^{\mu\nu}$ and its derivatives cannot, in general, be defined. This means that it is not clear what can be the role of Lorentz transformations, not only at very long distances, but also at very short distances.

We may illustrate the difficulties involved at small distances by considering a vector of finite length, V^μ, as made up of a sum of very small elements δx_i^μ, so that

$$V^\mu = \sum_i \delta x_i^\mu \qquad (8)$$

Under a Lorentz transformation, the δx_i^μ change according to

$$d\delta x_i^\mu = \varepsilon^{\mu\nu} g_{\nu\lambda}(x)\, \delta x_i^\lambda \qquad (9)$$

Because of the large relative quantum fluctuations of the $g_{\nu\lambda}(x)$ for elements at short distances from each other, the total vector dissolves into an undefinable "fog". Therefore this suggests that there should be some limit to the applicability of Lorentz invariance at very short distances. Lorentz invariance should therefore be restricted to vectors of intermediate size large enough to contain many micro-crystals, but small enough so that curvature may be neglected.

4. The Translational Gauge

What is the meaning of the translational gauge? To see this let us first consider a micro-crystal in space-time. Within it we may have various physical structures, such as vectors, triangles, loops, and strings. Within any particular crystal parameterised by an arbitarily chosen point, x^μ, we have to describe where these structures are located relative to x^μ. To do this we bring in fields $\phi^i(x^\mu)$ where the index i refers to internal coordinates which may be projections on the axes of the micro-crystal. The $\phi^i(x^\mu)$ then locate a point in the micro-crystal relative to x^μ. A single $\phi^i(x^\mu)$merely determines an arbitrary point in the micro-crystal. A vector is determined by two points, $\phi^i_1(x^\mu)$ and $\phi^i_2(x^\mu)$, or alternatively, by $\phi_M{}^i(x^\mu)$ and$V^i(x^\mu)$ where $\phi_M{}^i(x^\mu)$ represents the line from point x^μ to the mid point of the vector $V^i(x^\mu)$. (The specification of a vector along with its moment, is a simplification of the description.) A more general structure may be described by a whole set of points a, b, c, . . . whose locations relative to x^μ are given by $\phi^i_a, \phi^i_b, \phi^i_c$ . . . . (This set may be infinite.)

This approach may be regarded as an extension of tangent spaces as used in differential geometry. Each micro-crystal is in fact the local tangent space. A good image of what is meant here can be obtained by considering a polygonal space which is made up of a very large number of flat elements (where these elements meet, there will, of course, be dislocations.) The point x^μ corresponds to the point of support of the tangent space. But because the tangent space is finite rather than infintesimal, there is a certain arbitrariness in the location of this point. Clearly the physical laws should not depend on the relationship between the physical structures of interest and this point. In this way we are led to require invariance to an arbitrary translation $\lambda^i(x)$ which may, of course, be different for different micro-crystals specified by different values of x^μ. (But, of course, the translation has to be the same for each point ϕ^i_a of the structure in question).

The $\phi^i_a(x)$, $\phi^i_b(x)$, $\phi^i_c(x)$ may be regarded as a set of fields in this tangent space located at x^μ. For example a string would be described by an infinity of $\phi^i_a(x)$). But these fields now have a new kind of gauge invariance, for the physical state must be unchanged under the transformation.

$$\phi^i(x) \longrightarrow \phi^i(x) + \lambda^i(x) \tag{10}$$

If, for example, the ϕ^i_a satisfy a dynamical equation, their derivatives must also appear in the Lagrangian. This means that under the gauge transformation, the $\phi^i_a(x^\mu)$ undergo the change

$$\frac{\partial\phi^i_a(x)}{\partial x^\mu} \longrightarrow \frac{\partial\phi^i_a(x)}{\partial x^\mu} + \frac{\partial\lambda^i(x)}{\partial x^\mu} \tag{11}$$

To make a gauge invariant theory, we now have to assume a new kind of gauge potential which I shall call $A^i_\mu(x)$. The covariant derivative is then

$$D_\mu \phi^i_a(x) = \frac{\partial \phi^i_a(x)}{\partial x^\mu} + A^i_\mu(x) \tag{12}$$

Under a gauge transformation, the A^i_μ transform as

$$A'^i_\mu = A^i_\mu + \frac{\partial \lambda^i}{\partial x^\mu} \tag{13}$$

The A^i_μ may represent a new set of frames, in the sense that $A^i_\mu(x)\, dx^\mu = n_i$ defines a set of planes in each micro-crystal. The circuit integral of $\oint A^i_\mu dx^\mu$ will then respresent the change of position of these planes in going around a circuit and will thus correspond to a set of dislocations. It is the A^i_μ which will correspond to actual structures in space-time rather than being merely descriptive forms.

As with other gauge fields, the A^i_μ will also satisfy field equations. Their solutions will correspond to something physically unique except for the possibility of a translation along the axis resulting from the intersection of these planes. For example, consider a set of wave-like movements interfering to form surfaces of constant phase. With four of these, we have a possible frame. The basic idea is that some such process is going on in each region of space-time. It is this process that determines the physically relevant micro-crystal. I emphasise again that this micro-crystal is not an arbitrary descriptive form but rather corresponds to genuine physical state. The fields

$$F^i_{[\mu\nu]}(x) = \frac{\partial A^i_\mu}{\partial x^\nu} - \frac{\partial A^i_\nu}{\partial x^\mu} \tag{14}$$

then corresponds to the dislocations in this physically unique micro-crystal. And, as I have suggested already, the equations satisfied by these fields, along with their boundary conditions, will determine the A^i_μ up to an arbitrary relative translation $\partial \lambda^i(x)/\partial x^\mu$ of neighbouring elements.

Everything happening in a given neighbourhood may then be refered to the local frames A^i_μ which constitute a set of natural frames for the description of physical processes in this neighbourhood. Indeed the frames, A^i_μ, may be regarded as local absolute frames which are determined dynamically, however, by the field equations. What then is the role of Lorentz invariance? Let us recall that a Lorentz transformation of one frame relative to its neighbour is brought about by a set of dislocations. Clearly what happens in each frame need not, in general, be Lorentz invariant. But on a larger scale we have, as has been suggested, a very large number of micro-crystals, more or less randomly orientated. So on this scale it would be possible to have distributions of dislocations which allow for an effective Lorentz invariance (as in a glass we get effectively isotropic behaviour in the large scale).

We will then have a theory that is not Lorentz invariant e.g. on a small scale of the order of the Planck length. Lorentz invariance will arise as a kind of average. For example, consider

$$\oint_C A^i_\mu dx^\mu = n^i \tag{15}$$

which holds for an arbitrary curcuit. We then define an average value, $\bar{A}^i_\mu$,over a region containing many micro-crystals. We assume over a large circuit

$$\oint A^i_\mu dx^\mu = \oint \bar{A}^i_\mu dx^\mu \tag{16}$$

We can then define an average metric

$$g_{\mu\nu} = \sum_i \eta_i \bar{A}^i_\mu \bar{A}^i_\mu \tag{17}$$

and we can require approximate Lorentz invariance in terms of this metric even though in terms of the A^i_μ there is not even approximate Lorentz invariance. What this means is that I am assuming a kind of collective behaviour of the dislocations so that they restrict themselves to combinations giving rise to large scale statistical Lorentz invariance. The fundamental gauge-group is, however, translational, whereas the Lorentz group comes out from a particular distribution of the micro-crystals.

One can get some idea of what is meant here by considering how classical hydrodynamics comes out of a theory of the constituent particles of the fluid (which, of course, obey the laws of quantum theory). In a region large enougth to contain many particles, new macroscopic properties emerge such as density, local fluid velocity, etc, which will have certain invariance groups. These invariance groups are not , however, in any simple and direct relationship to those of the underlying constituent particles. To try to extrapolate the large scale classical properties by "quantizing" the equations of hydrodynamics will, clearly at best, be of limited validity and will not disclose the underlying structure. What I am suggesting is that to quantise gravitation as it is known on the large scale, may similarly not be the way to understanding the underlying structure on the small scale.

A related idea has in fact been considered by T.D. Lee [4] who has proposed a disordered underlying structure. Also M. Green [5] has said that a similar notion has been considered in string threory which begins with simple pseudo-Euclidean geometry which has string excitations of spin 2. These are connected with the metric by assuming some sort of collective average background with which the local oscillations of the string must fit. It is this background that is, in essence, the local gravitational field.

5. An Illustrative Example

I shall now illustrate these ideas in terms of a theory which treats the distribution of dislocations as continuous. The basic aim is to give a more mathematical example in which Lorentz invariance emerges in the large, while in the small, there is something like a field of dislocations in which there will be translational invariance, but no preservation of metric under parallel displacement. That is to say the general affine connection $\Gamma_{\mu\lambda}{}^{\nu}$ will not be equal to the Christoffel affinity $\{_{\mu}{}^{\nu}{}_{\lambda}\}$.

The general approach is to begin by assuming that there is in the Lagrangian, a term that tends to line up neighbouring frames so as to form something like a micro-crystal (though because of the continuity of the theory, it will only be approximately so). Such a term, for example, might be $\sum_i A^i{}_{[\mu\nu]}A^{i[\mu\nu]}$. This would tend to discourage situations in which neighbours had very different orientations (which would, of course, imply correspondingly strong dislocations at the boundary). In general, however, such terms would not be invariant to a local Lorentz transformation, so that they would not preserve metric under parallel displacement. Likewise one would expect that these terms would, in general, imply torsion. In accordance with the approach that has been descirbed here, there is no intrinsic reason to object to torsion and lack of preservation of metric in the small. However we must now add further terms to the Lagrangian which will elminate these in the large.

To do this we can begin by considering a special affine connection $\Gamma_{o\,\mu\lambda}{}^{\nu}$ which is such as to make the neighbouring frame vectors all parallel.

$$\frac{\partial A^i_\mu}{\partial x^\lambda} + \underset{o}{\Gamma}{}^{\nu}_{\mu\lambda} A^i_\nu = 0 \tag{18}$$

so that

$$\underset{o}{\Gamma}{}^{\nu}_{\mu\lambda} = -A^\nu_i \frac{\partial A^i_\mu}{\partial x^\lambda} \tag{19}$$

We will add to this affine connection, a tensor

$$T^\nu_{\mu\lambda} = T^\nu_{(\mu\lambda)} + T^\nu_{[\mu\lambda]} \tag{20}$$

This will give us a general connection

$$\Gamma^\nu_{\mu\lambda} = T^\nu_{(\mu\lambda)} + T^\nu_{[\mu\lambda]} - A^\nu_i \frac{\partial A^i_\mu}{\partial x^\lambda} \tag{21}$$

The net torsion tensor is

$$S^\lambda_{[\mu\nu]} = T^\lambda_{[\mu\nu]} - \frac{1}{2} A^\lambda_i \left[\frac{\partial A^i_\mu}{\partial x^\nu} - \frac{\partial A^i_\nu}{\partial x^\mu} \right] \tag{22}$$

We therefore introduce a term in the Lagrangian $aS^\lambda{}_{[\mu\nu]}S^{[\mu\nu]}{}_\lambda$, where a is a suitable coefficient. This term will tend to oppose large values of torsion. But since S has to be dynamic, we will need to introduce terms involving its derivatives into the Lagrangian. For example, such a term could be

$$D_\alpha T^\lambda_{[\mu\nu]} D^\alpha T^{[\mu\nu]}_\lambda \tag{23}$$

where D_α is the covariant derivative with the affine connection $\Gamma^\nu_{\mu\lambda}$.

The combination of these two terms will bring about the possibility of waves in which $T^{\lambda}{}_{[\mu\nu]}$ oscillates with a frequency dependent on a. The suggestion is to choose a large enough so that the minimum frequency will be of the order of the Planck time. For much lower frequencies, it is easily seen that this oscillation of $T^{\lambda}{}_{[\mu\nu]}$ will be such as to make $S^{\lambda}{}_{[\mu\nu]}$ very small. In this way we satisfy our requirement that effects of torsion are only significant near the Planck length and below.

However the term in the Lagranian $aS^{\lambda}{}_{[\mu\nu]}S^{[\mu\nu]}{}_{\lambda}$ will contain factors like

$$T^{\lambda}_{[\mu\nu]} - \frac{1}{2}A^{\lambda}_{i}\left[\frac{\partial A^{i}_{\mu}}{\partial x^{\nu}} - \frac{\partial A^{i}_{\nu}}{\partial x^{\mu}}\right] \tag{24}$$

[see equation 22.] But at higher frequencies it is clear that this term will introduce contributions like $\sum\eta_i A^i{}_{[\mu\nu]}A^{i[\mu\nu]}$ into the Lagrangian which will provide the tendency that we have mentioned for neighbouring frames to line up.

However the general affine connection that we have used still implies that metric is not preserved in a parallel displacement. That is to say the covariant derivature of $g_{\mu\nu} = \sum\eta_i A^i{}_{\mu}A^i{}_{\nu}$ does not approach zero in the large. To achieve this, we add yet another term to the Lagrangian which is

$$bD_{\lambda}g_{\mu\nu}D^{\lambda}g^{\mu\nu} \tag{25}$$

where b is a suitable coeffcient. This term will tend to oppose changes of metric in parallel displacement.

Because these fields are dynamic, we also have to add terms involving derivatives. These may be taken as $D_{\alpha}T^{\lambda}{}_{(\mu\nu)}D^{\alpha}T_{\lambda}{}^{(\mu\nu)}$. Since the covariant derivative of $g_{\mu\nu}$, which appears in equation 25, contains $T^{\lambda}{}_{[\mu\nu]}$, it follows that $T^{\lambda}{}_{[\mu\nu]}$ will oscillate. If we choose the coefficient b large enough so that the oscillation will have a period of the order to the Planck time, this will guarantee that for much lower frequencies the covariant derivative of $g_{\mu\nu}$ will also be small. Thus metric will be preserved approximately in parallel displacement in the large and the affine connection will reduce to its usual torsion-free form which implies a simple Lorentz transformation given by equations 1 and 3.

Of course to all these terms might be added the usual Lagrangian of the Einstein theory, which is proportional to the curvature scalar R. The energy momentum tensor is then obtained from an infintesmimal translation in the usual way as extended by P. Bergmann[2]. This tensor will contain, not only the frame fields $A^i{}_{\mu}$, but also the tensors $T^{\lambda}{}_{[\mu\nu]}$ and $T^{\lambda}{}_{(\mu\nu)}$ which contribute to the affine connection. These latter will constitute further fields of spin up to three, but their minimum frequencies will be so high that they will not show up at the scales of length that have thus far been investigated in phyiscs.

In this way it is possible to construct a theory in which the basic structure at small distances is characterised by a translational gauge, while the usual Einstein theory is obtained in the large.

6. Additonal illustration in terms of spinors

Thus far I have said very little about the physical structures within the micro-crystals. I shall now develop an illustrative example in which these physical structures are light waves within each micro-crystal (along the lines already suggested earlier.) The purpose is to show how translational invariance can be developed for such structures.

It is natural to discuss light waves (and rays) in terms of spinors. Given an arbitrary two-component spinor, ψ, we can define a light ray through the relationship

$$V^n = \Psi^* \sigma^n \Psi$$

where $\sigma^n = (1,\sigma)$ and where $n = 0, 1, 2, 3$. Of course we have in this example an internal pseudo-Euclidian metric in each micro-crystal. As is well known a pair of such rays given by two spinors, ψ^1 and ψ^2, determines a tetrad and therefore a Lorentz frame. Let us now consider spinors χ_1 and χ_2 defined by

$$\chi_1 = (\sigma_n Y^n)\psi_1, \quad \text{and} \quad \chi_2 = (\sigma_n Y^n)\psi_2$$

where Y^n is a four-vector within the frame of the micro-crystal. A pair (χ,ψ) form what Penrose calls a twistor [6]. We shall use this idea here but we will not use the full twistor theory. Consider then

$$\Psi^*{}_1\chi_1 = \Psi^*{}_1(\sigma_n Y^n)\chi_1 = \Phi_{11}$$

This tells us how many null hyper-planes of the first null-ray are crossed by the vector Y^n. Similarly for

$$\Psi^*{}_2\chi_2 = \Psi^*{}_2(\sigma_n Y^n)\chi_2 = \Phi_{22}$$

which tells how many null hyper-planes of the second null ray are crossed by the vector Y^n. Consider also

$$\Psi^*{}_1\chi_2 = \Psi^*{}_1(\sigma_n Y^n)\chi_2 = \Phi_{12}$$

(noting that $\Phi_{12} = \Phi^*{}_{21}$)

To see what this means, we transform to the special Lorentz frame in which the spinors are proportional to

$$\begin{bmatrix} 1 \\ 0 \end{bmatrix} \quad \text{and} \quad \begin{bmatrix} 0 \\ 1 \end{bmatrix}$$

Φ_{12} then gives us the projection of $Y^1 + iY^2$ of the vector Y^n into the hyper-planes normal to the light rays. The quantities Φ_{11}, Φ_{22}, Φ_{12} and Φ_{21}, can be represented by a matrix Φ_{AB}. They constitute a spinor expression of the vector Y^n projected into the local frame defined by the two spinors.

Let us go back to the description of structures given in section 3 in terms of vectors ϕ^i_a. It is clear that we can equally well describe these in terms of spinor structures Φ_{ABa}, along with the common spinor matrix formed from the components of the spinors ψ_1 and ψ_2, namely

$$\begin{bmatrix} \psi_{11} & \psi_{12} \\ \psi_{21} & \psi_{22} \end{bmatrix}$$

A translation on χ gives rise to a change

$$\delta\chi = \varepsilon = (\sigma_n \Delta Y^n)\Psi$$

where ε is a spinor. We then have

$$\chi \to \chi + \varepsilon$$

$$\frac{\partial \chi}{\partial x^\mu} \to \frac{\partial \chi}{\partial x^\mu} + \frac{\partial \varepsilon}{\partial x^\mu}$$

This suggests a further new gauge field Λ^μ which acts on the twistor. The covariant derivative would then be

$$D_\mu \chi = \frac{\partial \chi}{\partial x^\mu} + \Lambda_\mu$$

From this it follows that

$$D_\mu \Phi_{AB} = \left(D_\mu \psi^*_A\right)\chi_B + \psi^*_A\left(D_\mu \chi_B\right)$$

The Λ_μ would correspond to a field of spin three-half and is therefore reminscent of the Rarita-Schwinger theory [7], (though of course it should be recalled that the theory given here is classical).

As has already been pointed out, the Φ_{AB} are internal coordinates within a micro-crystal. They may be included in the Lagrangian, as may also their canonical conjugates Π_{AB}. We have an internal rotation operator, τ_{AB}, representing rotation around the internal axies and this commutes with the external rotations. In a quantized theory it could, for example, represent isospin. We could include terms like $\Pi_{AB}\tau^{AB}$ which are similar to those used by Skyme[8] to describe pions and kons. Other possibilities exist along these lines for further properties if we include a structure with many points (a special case of which is a string). It is essential for this approach, however, that the internal frames be determined by the dynamics of the theory in the way described earlier, rather than being arbitrary as in the usual approach to general relativity. For unless the frames are well defined, there is no meaning to attributing physical significance to properties that are projected into them.

Finally, of course, this has been thus far all classical. To make a quantum theory, we should first consider the wave function of a given micro-crystal.

$$\Psi(\ \Phi_{AB1}, \Phi_{AB2}, \ ....,A^{\mu}i,....)$$

The general wave functional involves all these fields in all the micro-crystals. In terms of such a wave function, we can quantise in the usual way. It may, however, be further remarked that since we do not require complete Lorentz invariance at the micro-level, we could bring about convergence of the resulting field theories by simply introducing a "cut off" in the rest frame of each micro-crystal. Becaue there is very nearly a random distribution of micro-crystals, it follows along the lines that have been discussed earlier, that the overall theory could be Lorentz invariant to a very high degree of approximation.

ᵗ. Conclusion

I have indicated several lines along which one could develop a new concept of ·anslation gauge invariance that would be important at short distances, while the usual orentz gauge invariance would emerge at longer distances. It seems worth while to :plore the meanings of such possible extensions of the notion of gauge invariance, pecially if we want to understand geometry from a quantum base.

7. Acknowledgments

I wish to thank Dr. B. Hiley for invaluable discussions and for further help preparation of this paper.

8. References

1. B.A. Bilby, R. Bullough and E. Smith. Proc. Roy. Soc. A231 263 (1955)
2. P.G. Bergmann Phys. Rev. 75 680 (1949)
3. F.W. Hehl, P.von der Heyde and G.D. Kerlick. Rev. Mod. Phys. 48 393
4. T.D. Lee, in How Far we from the Gauge Forces. Ed by A. Zichichi (Plen New York, 1984) p.15
5. M. Green. Private communication.
6. R. Penrose. J. Math. Phys. 8 345 (1967).
7. W. Rarita and J. Schwinger Phys. Rev. 60 61 (1941).
8. T.H.R. Skyme, Proc. Roy. Soc. A260 127 (1961).

FUNDAMENTAL ASPECTS OF QUANTUM THEORY RELATED TO THE PROBLEM OF QUANTIZING BLACK HOLES

GERARD 'T HOOFT

Institute for Theoretical Physics
Princetonplein 5, P.O. Box 80.006
3508 TA UTRECHT, The Netherlands

ABSTRACT

It is shown that the scattering amplitudes for elementary particles against a black hole can be deduced to some extent from the mere assumption that such amplitudes should exist and be in agreement with low energy field theory and general relativity. It automatically follows that the underlying field theory has a stringlike structure, albeit an unconventional one.

The fact that the black hole's entropy is finite implies that it has a discrete spectrum and this in turn suggests a discrete structure of Hilbert space, not only for the black hole but also for the field theory as a whole. Such a fundamental discreteness of Hilbert space allows a new interpretation of quantum mechanics that we call "evolutionary determinism": the local laws of physics could be deterministic while superposition of states and wave functions is still allowed and indeed necessary for deriving the low energy limit of these laws to obtain something resembling the standard model.

1. The Black Hole Spectrum

The standard Kerr-Newman solution[1] describing a black hole depends on three parameters, the mass M, charge Q and angular momentum L. All three are in principle continuous, but a particle physicist could convincingly argue that Q and L must be quantized. More speculatively one could suspect that also M is not continuous. Now, as is well-known[2], a black hole is expected to emit radiation. Therefore, once it is in a state (M, Q, L), it will not stay there very long. It is not difficult to estimate that because of Hawking radiation the masses M will be complex:

$$M^2 \Rightarrow M^2 - iB \quad , \tag{1}$$

where B is of the order of the Planck mass squared for all black holes.

But how dense is the spectrum of values (1)? The level density $\rho(M,Q,L)$ can be estimated the following way[3], where for simplicity we ignore the dependence on Q and L. Consider the reversible process where a black hole of mass M absorbs a particle with energy δM, becoming a black hole with mass $M + \delta M$. The absorption cross section σ is approximately

$$\sigma \cong \pi R^2 \ , \tag{2}$$

where R is the black hole radius, $R = 2M$. Conversely, a black hole with mass $M + \delta M$ can emit the same particle as Hawking radiation. The emission probability W is approximately

$$W \simeq \pi R^2 \ \rho_1(\delta M) \ e^{-\beta_H \delta M} \ , \tag{3}$$

where β_H is the inverse Hawking temperature, $\beta_H = 8\pi M$ (putting the gravitational constant κ and Boltzmann's constant k_B equal to one). $\rho_1(\delta M)$ is the density of states for the objects radiated out with energy δM.

If there were a quantum mechanical theory for the black hole, the same quantities could be expressed in terms of transition amplitudes, using the "golden rule":

$$\sigma = |\langle M+\delta M|\mathcal{T}|M,\delta M\rangle|^2 \rho(M+\delta M) \ ; \tag{4}$$

where $\mathcal{T}$ is the transition matrix, and

$$W = |\langle M,\delta M|\mathcal{T}|M+\delta M\rangle|^2 \rho(M)\rho_1(\delta M) \ . \tag{5}$$

By virtue of PCT invariance, the matrix elements in (4) and (5) should be each other's conjugates, and therefore we find

$$\rho(M+\delta M)/\rho(M) \ = \ \rho_1(\delta M) \ \sigma/W \ = \ e^{\beta_H \delta M} \ ; \tag{6}$$

this should hold for a range of values for δM as long as $\delta M \ll M$, and with $\beta_H = 8\pi M$ we find

$$\rho(M) = C \ e^{4\pi M^2} \ , \tag{7}$$

where the universal constant C is the only unknown. Note that the exponent is one quarter of the area of the horizon; this is what one also finds in more general cases.

C could be finite*, in which case we indeed have a finite spectrum density, or C is infinite, but in this case equations (4) and (5) could be considered at the lower end of the continuum. Let $|M+\delta M\rangle$ be in the continuum but $|M\rangle$ one of the discrete states directly underneath. Either $|M+\delta M\rangle$ would be an absolutely stable thing ($W=0$), in which case *virtual pair creation* of these things would give infinite contributions to graviton self-energy diagrams, or the cross section σ for collisions against $|M\rangle$ would tend to infinity. Neither of these latter options sound physically very attractive, which is why we suspect C to be finite. We must realize however that very large numbers are indigenous in quantum gravity. It could be that C is of the order of 10^{40} or 10^{-40}.

2. The Gravitational Back Reaction

Although Hawking radiation of black holes naturally leads to a picture of a discrete black hole spectrum, one does not find a discrete spectrum when one attempts to derive it more explicitly from Field Theory in the black hole background, in contrast with for instance the spectrum of magnetic monopoles that one could derive this way. The reason why one finds a continuous spectrum for black holes is technically that in any state of Hilbert space there seems to be no relation (in the form of some boundary condition at the horizon) between waves of incoming particles and waves of outgoing particles.

This cannot be right. If black holes can be seen as any form of matter there should be an S -matrix relating out states to in states. Now however we make the fundamental observation that it is possible to *assume* that an S -matrix *does* exist. To treat the black hole as a static background and subsequently rely on background field theory *ignoring gravitational interactions between in- and outgoing particles* may well be fundamentally incorrect, or at least imprecise. For any given in state $|in\rangle$, Hawking's derivation gives a *probabilistic distribution* of out states:

$$|in\rangle \Rightarrow \{|out\rangle, W_{out}\} \quad , \tag{8}$$

where W_{out} is the probability to find the state $|out\rangle$. Hawking's original interpretation of this result was that pure states may make transitions into mixed states, but it is more likely that the probabilities must be interpreted as "the probability that the calculation was right"; the calculation did not yield one single Hamiltonian for the black hole, but a probabilistic distribution of Hamiltonians. The true Hamiltonian is just one of these.

Can we devise a formalism that gives one single S-matrix, or one single Hamiltonian, in stead of a probabilistic distribution? We claim that it exists[4]. First, one must understand how infalling matter interacts gravitationally with outgoing Hawking radiation.

---

*It is unlikely that C is exactly constant. There may well be subdominant corrections either in the exponent, or in the form of powers in front of the exponent.

Consider a representation of the Schwartzschild solution in terms of its Kruskal coordinates[5]:

$$xy = -(r/2M - 1)e^{r/2M} \quad ; \qquad (9)$$

$$x/y = e^{(t-t_0)/2M} \quad , \qquad (10)$$

where t_0 is some arbitrary reference time. In these coordinates space-time is regular at the future and past horizons, which are given by $y = 0$ and $x = 0$, respectively. A boost in time t corresponds to a Lorentz transformation in x and y around the origin $(x=y=0)$. Suppose a particle falls in around the time t_0, and a Hawking particle leaves the hole around time $t \gg t_0$. We see that at the horizon these two particles cross each other with a tremendous center-of-mass energy, given by the Lorentz boost factor $\exp((t-t_0)/M)$. Because of this large relative energy the exchange of gravitational interactions may not be ignored.

This interaction is easy to compute in the limit where we neglect the particles' rest masses. An infalling particle causes a *shift* in the metric: two ordinary Schwartzschild metrics are glued together along the surface $x = 0$ with a relative shift δy in the y coordinate[6], which depends on the angles $\Omega = (\vartheta, \varphi)$. This is not unlike a "sonic boom" or Čerenkov radiation. The functional dependence of δy on Ω is determined by the equation

$$(1-\Delta_\Omega)\ \delta y(\Omega) = 4\pi\kappa p_{in}\ \delta^2(\Omega,\Omega_1) \quad , \qquad (11)$$

where Δ_Ω is the angular Laplacian and Ω_1 is the set of angles at which the particle dropped in, with momentum p_{in} with respect to the Kruskal coordinates.

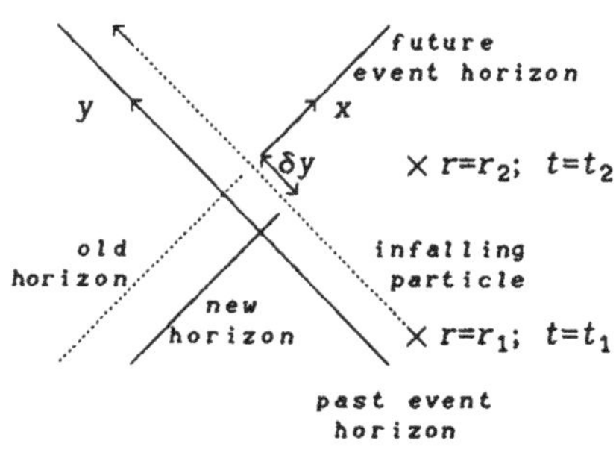

Fig. 1. The horizon displacement.

The solution of Eq. (11) can be written as

$$\delta y(\Omega) = f(\Omega, \Omega_1)\, p_{in} \quad , \tag{12}$$

where $f(\Omega,\Omega_1)$ is a Green function. It is uniquely determined by Eq. (11) because $1-\Delta_\Omega$ has a unique inverse.

This solution of Einstein's equations is found by writing the space-time metric obtained from gluing the two Schwarzschild solutions together as an *Ansatz* and then imposing Einstein's equations. The Green function f can be given in an integral form but this is not very illuminating. One finds that $f(\Omega,\Omega_1) > 0$ for all Ω, Ω_1, and f diverges logarithmically when $\Omega \to \Omega_1$.

A consequence of these observations is that if we drop a particle into the black hole, the position of the horizon at times t *before* the particle fell in, changes, as drawn in Fig. 1. This change is barely perceptible at times $t \lesssim t_1$, but at times $t_2 \gg t_1$ the change is large. An observer there sees Hawking radiation that now originated in a *different* region of space-time than it would have if the first particle had not been thrown in.

Is this consequence of any importance? What does it matter if the Hawking radiation originated somewhere else? It will certainly look the same as before.

We will argue in the next section that only in a quantum theory that is detailed enough to give us a scattering matrix instead of a density matrix, shifting horizons will be relevant. Indeed, important constraints on the scattering matrix will be found. As was argued in ref[7] the no-hair theorem is *not* true for the quantum black hole. In ref[7] it was found that finite, discrete quantum numbers may exist for black holes; we will go one step further: we expect one 'quantum hair follicle' (one Boolian degree of freedom) per Planckian unit of surface on the horizon; this would give the right spectral density.

3. Constructing the S-matrix

Suppose a scattering matrix exists. This means that if we have completely specified the state $\{p_1, p_2, \dots\}$ of all particles that ever went into the black hole, the outgoing matter should be in one well-specified state $|\psi\rangle_{out}$. A basis for $|\psi\rangle_{out}$ is the set of states where all outgoing particles have well-specified momenta at a certain time $t=t_0$. Now at $t=t_1>t_0$ we drop a light particle into the hole, with momentum p_{in} (in Kruskal coordinates) at solid angle Ω_1. The change this induces for the outgoing waves is now determined primarily by the horizon shift (if other, non-gravitational interactions may be ignored). Thus, the new state will now be

$$|\psi\rangle_{out} \to e^{-i\int P_{out}(\Omega)\delta y(\Omega) d^3\Omega}\, |\psi\rangle_{out} \quad , \tag{13}$$

where $P_{out}(\Omega)$ is the operator that generates a shift in the configurations at the solid angle Ω. It is, of course, also the total momentum emerging at solid angle Ω. In here, we can now substitute

Eq. (12) for δy .

Now this means that if we know $|\psi\rangle_{out}$ at one stage, then $|\psi\rangle_{out}$ can, in principle, be determined after allowing any number of particles to fall in. If we may ignore non-gravitational interactions, we see that all states $|\psi\rangle_{out}$ ever to be produced by the black hole are generated by the operator $P_{out}(\Omega)$ from one single state. Therefore, $|\psi\rangle_{out}$ must be generated by the *algebra* of these operators. Similarly, the *ingoing* particles are only distinguished by the *total* momentum $p_1(\Omega_1)$ at each solid angle Ω_1 .

We find the following important result[4,8]. For the incoming wave functions one may diagonalize the operators $P_{in}(\Omega_1) = p_{in}(\Omega_1)$, and for the outgoing states we diagonalize $P_{out}(\Omega)$. Eq. (13) then tells us how a change in p_{in} affects the outgoing state. Up to a proportionality factor, the complete transformation rule for ingoing states into outgoing states should be generated by this equation. This rule is not difficult to find:

$$\langle\{p_{out}(\Omega)\}|\{p_{in}(\Omega')\}\rangle = N\, e^{-i\int d^2\Omega\, d^2\Omega'\, p_{out}(\Omega)f(\Omega,\Omega')p_{in}(\Omega')} , \tag{14}$$

where N is a normalization factor†.

Eq. (14) is the *S*-matrix we wanted. *If* an *S*-matrix exists, and if we may ignore other than the longitudinal gravitational forces, it must be this one. It can be rewritten as (neglecting the 1 and the $4\pi\kappa$ in Eq. (11)),

$$\langle out|in\rangle = \int \mathcal{D}x^+(\Omega)\mathcal{D}x^-(\Omega)e^{i\int d^2\Omega(-\partial_\Omega x^+\partial_\Omega x^- + p_{out}x^- - p_{in}x^+)} . \tag{15}$$

It is here that we see a strikingly close resemblance to string theory[4]. As in string theory, Eq. (15) should be universal. The amplitude is nearly the same as the one used as a starting point in string theory; only the string coupling constant comes out being imaginary[4,8]. If we replace the usual string amplitudes (for which after all no direct physical motivation can be found) by Eq. (15) or possible refinements[8] of Eq. (15), then the *spectrum of massless states at the zero-slope limit* will remain the same. Thus we imagine that the qualitative successes of string phenomenology can also be attributed to this amplitude.

However, our amplitudes were directly motivated by consistency requirements for black holes, and in our implementation of these requirements a number of approximations were made. And it is not hard to argue that Eq. (15) cannot be exactly correct.

4. Horizon operator algebra

The problem with it is the algebra that generated the basis in which it is defined. We have the following commutation rules

---

†If other properties of the in- and outgoing particles are taken into account besides their momenta, (e.g. electric charge) then N becomes a unitary matrix. In the case of electric charge this matrix represents the contribution of a fifth, compactified, dimension[8].

$$[p_{in}(\Omega), p_{in}(\Omega')] = 0 \quad ; \quad [p_{in}(\Omega),\ x_{in}(\Omega')] = -i\delta^2(\Omega,\Omega') \quad ;$$
$$[p_{out}(\Omega), p_{out}(\Omega')] = 0 \ ; \ [p_{out}(\Omega),\ y_{out}(\Omega')] = -i\delta^2(\Omega,\Omega') \quad , \tag{16}$$

and we have the relation

$$y_{out}(\Omega) \quad = \quad \int d^2\Omega_1 \ f(\Omega,\Omega_1) \ p_{in}(\Omega_1) \quad . \tag{17}$$

This implies

$$[x_{in}(\Omega), y_{out}(\Omega')] \quad = \quad if(\Omega,\Omega') \quad , \tag{18}$$

so that we have also

$$x_{in}(\Omega) \quad = \quad -\int d^2\Omega' \ f(\Omega,\Omega') \ p_{out}(\Omega') \quad . \tag{19}$$

The operators x_{in} and y_{out} could be interpreted as coordinates of "particles", but then there should be *exactly one particle at every value of* Ω . This is where this Hilbert space differs from ordinary Fock space, where we may have any number of particles (mostly this will be zero) at every mode. This is also why it will be difficult to interpret our *S*-matrix directly as a matrix describing scattering of familiar particles.

But in spite of the unusual way in which the dynamical variables are represented we do believe that this description of the Hilbert space surrounding the black hole can be defended. Imagine a lattice-like cut-off on the horizon, where the lattice length is of the order of the Planck length. At every lattice site Ω there is exactly one particle. This leaves us more than enough particles to reconstruct ordinary Hilbert space. Ordinary particle physics is at the low-energy limit, where we never need to know what happens when two or more particles sit at exactly the same site Ω .

Thus, on the one hand we have ordinary particles, but alternatively, x_{in} and y_{out} can be seen as the position operators for the *past* and the *future* horizon. We then recognise an important consequence of our description of Hilbert space: past and future horizons cannot both be localized accurately; these obey an uncertainty relation. Indeed, they are each other's dual conjungates, much in the same way as coordinates are dual to momenta.

We claim that this also does away with a question considered often in the literature: does the time-reversed black hole (called "white hole") exist? Does the "eternal black hole" (one with a past white hole that was already there before the universe began) exist? Our discovery is that these questions are not appropriate if our *S*-matrix exists: white hole coordinates are ill-specified once a black hole was localised.

The difficult but perhaps exciting picture that emerges is that the exact shape of either the past or the future horizon may completely determine the particle content of the black hole's vicinity.

It should be possible to refine this picture by incorporating gravitational forces in the transverse direction, and non-gravitational forces. It is not hard to take the electromagnetic force into account. Here, the electric charge density operator $\rho(\Omega)$ and the gauge phase operator $\phi(\Omega)$ are each other's duals. Electromagnetic shock waves (Čerenkov radiation) surrounding charged massless particles are very similar to gravitational shock waves[9].

As is discussed further in Ref.[8], we expect a cut-off in Ω space. Probably the transverse forces are responsible for that. Surely, if coordinates in the transverse direction would be specified with accuracies better than the Planck length (implying $\delta\Omega \lesssim M_{Pl}/M$, where M is the black hole mass), then momenta in the transverse direction exceed the Planck energy so that also shifts in the transverse direction will arise that are bigger than $\delta\Omega$. The simplest cut-off would be a lattice in Ω space, but reality will be more complicated. What we expect actually to happen is that the Hilbert space algebra itself will produce a cut-off. The operator algebra is worked out further in Ref.[8].

We should not expect to get discrete representations of this algebra directly, because our Hilbert space still includes particles at infinite distances from the black hole, which certainly form a continuum[♪]. But if we erect a laboratory wall at several black hole radii then the bulk of all entropy resides near the hole, and the total number of states then *is* determined by the hole's surface: one flip-flop degree of freedom par Planckian unit of surface area. The necessity of a laboratory wall presumably implies that we should impose further boundary restrictions at infinity. This may explain why we actually do find a discrete representation, which however is not unitary[8].

Let us summarise what was found in Ref.[8]. The algebra is replaced by a Lorentz covariant algebra, assuming that this way also gravitational shifts in the transverse direction will be incorporated. One then discovers representations of this algebra as follows. On the black hole's horizon surface area we imagine a lattice built from surface elements Σ_i , $i=1, \ldots , N$. Any such lattice corresponds to a representation. At each surface element Σ_i a 3-vector operator $L^a{}_{(i)}$, $a=1,2,3$, is defined, satisfying the commutation rules of angular momenta:

$$[L^a{}_{(i)} \ , \ L^b{}_{(j)}] = i\delta_{ij}\varepsilon^{abc}L^c{}_{(i)} \quad . \tag{20}$$

Further,

$$L_{(i)}{}^2 = \tfrac{1}{2} \ \text{ or } \ 1 \quad , \tag{21}$$

to guarantee that they represent minimal surface elements. Angular momenta corresponding to larger agglomerations of surface segments are obtained by adding the corresponding L operators in the usual way. A representation would then be given by the set of numbers

$$|\{m_{(i)}\}\rangle \quad , \tag{22}$$

which of course is discrete.

Unfortunately, because the operators $L^a{}_{(i)}$, from their definition, are not Hermitean, the states (22) are not orthonormal.

[♪] I thank S. Coleman for a discussion on this point.

5. Discrete Quantum Mechanics

It is very interesting to study the quantum mechanics of systems that are entirely discrete such as the sets of numbers (22). Although (22) is a Heisenberg state, not evolving in time, we could imagine a similar configuration obeying a Schrödinger equation,

$$\mathrm{d}|\psi\rangle/\mathrm{d}t = -iH|\psi\rangle \quad . \tag{23}$$

At first sight one might think that this is fundamentally different from a "classical" discrete system, evolving in a jumpy way,

$$\{m_{(i)}\}_{t_1} \Rightarrow \{m'_{(i)}\}_{t_2} \Rightarrow \{m''_{(i)}\}_{t_3} \Rightarrow \ldots , \tag{24}$$

but the transition "classical" ⇒ "quantum mechanical" for discrete systems is very different from the continuous case.

The evolution (24) is *a special case* of the *more general* quantum mechanical law of evolution (23). There is no need for a limiting procedure $\hbar \to 0$.

Even if a discrete system is classical, one can formally introduce a Hilbert space of states, $|\{m_{(i)}\}\rangle$, defined to be orthonormal. In terms of these states the law of evolution (24) is

$$|\{m_{(i)}\}\rangle_{t_1} \Rightarrow |\{m'_{(i)}\}\rangle_{t_2} = U(t_2,t_1)|\{m_{(i)}\}\rangle \quad , \tag{25}$$

where $U(t_2,t_1)$ is a matrix containing ones and zeros, corresponding to the permutation $m_{(i)} \Rightarrow m'_{(i)}$. If we are dealing with a true permutation U is unitary and an operator H can be found so that

$$U(t_2,t_1) = e^{-iH(t_2-t_1)} \quad . \tag{26}$$

Not all operators H satisfy an equation of the form (26) such that U is a permutation operator. These Hamiltonians form a subclass of all Hamiltonians.

If a basis $\{|n\rangle\}$ *and a discrete set of time variables* t_i *exists such that* $U(t_i-t_j)$ *in this basis are genuine permutation operators we define this system to have "evolutionary determinism".*

In practice we will also require that the definition of this special basis (called "primordial basis" in Ref.[10]) is in some sense local, so that projection operators selecting out certain basis elements are physically observable.

Our evolutionary determinism is to be distinguished from other forms of determinism in the sense that we do not object against transformations to some other basis in Hilbert space in terms of which U can no longer be seen to be an obvious permutator. The initial state may well be chosen to be some linear superposition. Actually we believe that linear superpositions are absolutely necessary if we want to deduce macroscopic, low energy laws of physics from the microscopic ones.

6. Examples

Consider an atom with a large angular momentum ℓ . Its Hilbert space is finite dimensional, and usually we chose as a basis the states

$$|m\rangle \quad , \quad m = -\ell, \; \ldots \; , +\ell \; . \tag{27}$$

Let's have a magnetic field B in the z-direction so that the Hamiltonian is given by

$$H|m\rangle \;=\; \mu B \; m|m\rangle \; ; \tag{28}$$

where for simplicity we will take $\mu B = 1$. Of course the evolution operator U is

$$U(t)|m\rangle \;=\; e^{-imt}|m\rangle \quad . \tag{29}$$

Now choose the new basis $\{|g\rangle\}$, with

$$|g\rangle = \frac{1}{\sqrt{2\ell+1}} \sum_m e^{-\frac{2\pi i m g}{2\ell+1}} |m\rangle \quad , \quad g = 0, \; \ldots, \; 2\ell \; . \tag{30}$$

Writing $2\ell+1 = N$ we see

$$U(t)|g\rangle = \frac{1}{\sqrt{N}} \sum_m e^{-2\pi i m g/N \, - i m t}|m\rangle \quad , \tag{31}$$

and if we limit ourselves to time intervals $t = 2\pi k/N$, k integer, then we see

$$U(t)|g\rangle \;=\; |g + k \; (\mathrm{mod}\; N)\rangle \; , \tag{32}$$

and this is exactly a (cyclic) permutation. We observe that the spinning atom in a homogeneous magnetic field has evolutionary determinism.

Evolutionary determinism does not forbid us to introduce non commuting operators such as L_x and L_y . Indeed, they may be called "observables" in the usual sense. In fact, the Hamiltonian H itself is non diagonal in $\{|g\rangle\}$, and we accept it as a truly observable variable. Treating the spinning atom as a model of the Universe, our philosophy will be that at the beginning of the Universe a single state $|g\rangle$ was selected, so that the evolution is completely deterministic. But we may decide to select eigenstates of L_x, L_y, or H if that comes out handy to formulate the (apparently quantum mechanical) "macroscopic" laws of physics.

A slightly less trivial example is a one space - one time dimensional model. Imagine an infinite series of cells x , $-\infty < x < \infty$, each of which may or may not contain one particle going to the right and/or one particle going to the left. These states can be characterized as

$$|\psi\rangle \;=\; \prod_{x=-\infty}^{\infty} |r_x, \, \ell_x\rangle \quad ; \qquad r_x = 0 \text{ or } 1 \; ; \; \ell_x = 0 \text{ or } 1 \quad . \tag{33}$$

There is a clock ticking in definite time intervals. The law of evolution is simple: at each tick of the clock the left movers make one step to the left and the right movers make one step to the right.

To write the evolution operator in a quantum mechanical way we introduce annihilation operators:

$$\sigma_\ell^-(x)|r_x,\ 1\rangle = |r_x,\ 0\rangle \quad ; \quad \sigma_\ell^-(x)|r_x,\ 0\rangle = 0 \ ;$$
$$\sigma_r^-(x)|1,\ \ell_x\rangle = |0,\ \ell_x\rangle \quad ; \quad \sigma_r^-(x)|0,\ \ell_x\rangle = 0 \ . \tag{34}$$

These operators do not satisfy nice (anti-)commutation rules. Therefore we introduce the Jordan Wigner transformation

$$\psi_\ell(x_1) = (-1)^{\sum_{x<x_1} \ell_x} \cdot \sigma_\ell^-(x_1) \quad ;$$
$$\psi_r(x_1) = (-1)^{\sum_x \ell_x + \sum_{x<x_1} r_x} \cdot \sigma_r^-(x_1) \quad . \tag{35}$$

These satisfy

$$\{\psi_i(x), \psi_j(x')\} = 0 \quad ; \quad \{\psi_i(x), \psi_j^\dagger(x')\} = \delta_{ij}\delta_{xx'} \quad . \tag{36}$$

They have been constructed carefully in order to have

$$\psi_\ell(x,t) = \psi_\ell(x+t,0) \quad ; \quad \psi_r(x,t) = \psi_r(x-t,0) \ . \tag{37}$$

Let us now introduce their Fourier transforms,

$$\psi_i(x) \ = \ (2\pi)^{-\frac{1}{2}} \int_{-\pi}^{\pi} \mathrm{d}k \ \mathrm{e}^{-ikx} \ \hat{\psi}_i(k) \ . \tag{38}$$

One finds

$$\hat{\psi}_i(k,t) = \mathrm{e}^{ik\sigma_3 t} \ \hat{\psi}_i(k,0) \quad ; \quad \sigma_3\psi_{\substack{\ell\\r}} = \pm\ \psi_{\substack{\ell\\r}} \ . \tag{39}$$

Using the commutation rules we find that this evolution is generated by the evolution operator

$$U(t) = \mathrm{e}^{-iHt} \quad ; \quad H = -\int_{-\pi}^{\pi} k \ \hat{\psi}^\dagger(-k)\sigma_3\hat{\psi}(k) \ \mathrm{d}k \ . \tag{40}$$

Notice that in the continuum limit this Hamiltonian tends to

$$H \to \int \mathrm{d}x \ \psi^\dagger(x) \ i\sigma_3 \frac{\partial}{\partial x} \ \psi(x) \ , \tag{41}$$

which is Dirac's Hamiltonian for a one-dimensional massless fermion gas.

This gas clearly also has evolutionary determinism. The model is less trivial than the previous example because we can also consider its more usual description. Filling up the Dirac sea we can describe the lowest energy levels as left or right moving "neutrinos". The field $\psi(x)$ however creates a superposition of a neutrino and an antineutrino, so the "macroscopic" law shows quantum mechanical interference effects. On the other hand, there is no "Zitterbewegung" because the neutrinos naturally move with the speed of light; most of the usual quantum paradoxes are still absent in this model.

To construct less trivial models, in particular models that can show how the fundamental paradoxes of quantum mechanics such as the EPR paradox can emerge, seems to be difficult. *Formally*, every deterministic discrete model *corresponds to* a quantum system. We imagine that extremely non-trivial cellular automata[11] are equivalent with equally non-trivial quantum field theories, but we have not been able to show this in a detailed construction. A difficulty is that the quantum mechanical Hamiltonian H does not follow uniquely from the evolution operator U if we only know it at discrete time intervals δt . All Eigenvalues of H are only determined modulo $2\pi/\delta t$. This has two difficult consequences:

i) One discrete classical model may correspond to many apparently different quantum field theories, and

ii) The vacuum state, defined to be the state with lowest energy, depends on how many times we choose to add $2\pi/\delta t$ to the various eigenvalues.

7. Discussion

We believe that black holes shed a new light upon the question what to think about the quantum mechanical nature of our laws of physics. They suggest namely that at the Planck level everything is discrete. The set of discrete quantum mechanical models has a subset featuring evolutionary determinism. This is a form of determinism fundamentally different from the "hidden variables" theory, because none of the variables in our states are hidden from direct observation. There is determinism at the microscopic level which however needs not lead to determinism at a macroscopic level. The reason why one can say this is that the state we call "vacuum state" needs not be one of the "primordial" basis elements but may well be a superposition of these. If this is the case then quantum mechanical superposition will be a natural phenomenon in macroscopic physics, since all experiments we do are surrounded by vacuum.

The way to visualize what is actually happening is that our world could be something like a cellular automaton, such that the state we call "vacuum" is a "chaotic" solution. It should be called a miracle that such a system should show any regular effective laws of physics at all at macroscopic scales. At best one would expect certain forms of statistical regularities. This may be what we call quantum mechanics today.

Physicists have learned to work with these quantum mechanical rules in order to be able to make predictions concerning the future behaviour of a system. The truth is (could be) that these predictions cannot be better than statistical in nature because of the ubiquitous chaos in all but a few of our physical variables.

The macroscopic "effective" laws of quantum mechanics can only be formulated in terms of observables that are non diagonal in the primordial basis. This includes the laws according to which our accelerator devices and measuring apparatus, including our own eyes and our own brains, work. We can only say "there is an electron here, with spin up" when we assume these laws to describe absolute truths. In reality we may be talking about some minute statistical fluctuations around the chaotic oscillations of the vacuum. If we say "it is sent towards the moon", we really mean that we know from experience that the chaotic vacuum oscillations bombarding us from all directions in outer space will propagate this fluctuation towards the moon. The best way for us to describe these oscillations is the remarkable discovery that they are nearly perfectly described by the zero energy eigenstate of the Hamiltonian, which is a superposition of all basis elements.

We do not claim that all or even some of the usual paradoxes and other difficulties with the quantum mechanical nature of our world are resolved this way. One problem, among many others, is the question why our Universe should be so close to the vacuum state. Our main contribution to the discussion is this one proposal, namely that the microscopic laws of physics show some form of evolutionary determinism.

References

1. E.T. Newman *et al*, *J. Math. Phys.* **6** (1965) 918; B. Carter, *Phys. Rev.* 174 (1968) 1559.
2. S.W. Hawking, *Nature*, **248** (1974) 30; *Comm. Math. Phys.* **43** (1975) 199; R.M.Wald, *Comm. Math. Phys.* **45** (1975) 9; L. Parker, *Phys. Rev.* **D12** (1975) 1519; J.B. Hartle and S.W. Hawking, *Phys. Rev.* **D13** (1976) 2188; W.G. Unruh, *Phys. Rev.* **D14** (1976) 870; S.W. Hawking and G. Gibbons, *Phys. Rev.* **D15** (1977) 2738; S.W. Hawking, *Comm. Math. Phys.* **87** (1982) 395.
3. G. 't Hooft, *Nucl. Phys.* **B256** (1985) 727.
4. G. 't Hooft, *Phys. Scripta* **T15** (1987) 143.
5. S.W. Hawking and G.F.R. Ellis, *The large-scale structure of space-time* (Cambridge Univ. Press 1973); C.W. Misner, K.S. Thorne and J.A. Wheeler, *Gravitation* (Freeman, San Francisco, 1973)
6. T. Dray and G. 't Hooft, *Nucl Phys.* **B253** (1985) 173
7. F. Wilczek, contribution at this Conference; J. Preskill and L.M. Krauss, Caltech Preprint CALT-68-1601 (1990)
8. G. 't Hooft, *to appear in Proceedings of the Banff Summer School in Theoretical Physics on "Physics, Geometry and Topology"* (Aug. 14-25, 1989), Ed. H.C. Lee (Plenum, 1990); G. 't Hooft, *Nucl. Phys.* **B** (1990), to be publ.
9. G. 't Hooft, *Phys. Lett.* **B 198** (1987) 61; *Nucl. Phys.* **B 304** (1988) 867.
10. G.'t Hooft, *J. Stat. Phys.* **53** (1988) 323.
11. E. Fredkin, private communication.

QUANTUM MECHANICS AND QUANTUM COSMOLOGY

JAMES B. HARTLE

Department of Physics
University of California
Santa Barbara, CA 93106

ABSTRACT

In the search for a theory of the initial condition of the universe, quantum mechanics must be applied to the universe as a whole. For this the "Copenhagen interpretations" of quantum mechanics are insufficiently general on at least two counts. The first concerns the characteristic assumptions of these interpretations of the existence of a separate "classical domain" or its equivalent. However, in a unified theory of the cosmology there is no fundamental basis for a separate classical physics. The second count concerns the preferred role of time in Hamiltonian quantum mechanics. However, in the early universe, when spacetime geometry is fluctuating quantum mechanically, there is no natural preferred family of spacelike surfaces to supply the preferred time of Hamiltonian quantum mechanics.

This talk sketched a quantum mechanical framework adequate for cosmology. The first part described joint work with Murray Gell-Mann[1] in which the complications introduced in the Planck era by gross quantum uncertainties in the structure of spacetime were ignored. The familiar notions of time, Hamiltonian, and Hilbert space of states could then be employed. The framework described is an extension and clarification of that of Everett and incorporates the ideas of the post-Everett development, especially the work of Zeh, Zurek, and Joos and Zeh. In the discussion of history and other points it is consistent with the work of Griffiths and Omnès. The second part of the talk described work by the author[2] on proposed generalizations of the familiar framework to deal with quantum spacetime and the problem of time.

Three forms of information are necessary for prediction in the quantum mechanics of a closed system. When quantum spacetime is ignored these are the Hamiltonian, the initial quantum state of the universe, and the information specifying a particular class of histories. The roles played by the different types of information were discussed for physics and chemistry on the one hand and for the enviromental sciences on the other. The general objects of quantum-mechanical prediction are time histories. However, quantum mechanics prohibits the assignment of probabilities to arbitrary sets of alternative histories. Probabilities can be assigned only to classes of histories that decohere to a good approximation, given a particular quantum initial condition for the universe. Histories described at an arbitrarily fine grained level do not decohere; some coarse graining is necessary for decoherence. The decoherence functional was introduced. With it decoherence was defined and the mechanism of decoherence was reviewed. For projection operators onto ranges of values of certain "quasi-classical" operators, habitual decoherence is widespread in the universe with its initial condition. A "classical domain" consists of classes of alternative histories described as a coarse graining that is as refined as is consistent with decoherence having individual branches that are correlated in time mostly according to classical laws.

A measurement situation is one in which a variable becomes correlated with a quasi-classical operator of the "classical domain". The theory of measurements in quantum mechanics was discussed from this point of view. This work suggests a view point toward quantum mechanics based on a "classical domain" that is emergent from the Hamiltonian of the elementary particles, the initial condition of the universe, and certain features of history. An "observer" (or information gathering and utilizing system) was treated as a complex adaptive system that evolves to utilize the relative predictability of such a "classical world". Resolution of many of the problems of interpretation presented by quantum mechanics is not to be found within the theory *in general* but rather through an examination of the origin of the universe and its subsequent history.

The second part of this talk addressed the possible generalizations of quantum mechanics necessary to deal with generally covariant theories of quantum spacetime and the consequent absence of a preferred time for Hamiltonian quantum mechanics. Most generally the essential elements of quantum mechanics are: 1) a set of fine grained alternative histories which are the most refined descriptions to which one can contemplate assigning probabilities, 2) a notion of coarse graining of these histories, and 3) a decoherence functional consistent with the principle of superposition. With these three elements decoherence can be defined and probabilities assigned to alternative sets of coarse grained histories of the universe. Hamiltonian quantum mechanics is one specific implementation of these three elements, but more general ones can be contemplated. A sum-over-histories generalized quantum mechanics for cosmological spacetimes was sketched which had no equivalent Hamiltonian formulation. In such a framework, Hamiltonian quantum mechanics with its preferred time emerges not generally and exactly, but rather as an approximation appropriate to the late universe for those special initial conditions which imply classical spacetime there.

Acknowledgement

Preparation of this abstract was supported in part by the National Science Foundation under NSF grant PHY90-08502.

References

1. M. Gell-Mann and J.B. Hartle, "Quantum Mechanics in the Light of Quantum Cosmology" in *Complexity, Entropy and the Physics of Information, Santa Fe Institute Studies in the Sciences of Complexity*, Volume VIII, edited by W. Zurek (Addison Wesley, Reading, 1990) or in *Proceedings of the 3rd International Symposium on the Fundations of Quantum Mechanics in the Light of New Technology*, edited by S. Kobayashi, H. Ezawa, Y. Murayama, S. Nomura (Physical Society of Japan, Tokyo, 1990).

2. See, e.g. J.B. Hartle, "The Quantum Mechanics of Cosmology" in *Proceedings of the Jerusalem Winter School on Quantum Cosmology and Baby Universes*, edited by S. Coleman, J.B. Hartle, and T. Piran (World Scientific, Singapore, 1990).

CLOSING IN ON A RENORMALIZABLE AND UNITARY POINT-LOCAL QUANTUM FIELD THEORY OF GRAVITY

YUVAL NE'EMAN
Wolfson Chair Extraordinary,
Sackler Faculty of Exact Sciences
Tel-Aviv University, Tel-Aviv, Israel
and
Center for Particle Theory, University of Texas
Austin, Texas 78712, USA

and

CHANG-YEONG LEE
Center for Particle Theory, University of Texas
Austin, Texas 78712, USA

ABSTRACT

We discuss a conjecture assigning a geometrical origin to all non-local features in Quantum Mechanics - consistent with the fact that all basic interactions are described by relativistic quantum gauge field theories. Gravity, a gauge theory classically, has not yet been quantized. We describe the proof of renormalizability for a point- (as against a string) local $\overline{GL}(4,R)$ gauge theory of gravity, spontaneously broken so as to yield Einstein's theory as the long-range or low-energy theory. Unitarity has not yet been proven.

1. Quantum Non-Locality and Gauge Theories

1.1. One of the important issues in the interpretation of non-relativistic quantum mechanics relates to the theory's features of non-locality. These are of two types topologically:

1) Effects such as the Aharonov-Bohm phase, the magnetic monopole appearing in the solutions of a spontaneously broken Yang-Mills gauge theory, instantons - all of these are global

features characterizing specific geometrical manifolds, and in particular fiber-bundles.

2) The EPR-type long-range correlations. These all presuppose some measure of parallel transport for the relevant observable. In the non-relativistic theory, the connections are trivial, but in the fully (special) relativistic quantum field theory, they are given by gauge 4-potentials. Their role is to provide for an infinity of long-range correlations; taking a horizontal section, we get a local differential operator (the generalized "covariant derivative") for the parallel-transport of the relevant variable.

1.2. One of us has discussed these features elsewhere[1]) and formulated the conjecture that quantum non-locality as a whole, in the non-relativistic theory, is a residual effect of the $c \to \infty$ contraction, together with the fact (an "experimental" discovery) that all fundamental physical dynamical theories are gauge theories ("the standard model" in the parlance of this end-of-a-century).

1.3. In the quantum field theory, we have in addition the essential role played by off-mass-shell contributions. This is of course yet another non-local feature, an apparent violation of Lorentz invariance. However, these violations are completely hidden, being confined as they are to the virtual (off-mass-shell) domain. Relativistic Quantum Field Theory thus manages to reconcile Lorentz Invariance with the intrinsic geometrical non-locality of Quantum Mechanics by ensuring that signals travelling faster than light exist only in the unphysical regions. We well know that these off-mass-shell contributions are an essential constituent of the theory: witness the prediction of the mass of the W meson at 87 GeV, indicated by this meson's off-mass-shell role in neutron beta decay - where the available physical mass is of the order of 1 Mev! It is edifying to learn that this virtual region is also essential for the co-existence of relativity and quantum mechanics in this deeper sense: extracting a Lorentz invariant physical world from a non-local framework.

1.4. Phenomenologically, Gravity is the main interaction which is not covered by the Standard Model - although there are indications for the existence of several additional forces, either as extensions-unifications of $SU(3) \times SU(2) \tilde{\times} U(1)$ or as separate interactions. Gravity does contain many of the features of a gauge theory at the classical level already - which is why it inspired Hermann Weyl and later C.N. Yang and R.L. Mills in conceiving the modern quantum gauge theories.

Our above discussion just adds one more reason for our interest in a Field Theory of Quantum Gravity. It was shown years ago that the proposal that Gravity stay as a classical theory would have created inconsistencies in terms of the Uncertainty Principle. We now also need Quantum Gravity in order to understand how EPR experiments in which the distance-correlated variable is momentum (as in the original EPR article) or spin (as in most present experiments) fit with a Quantum connection (the gravitational potential) that is there to ensure that these variables can indeed be parallel-transported, and that the corresponding faster-than-light signals can be "buried" in the virtual domain..

2. The Search for Quantum Gravity

2.1. Quantum Gravity has evaded the theoretical physicists for almost half a century. Since 1974, there is a possibility that the answer might be provided by the Quantum String or Superstring[2]. In particular, when in 1984 it was shown that some specific models of the string are distinguished by having their divergence-causing quantum anomalies vanish, it was generally assumed that there is no (point-) local quantum field theory of gravity and that the spatial extension of the string (dimensionally minimal as it is) is essential to the finiteness of the theory. The superstring amplitudes indeed appear finite, though there is as yet no proof that it is so to all orders. The string diagrams are not Borel-summable, but this is also true e.g. in QED, indicating the presence of (compensating) non-perturbative solutions. In strings, this requires the construction of a quantum string-local field theory, also vital for the removal of the tremendous arbitrariness in the reduction to the four-dimensional theory (from 10 or 26 dimensions).

2.2. Casher[3] has argued that the interpretation of the sub-Planck-energy truncated massless sector of the string as representing the phenomenologically fitting point-local field theory of the Standard Model would be inconsistent without a complementary renormalizable point-local theory of gravity. As a "theory of everything" (TOE) the string must have an eigenstate representing the bound state of positronium, for instance. The field theory of the truncation includes QED, which predicts the energy of positronium very precisely. However, this accounts for the electromagnetic binding only. There is, we know, a very tiny contribution due to gravitational binding, and it represents the difference between the TOE eigenstate and the QED result. Its finiteness cannot be due to the higher massive excitations of the graviton (with masses of the order of the Planck mass), from considerations similar to the "decoupling" theorem[4]. The truncated theory thus has to include a renormalized local field theory of gravity.

In the following sections, we sketch the outline of a renormalizable theory of gravity, modeled on those of the Standard Model: a $\overline{GL}(4,R)$ gauge theory with a quadratic Lagrangian, spontaneously broken so as to leave only the Lorentz subgroup as the low energy symmetry for local frames[5]. The theory is BRST invariant[6] and has been shown to be renormalizable[7]; it has yet to be cast in a mold guaranteeing unitarity, considering the noncompact nature of the entire $SL(2,C) \approx \overline{SO}(1,3) \subset \overline{SL}(4,R)$ sequence (the upper bar over the name of a group denotes its double-covering).

Should this program succeed, and should it turn out that the field theory's predictions differ, above Planck energies, from those of the superstring, we might find it difficult to decide between the two theories through experiments. Conceptually, two interpretations would be available: a) our field theory is Casher's field theory for the truncated low-energy region of the string, with the string TOE ruling the higher energy region; b) our theory is the complete theory of quantum gravity at all energies, and the string is just another "regulator" method, like Pauli-Villars' or dimensional regularization.

2.3. Why is gravity difficult to renormalize? We list 5 basic problems (for a lagrangian treatment):

a) the linearity of Einstein's lagrangian [dimension(R) = 2] also forcing a dimensionality d=2 on Newton's constant, and therefore an infinite number of different counter-terms in the renormalization.

b) the non-compactness of the gauge symmetry group SL(2,C) of the local frames; replacing the linear lagrangian by a Yang-Mills-like quadratic $R \wedge {}^{*}R$ puts a minus sign in front of the squared Lie-algebra-valued curvatures for the Lorentz boost components and makes the relevant components of the connection into ghosts (or tachyons, after a certain transformation, which are no better than ghosts).

c) the Riemannian condition

$$D_\mu \, g_{\rho\sigma} = 0 \tag{1}$$

relating metric $g_{\rho\sigma}$ and connection $\Gamma^\lambda{}_{\mu\nu}$ or $\omega_\mu{}^{ab}$. This causes quartic p^{-4} propagators, indicating a violation of unitarity, since the identity

$$\frac{1}{p^2\,(p^2 - m^2)} = \frac{1}{m^2}\left[\frac{1}{p^2 - m^2} - \frac{1}{p^2}\right] \tag{2}$$

implies that *one of the two poles must have the wrong sign* for its residue and is a ghost.

d) the dimensionlessness of the metric field, due to the Equivalence-Principle-imposed smooth transition to a Minkowski constant metric (as against d=1 for a Yang-Mills field or for the connection in gravity).
e) the nature of the infinite group of diffeomorphisms $\overline{Diff}(4,R)$ as yet another gauge group, with its algebra acting as a system of Lie-derivatives rather than as a Lie-algebra and the fact that it is represented non-linearly over its sl(4,R) subalgebra; the role of the sl(4,R) parameter is then fulfilled by the gradient term $\partial_\rho\xi^\sigma$ which therefore appears in all field variations for the "alias" part, and not just in those of the relevant gauge field (metric, tetrad or connection; e.g. for a tensor field $\Xi^{\mu\nu\cdot\cdot}$

$$\delta\Xi^{\mu\nu\cdot\cdot} = -\partial_\rho\xi^\sigma \cdot \{\Omega^\rho{}_\sigma\}\,\Xi + \xi^\sigma\,\partial_\sigma\Xi \tag{3}$$

where $\{\Omega\}$ is the sl(4,R) matrix for the representation corresponding to the tensor $\Xi^{\mu\nu\cdot\cdot}$ and the second term is the "alibi" part of the variation.

2.4. As a result, we have to modify the theory so as to make the following improvements:
a) Provide a "soft" derivation for Newton's constant; a model of that type was proposed in a revived version[8] of Weyl's theory.
b) The high energy region should be determined by d=4 terms, like in an ordinary "internal" Yang-Mills theory (with a dimensionless coupling and natural cut-off independence).
c) The Riemannian condition (Eq. 1) should be "weak", holding only for low-energy (i.e. lower than Planck energies) matrix elements.
d) Find ways of dealing with the $\overline{Diff}(4,R)$ gauge through appropriate BRST transformations. This was achieved in the seventies[9].
e) Find ways of coping with the dimensionlessness of the metric (and tetrads) by an appropriate selection of the gauge fixing and ghost terms in the quantum Lagrangian.
f) The effects of the non-hermiticity of the finite representations for non-compact generators' curvature bilinears have to be cancelled.

The model we present already incorporates - at this stage of our on-going program[5,6,7] - the first five modifications and thereby achieves renormalizability, has no higher-derivative terms (nor quartic propagators) in the invariant part of the lagrangian and manages to cope with the $\overline{Diff}(4,R)$ variations and with the metric's dimensionlessness. Only problem f), the cure for non-compactness-caused diseases, remains to be devised, hopefully in the next stages of this program.

2.5. Our invariant lagrangian is a Yang-Mills-like quadratic polynomial in $\overline{GL}(4,R)$ curvatures. In this discussion, we deal with the expression resulting from the use of the gl(4,R) Killing metric in the adjoint representation, leaving the base-space (holonomic) greek indices to be contracted separately. This contraction involves the symmetric tensor $g_{\mu\nu}(x)$ which continues to make world-scalars under the passive general co-ordinate transformations. On the other hand, writing

$$g_{\mu\nu}(x) = e_\mu{}^a(x) \cdot e_\nu{}^b(x) \cdot g_{ab}(x) \qquad (4)$$

we have by definition of the tetrad frames, as an identity

$$D_\rho e_\mu{}^a(x) \cong 0 \qquad (5)$$

$$D_\rho g_{\mu\nu}(x) = e_\mu{}^a(x) . e_\nu{}^b(x) . D_\rho g_{ab}(x) \neq 0 = K_{\rho\mu\nu}(x) \qquad (6)$$

The field $g_{ab}(x)$ is a symmetric tensor under the anholonomic (gauge) group $\overline{GL}(4,R)$ but is not yet an invariant metric. We may use it, though, to contract sl(4,r) indices, to produce sl(4,R) global invariance. The geometry in the high energy region (HER) is thus non-Riemannian. Such lagrangians have been proposed before[10] ; they do not obey the Principle of Equivalence. We manage to correct this feature through a mechanism of spontaneous symmetry breakdown: the Higgs field is a bosonic infinite-component field ("manifield"[11]) representation of $\overline{SL}(4,R)$, $\Phi(x)$, with a Lorentz (0,0) lowest state with non-vanishing vacuum expectation value,

$$v = \langle 0 | \Phi_{00} | 0 \rangle \neq 0 \qquad (7)$$

As a result, the low energy region (LER) has a pseudo-orthogonal local isotropy group and is "weakly" Riemannian,

$$\langle LER | D_\rho g_{ab} | LER \rangle = 0 \qquad (8)$$

Note that the connection $\omega_\mu{}^{ab}(x)$ has 64 independent components. In the LER, Eq. (7), resulting in (8), fixes 40 components in terms of the "effective" LER metric's components. The other 24 components will be given by the equations of motion in terms of the torsion sources. We may or may not include torsion-squared terms in the invariant lagrangian, depending on the method we shall use to reproduce Einstein's theory for the LER, since a certain combination of torsion-squared terms has the same solutions as Einstein's lagrangian up to 5th ppn order. In any case, whether we use Einstein's curvature scalar or this torsion-squared term, we replace Newton's constant by ϕ^2 (an sl(4,R) scalar "dilaton" with d=1) or Φ^2.

The ϕ field also acquires a non-vanishing v.e.v., through a "$-\lambda_{mix}\,\phi^2.\Phi^2$" term in the potential, aside from "$-\lambda^0\,\phi^4$" and "$-\lambda\,(\Phi^*.\,\Phi)^2$". Matter is represented by world-spinor manifields[11], also coupled to the Higgs field $\alpha\,\Psi^*.\Phi.\Psi$. The Φ components "above" (0,0) include (2,0),(1,1) and (0,2); the (1,1) transmute into longitudinal components for the (now massive) connections $\omega_\mu{}^{(ab)}$. All other Higgs field components acquire masses, and so does the $\phi(x)$. The scale for these masses is fixed by the phenomenological identification

$$G_N = 1/(16\Pi\,\hat{v}^2), \quad \hat{v} = \langle 0\,|\,\phi\,|\,0\rangle = -\lambda_{mix}/(2\,\lambda^0)^{1/2}\,\hat{v} \qquad (9)$$

if we adopt the version in which Einstein gravity is reproduced in the LER through the term

$$\sqrt{g}\,.\,\phi^2.\,R \qquad (10)$$

(the $\sqrt{g}$ multiplies all terms, including the above matter and spontaneous symmetry breakdown elements).

In addition to the 64 components of the connection we start out with 10 components of g_{ab} and 16 tetrads. The spontaneous symmetry breakdown makes $g_{ab} \to \eta_{ab}$ and leaves us with the 10 components of the Riemannian $g_{\mu\nu}$, with the residual 6 degrees of freedom corresponding to local Lorentz invariance.

The HER symmetric Yang-Mills-like curvature-squared lagrangian is not irreducible, since both anholonomic (gauge) indices and base-space holonomic indices refer to the same variables, up to a tetrad-induced transvection. In principle, all irreducible components will appear anyhow in the counter terms generated by the renormalization process; it is however possible to discard some of them through hermiticity or some similar condition. We shall deal with this aspect when we treat the question of unitarity. For a general discussion of "Metric Affine" models of Gravity, see reference [12].

3. BRST and the "Background Field" Method

3.1. The BRST transformations for the fields and ghosts[6] are derived by the usual method of replacing the parameters $\lambda^a{}_b$ (of sl(4,R)) and ξ^ρ (of Diff(4,R)) by $C^a{}_b\,\Lambda$ and $C^\rho\,\Lambda$, where Λ is a "constant anticommuting element" and $C^a{}_b$ and C^ρ are the appropriate ghosts:

$$\delta_B e^a{}_\mu = (C^\lambda e^a{}_{\mu,\lambda} + C^\lambda{}_{,\mu} e^a{}_\lambda + C^a{}_b e^b{}_\mu).\Lambda \tag{11a}$$

$$\delta_B \omega^a{}_{b\mu} = (C^\lambda \omega^a{}_{b\mu,\lambda} + C^\lambda{}_{,\mu} \omega^a{}_{b\lambda} - C^a{}_{b,\mu} + C^a{}_d \omega^d{}_{b\mu} - C^d{}_b \omega^a{}_{d\mu}).\Lambda \tag{11b}$$

$$\delta_B C^\mu = - C^\lambda C^\mu{}_{,\lambda} . \Lambda \tag{11c}$$

$$\delta_B C^a{}_b = (- C^\lambda C^a{}_{b,\lambda} - C^a{}_d C^d{}_b).\Lambda \tag{11d}$$

and $\delta^2_B = 0$ for all fields and ghosts. The last three terms of (11b) form $D_{sl(4R)}$ C.Λ, i.e. the sl(4,R) covariant derivative of the appropriate gauge parameter, here C.Λ .

For matter fields, there are alibi and alias terms in C^λ and $C^\lambda{}_{,\mu}$ respectively, for every "world" (or holonomic) index[9] - and a term $C^a{}_b \{\Omega^b{}_a\} \Xi$, where $\{\Omega\}$ is the sl(4,R) unirrep acting on the field Ξ (finite or infinite). The antighosts transform according to

$$\delta_B \bar{C} = F(e^a{}_\mu , \omega^a{}_{b\mu} , \Xi).\Lambda \tag{11e}$$

where F is defined by the explicit construction of the gauge fixing term in the effective quantum lagrangian. The latter thus includes gauge-fixing and ghost terms, aside from the gauge-invariant terms that we dealt with up to this point (Y represents gauge fields, Ξ are the various matter fields - including Higgs-Goldstone etc and C are the ghost fields):

$$\mathcal{L}_{eff} = \mathcal{L}_{inv} + \mathcal{L}_{g.f.} + \mathcal{L}_{ghost} \tag{12a}$$

$$\mathcal{L}_{g.f.} = - 1/2\, F(Y, \Xi)^2 \tag{12b}$$

$$\delta_B F(Y, \Xi) = \underline{M}\, (Y, \Xi)\, C.\Lambda \tag{11f}$$

$$\mathcal{L}_{ghost} = - \bar{C}\, \underline{M}\, C \tag{12c}$$

3.2. In the "Background Field" method[13,14] the gauge fields Y are split into a "background" part A and a "quantum" piece q, Y = A + q . The above (11b) variations δ_B Y = DC.Λ are then reproduced in two different manners, which we denote as δ' and δ" and whose action is given by (D' is ∂ + A)

$$\text{type I : } \delta'A = D'C\Lambda\ ,\ \delta'q = C\Lambda f q\ ,\ \delta'(Y+q) = DC.\Lambda \tag{13a}$$

$$\text{type II: } \delta''A = 0\ ,\ \delta''q = DC.\Lambda \tag{13b}$$

where f (for sl(4,R)) are the structure constants

$$f^{a}{}_{b}{}^{c}{}_{d}{}^{e}{}_{h} = \delta^{a}{}_{d}\,\delta^{c}{}_{h}\,\delta^{e}{}_{b} - \delta^{c}{}_{b}\,\delta^{e}{}_{d}\,\delta^{a}{}_{h} \tag{13c}$$

Here again, with sl(4,R) treated precisely as a Yang-Mills gauge, it is in selecting the diff(4,R) variations that we have to make new decisions.

Note that type I variations leave $\mathcal{L}_{g.f.}$ and $\mathcal{L}_{ghost}$ separately invariant, a useful feature. It is a gauge-invariance obeyed by the quantum lagrangian and thus propagated throughout the renormalization counter-terms, for instance. For gravity and supergravity, this is an important tool[13].

First, we have to select the Y fields themselves, out of the set $g_{\mu\nu}$, $g^{\mu\nu}$, $e^{a}{}_{\mu}$, $e_{a}{}^{\mu}$; we also have to take into account the measure $\sqrt{g}$, $g = \det g_{\mu\nu}$. Note that we have already chosen $\sqrt{g}$ rather than e in order to avoid the effect of the anholonomic part of the variation, which would yield an undesirable trace term[6]. Note also that using the inverse metric $g^{\mu\nu}$ we can also define $g^{ab} = e^{a}{}_{\mu}\, e^{b}{}_{\nu}\, g^{\mu\nu}$ as in (4). We redefine our fields so as to include the measure

$$G^{\mu\nu} \cong g^{1/4}\, g^{\mu\nu}\ , \quad G_{\mu\nu} \cong g^{-1/4}\, g_{\mu\nu} \tag{14}$$

The new fields thus transform as tensor densities of weight $\delta = 1/2$ and $\delta = -1/2$ ($\delta = 1$ for $\sqrt{g}$). In the Appendix, we list the dimension, density weight and ghost number of all fields and their functionals.

Our choice of fields Y = A + q for diff(4,R) is now given by the definitions (H,T and U are the background "A" fields, h, e and u are the quantum fields "q")

$$G^{\mu\nu} = H^{\mu\nu} + h^{\mu\nu} \tag{15a}$$

$$e^{a}{}_{\mu} = T^{a}{}_{\mu} + t^{a}{}_{\mu} \tag{15b}$$

and more conventionally for sl(4,R),

$$w^{a}{}_{b\mu} = U^{a}{}_{b\mu} + u^{a}{}_{b\mu} \tag{15c}$$

The inverse fields for h and t are then defined through the relevant Taylor expansions,

$$G_{\mu\nu} = H_{\mu\nu} + h'_{\mu\nu}\,(h^{\rho\sigma}) \tag{15a'}$$

$$e_{a}{}^{\mu} = T_{a}{}^{\mu} + t'_{a}{}^{\mu}\,(t^{b}{}_{\nu}) \tag{15b'}$$

The functionals $h'_{\mu\nu}$ and $t'_a{}^{\mu}$ are determined by the relations $H^{\mu\nu} H_{\nu\rho} = \delta^{\mu}{}_{\rho}$ together with $G^{\mu\nu} G_{\nu\rho} = \delta^{\mu}{}_{\rho}$ etc.

We choose the variations (13) for diff(4,R) according to the following scheme: to leave $\mathcal{L}_{inv}$ invariant under typeI, we are forced to use ordinary general coordinate transformations (or the corresponding BRST) for both A and q fields, whereas in type II we have all $\delta''A$ vanishing and assign the GCT (or corresponding BRST) of the complete Y to the variations of q:

$$\delta'_{diff} = \delta_{Bdiff} \; ; \tag{16a}$$

$$\delta''_{diff} (H,T,U) = 0 \; ; \; \delta''_{diff} (h,t,u) = \delta''_{Bdiff} (G,e,\omega) \tag{16b}$$

3.3. We now proceed to construct $\mathcal{L}_{eff}$. The dimensionality of the entire lagrangian should maintain d = 4 as for $\mathcal{L}_{inv}$. The gauge-fixing is required for the "q" fields. Out of the set (15) we select $h^{\mu\nu}$; for sl(4,R) this is $u^a{}_{b\mu}$. We write

$$\mathcal{L}_{g.f.} = -(1/2\alpha) \, (a^{\nu} u^a{}_{b\nu})^2 - (1/2\beta) \, \mu^{-2} (G_{\sigma\tau} \; b_{\lambda} \, h^{\lambda\sigma})^2 \tag{17}$$

where a and b are differential operators. For sl(4,R) this is

$$a_{\nu} = g^{1/4} \; g^{\mu\nu} D_{\mu}' \; = \; G^{\mu\nu} D_{\mu}' \tag{18a}$$

Here $D' = \partial + U$ and d(a) = 1. The structure of the diff(4,R) gauge-fixing term is determined by the constraints imposed by the dimensional structure of the corresponding ghost lagrangian, in view of (12b), (11f) and (12c). Since $d(C^{\lambda}) = 0$ and $d(\bar{C}_{\lambda}) = 0$, from (12c) we get $d(\underline{M}) = 4$. From (17) we have $F \approx b\, h$ and thus (11f) $\delta_B F = \underline{M}\, C.\Lambda \approx b\, \delta_B h \approx b\, D'C.\Lambda$, we get d(b) = 3. We thus select (here $D' = \partial + \Gamma$)

$$b_{\lambda} = g^{1/4} \; g^{\mu\rho} D_{\mu}' \, D_{\rho}' \, D_{\lambda}' = G^{\mu\rho} D_{\mu}' \, D_{\rho}' \, D_{\lambda}' \tag{18b}$$

The density weight $\delta=1$ is supplied by $\delta=1/2$ for a^{ν} (see 18a) for the first term in (17); for the second term it is contained in the $\delta=1/2$ of the h field (see 14 and 15a) and in $\delta(b_{\lambda})=1/2$ (see 18b) so that it has to be compensated by $G_{\sigma\tau}$ with $\delta=-1/2$. The μ^{-2} factor is a mass-scale parameter, necessary in order to ensure d=4 for (17), considering that d(b) = 3, d(b²) = 6.

We can now construct the ghost lagrangian,

$$\mathcal{L}_{ghost} = - \bar{C}^b{}_a \; a^\nu \; \delta u^a{}_{b\nu} \; - \; \bar{C}_\mu \; b_\nu \; \delta h^{\mu\nu} \tag{19}$$

where δq stands for $\delta_B\, q.\Lambda^{-1}$. Note that $\bar{C}^b{}_a$ will have to transform as a density with weight $\delta=1/2$, to supplement the weight 1/2 of a^ν - whereas in the second term, the weight 1 is supplied by both b_ν and $h^{\mu\nu}$. We can now write the explicit expressions for the antighost variations

$$\delta_B\, \bar{C}^b{}_a = -(1/\alpha)\; a^\nu\; u^b{}_{a\nu}\; \Lambda \tag{20a}$$

$$\delta_B\, \bar{C}_\nu \;\; = -(1/\beta)\; \mu^{-2}\; G_{\nu\sigma}\; g^{-1/4}\; b_\lambda\; h^{\lambda\sigma}\; \Lambda \tag{20b}$$

4. The Effective Action and the Slavnov-Taylor Identities

4.1. We sketch the general construction of the effective quantum action. The logarithm W[J] of the generating functional Z[J] produces the normalized vacuum-expectation-values for the connected n-point functions:

$$Y = \langle 0| \, Y \, |0\rangle \;\; \langle 0|0\rangle = \;\; \delta W/\delta J \tag{21a}$$

$$\langle 0| \, T\{Y...Y\} \, |0\rangle \;\; \langle 0|0\rangle \; = (1/i)^{n-1}\; \delta^n\; W/\delta J^n \tag{21b}$$

$$Z[J] = \int \delta Y \exp\{i\int d^4x\; (\mathcal{L}[Y] + J.Y)\} \cong \exp\{iW[J]\} \tag{21c}$$

and the effective action is given by $\Gamma[\tilde{Y}] = W[J] - J.\tilde{Y}$ (21d)

yielding the inverse propagator for one-particle-irreducible ("1PI") Green's functions

$$D^{-1} = (1/i)\; \delta^2\Gamma[\tilde{Y}] \; / \; \delta\tilde{Y}^2 \tag{21e}$$

and the full connected n-point functions can be constructed from the 1PI through (21b) or $\delta^n\; \Gamma\; /\delta\tilde{Y}^n$.

Using the Background-Field method $Y = A + q$, we have $\tilde{q}$ replacing $\tilde{Y}$ and q replacing Y in (21a) and (21b). We have

$$Z'[J,A] = Z[J]\; e^{-iJ.A} = \int\delta q\; \exp i\int d^4x\; (\mathcal{L}[q+A]+J.q) \cong e^{iW'[J,A]} \tag{22a}$$

$$q \to \tilde{q} = \delta W'/\delta J \;,\; \Gamma'[\tilde{q},A] = W'[J,A] - J.\tilde{q} = \Gamma[\tilde{q} + A] \tag{22b}$$

$\Gamma'[0,A]$ generates no graph with external lines $\tilde{q}$. It is the sum of all 1PI vacuum graphs, in the presence of the A field.

The effective action Γ' is thus calculated by summing only vacuum graphs, with the A appearing as the external lines and treated perturbatively.

4.2. We now apply the formalism to our results. The generating functional of the fields (A represents the three background fields H,U,T), sources J, ghost-sources $\bar{\zeta}$, anti-ghost sources ζ, field-BRST-sources K, ghost-BRST-sources L - is given by the expression ($\zeta,\bar{\zeta}$,K anti-commute)

$$Z'(A,J,\bar{\zeta},\zeta,K,L) = \int D(u,h,t,\bar{C},C) \exp i\int d^4x\{\Sigma'(U,H,T;u,h,t; C^a{}_b,C^\mu;\bar{C}^b{}_a,\bar{C}_\nu;K,L) + J_u.u + J_h.h + J_t.t + \bar{\zeta}^b{}_a.C^a{}_b + \bar{\zeta}_\nu.C^\nu + \bar{C}^b{}_a.\zeta^a{}_b + \bar{C}_\nu.\zeta^\nu\} \quad (23)$$

$$\Sigma'(A,q,\bar{C},C,K,L) = I_{inv}(U+u,H+h,T+t) -(1/2\alpha)\,(a^\nu.u^a{}_{b\nu})^2 - -(1/2\beta)\,\mu^{-2}(G_{\sigma\tau}.b_\lambda\, h^{\lambda\sigma})^2 + \bar{C}^b{}_a.a^\nu\ \delta u^a{}_{b\nu} + \bar{C}_\nu\ b_\nu.\delta h^{\mu\nu} + (K_u)^{\nu b}{}_a.\delta u^a{}_{b\nu} + (K_h)_{\mu\nu}.\delta h^{\mu\nu} +(K_t)^\mu{}_a.\delta t^a{}_\mu - L^b{}_a.\delta C^a{}_b - L_\nu.\delta C^\nu\,. \quad (24)$$

K_h, K_t and L_ν have dimensionality d=3, K_u and $L^b{}_a$ have d=2. J_u, J_t, $\bar{\zeta}^b{}_a$, $\bar{\zeta}_\nu$, ζ^ν, K_u, K_t, $L^b{}_a$, L_ν transform as tensor densities of weight 1; J_h, K_h, $\zeta^a{}_b$ have weight w=1/2. Z' is invariant under type I transformations, since we have set the weight densities accordingly and we have ensured global sl(4,R) invariance for the J, $\bar{\zeta}$, ζ, K, L source terms. Z' defines W' according to (22a) and this defines $\Gamma'[A,\bar{q}']$ as in (22b), i.e.

$$\Gamma'(A,\tilde{q}\,,\tilde{\bar{C}}\,,\tilde{C}\,,K,L) = W'(A,J,\zeta,\bar{\zeta},K,L) - J.\tilde{q} - \bar{C}.\zeta - \bar{\zeta}.C \quad (25)$$

4.3. The Slavnov-Taylor identities state the invariance of Γ' under the two types of Background-Field BRST transformations, i.e. $I[\Omega+\epsilon\Delta] = I[\Omega]$, where $\Omega\to\Omega'=\Omega+\Delta$; since $D\Omega'=D\Omega$, we write

$$\int\{\delta\Gamma'[\Omega]/\delta\Omega\}\Delta(x;\Omega) = 0 \quad (26)$$

At this stage one goes over to the reduced effective action $\Gamma^*[A,\tilde{q}\,]$ in which the two gauge-fixing terms in $1/2\alpha$ and $1/2\beta$ in (24) are resubtracted. Applying the two types of BF BRST transformations yields the four identities, the first from typeI, the others from typeII (varying fields and ghosts in b and the antighosts in c and d) and making use of the fact that the K are coupled to the variations of the anti-ghosts:

$$\delta' A.\delta\Gamma^*/\delta A + \delta'\tilde{q}\,.\delta\Gamma^*/\delta\tilde{q} + \delta'\tilde{C}\,.\delta\Gamma^*/\delta\tilde{C} + \delta'\bar{C}\,.\delta\Gamma^*/\delta\bar{C} = 0 \quad (27a)$$

$$(\delta\Gamma^*/\delta\tilde{q})(\delta\Gamma^*/\delta K) + (\delta\Gamma^*/\delta\tilde{C})(\delta\Gamma^*/\delta L) = 0 \quad (27b)$$

$$a_\nu\,(\delta\Gamma^*/\delta K_{h\mu\nu}) - \delta\Gamma^*/\delta\bar{C}_\nu = 0 \quad (27c)$$

$$b^{\mu b}{}_a\,(\delta\Gamma^*/\delta K_u{}^{\mu b}{}_a) - \delta\Gamma^*/\delta\bar{C}^b{}_a = 0 \quad (27d)$$

These symmetry properties have to survive the renormalization process.

5. Proof of Renormalizability

5.1. Assume that Γ^* has been successfully renormalized up to (n-1) loops, i.e. all divergences have been removed by a redefinition of the fields, parameters and transformations of the theory. We then proceed to n-loop order and have to prove that the divergent parts in $\Gamma^*_{(n)}$ can be similarly removed. $\Gamma^*_{(n)}$ is split into finite and divergent parts:

$$\Gamma^*_{(n)} = \Gamma^*_{(n)}{}^{\text{finite}} + \Gamma^*_{(n)}{}^{\text{diverg.}} \quad (28)$$

Eqns. (27) have to be satisfied to n-loop order; the new divergent contributions from n-loops have a pole in ϵ in the dimensional reduction regularization method (working in 4-ϵ dimensions) and should thus obey (27) separately . We thus rewrite (27) for $\Gamma^*_{(n)}{}^{\text{diverg.}}$, which we denote $\Gamma^d{}_{(n)}$:

$$\delta' A.\delta\Gamma^d{}_{(n)}/\delta A + \delta'\tilde{q}\,.\delta\Gamma^d{}_{(n)}/\delta\tilde{q} + \delta'\tilde{C}\,.\delta\Gamma^d{}_{(n)}/\delta\tilde{C} +$$

$$\delta'\bar{C}\,.\delta\Gamma^d{}_{(n)}/\delta\bar{C} = 0 \quad (29a)$$

$$(\delta\Gamma^d{}_{(n)}/\delta\tilde{q})(\delta\Gamma^\circ/\delta K) + (\delta\Gamma^\circ/\delta\tilde{q})(\delta\Gamma^d{}_{(n)}/\delta K) +$$

$$+ (\delta\Gamma^d{}_{(n)}/\delta\tilde{C})(\delta\Gamma^\circ/\delta L) + (\delta\Gamma^\circ/\delta\tilde{C})(\delta\Gamma^d{}_{(n)}/\delta L) = 0 \quad (29b)$$

$$a_\nu\,(\delta\Gamma^d{}_{(n)}/\delta K_{h\mu\nu}) - \delta\Gamma^d{}_{(n)}/\delta\bar{C}_\nu = 0 \quad (29c)$$

$$b^{\mu b}{}_a\,(\delta\Gamma^d{}_{(n)}/\delta K_u{}^{\mu b}{}_a) - \delta\Gamma^d{}_{(n)}/\delta\bar{C}^b{}_a = 0 \quad (29d)$$

where $\Gamma^\circ = \Sigma' + 1/2\alpha \ldots + 1/2\beta \ldots = \Sigma''$
Eqn. (29a) states typeI invariance for $\Gamma^d{}_{(n)}$. Equation b is the "renormalization equation". It can be written as

$$X\Gamma^d{}_{(n)} = 0, \tag{30}$$

where X contains the variations of Γ^0 as coefficients of the variations of $\Gamma^d{}_{(n)}$. One checks that $X^2 = 0$.
As a result we can write,

$$\Gamma^d{}_{(n)} = G(A,\tilde{q}\,) + X\,P(A,\tilde{q}\,,\tilde{C}\,,\bar{C}\,,K,L) \tag{31}$$

G should have d=4, and to satisfy typeII invariance $G(A,\tilde{q}\,) = G(A+\tilde{q})$. To satisfy typeI invariance too, it has to coincide with I_{inv} in (12), up to changes in the coefficients of the various irreducible parts. By inspection (compare with the Appendix) we see that X has d=1 and ghost number 1, so that P should have d=3 and ghost number -1. From (29c)(29d) K and $\bar{C}$ in P can only appear in the combinations $(K_h)_{\mu\nu} + \bar{C}_\nu\, a_\mu$ and $(K_u)^{\nu b}{}_a + \bar{C}^b{}_a\, b^\nu$. This enables us to write the most general expression for P (M,M',N,N' are arbitrary scalar functionals of the t,T fields and F a vector in both tangent and world indices)

$$\begin{aligned} P = {} & \kappa_1 (K_t)^\mu{}_a \cdot F^a{}_\mu(t^b{}_\nu, T^b{}_\nu) + \\ & + \kappa_2\{(K_u)^{\nu b}{}_a + \bar{C}^b{}_a \cdot b^\nu\} \cdot (\tilde{u}^a{}_{b\nu} + U^a{}_{b\nu})\, M(t^d{}_\rho, T^d{}_\rho) \\ & + \kappa_3\{(K_h)_{\mu\nu} + \bar{C}_\nu \cdot a_\mu\} \cdot (h^{\mu\nu} + H^{\mu\nu})\, M'(t^d{}_\rho, T^d{}_\rho) \\ & + \lambda_1\, L_\nu\ \tilde{C}^{\,\nu}\ N(t^d{}_\rho, T^d{}_\rho) \\ & + \lambda_2\, L^b{}_a\ \tilde{C}^{\,a}{}_b\ N'(t^d{}_\rho, T^d{}_\rho) \end{aligned} \tag{32}$$

5.2. The degree of divergence of an arbitrary diagram is

$$D = 4\Lambda + \textstyle\sum_v \delta_v - \sum_j \Pi_j\, \theta_j \tag{33}$$

where Λ is the number of independent loops, δ_v is the number of derivatives acting on internal lines at the vertex "v", Π_j the number of internal lines in the diagram corresponding to the field "j" and θ_j is the negative power of the momenta in the relevant propagator for "j". Topologically, with V the number of vertices, we have the identity $\Lambda = \sum_j \Pi_j + 1 - V$. We can thus reexpress D as

$$D = 4 + \textstyle\sum_j (4 - \theta_j)\, \Pi_j - \sum_v (4 - \delta_v) \tag{34}$$

in the Background Field method, the 1PI diagrams with external "q" lines and those with internal "A" lines (see section 3.2) will not contribute to the effective action. Hence, $\Pi_j = (1/2) \sum_j N^v{}_j$, where $N^v{}_j$ is the number of internal lines j leaving the vertex v. As a result,

$$D = 4 - \sum_v \{4 - \delta_v - (1/2) \sum_j (4 - \theta_j) N^v{}_j\} \tag{35}$$

Our effective action yields values of θ_j as listed in our Appendix. Inserting these values yields

$$D = 4 + \sum_v d_v$$

$$d_v \cong \delta_v + N^v{}_u + N^v{}_{Cu} - N^v{}_h - 4 \tag{36}$$

where "Cu" denotes the $C^b{}_a$ propagators. The theory is **renormalizable if all vertices ("v") in the action Σ have non-positive d_v**; this ensures that the divergences can be removed, since higher order diagrams will not generate higher degree divergences.

5.3. In the case of our Σ, d_v is indeed non-positive. In particular, vertices involving K_h (the κ_3 term in Eq. (32)) have $d_v = -4$ and do not yield divergent contributions at all. Because of (27c), this also ensures that there is no divergent contribution due to the the anti-ghost $\bar{C}_\nu$, i.e. it does not appear in $\Gamma^*{}_{(n)}{}^{\text{diverg.}}$, which implies that any contribution related to the h' field in $\Gamma^*{}_{(n)}{}^{\text{diverg.}}$ is gauge invariant.

The divergences in XP of (31) can all be removed through field renormalizations or by renormalizing the BRST transformations. For example, the κ_1 term in P of (32) will yield the divergent contribution (we read out X from eqn. 29b, inserted in eqn. 30),

$$\kappa_1\, \delta\xi^b{}_\nu \cdot (K_t)_a{}^\mu\, (\delta F^a{}_\mu / \delta\xi^b{}_\nu) + \kappa_1\, (\delta\Gamma^\circ / \delta\xi^a{}_\mu)\, F^a{}_\mu\, . \tag{37a}$$

which can be removed by the redefinitions

$$\xi^a{}_\mu \to \xi^a{}_\mu - \kappa_1\, F^a{}_\mu(\xi^b{}_\nu, T^b{}_\nu) \tag{37b}$$

$$\delta\xi^a{}_\mu \to \delta\xi^a{}_\mu - \kappa_1\, (\delta F^a{}_\mu / \delta\xi^b{}_\nu)\, \delta\xi^b{}_\nu\, . \tag{37c}$$

All divergent contributions in P can be removed similarly, using the redefinitions

$$\tilde{u} \rightarrow \tilde{u} - \kappa_2 M (\tilde{u} + U) \quad (38a)$$

$$\delta\tilde{u} \rightarrow \delta\tilde{u} - \kappa_2 (\delta M/\delta\xi^d{}_\rho)\, \delta\xi^d{}_\rho (\tilde{u} + U) - \kappa_2 M\, \delta\tilde{u} \quad (38b)$$

for the 2nd term in P. The third term does not diverge, as can be seen from our argument regarding the anti-ghost $\bar{C}_\nu$. The λ terms are renormalized through the redefinitions,

$$\tilde{C}^\nu \rightarrow \tilde{C}^\nu - \lambda_1 N \tilde{C}^\nu \quad (38c)$$

$$\delta\tilde{C}^\nu \rightarrow \delta\tilde{C}^\nu - \lambda_1 (\delta N/\delta\xi^d{}_\rho)\, \delta\xi^d{}_\rho\, \tilde{C}^\nu \quad (38d)$$

$$\tilde{C}^a{}_b \rightarrow \tilde{C}^a{}_b - \lambda_2 N' \tilde{C}^a{}_b \quad (38e)$$

$$\delta\tilde{C}^a{}_b \rightarrow \delta\tilde{C}^a{}_b - \lambda_2 (\delta N'/\delta\xi^d{}_\rho)\, \delta\xi^d{}_\rho\, \tilde{C}^a{}_b - \\ - \lambda_2 N'\, \delta\tilde{C}^a{}_b \quad (38f)$$

The divergences contributed in $\Gamma^*{}_{(n)}{}^{\text{diverg.}}$ by $I_{inv}\,(A+\tilde{q})$ can be removed by a redefinition of the coefficients of the irreducible parts. As a result, the entire divergent piece of $\Gamma^*{}_{(n)}$ has been removed, and our proof by induction from (n-1) to n is complete. We are left with the task of studying unitarity, which we hope to address in the next phase of this program.

6. References

1.Y. Ne'eman, Proc. Nat. Acad. Sci., 80, 7051; same author, in Foundations of Modern Physics, Proc. Joensuu (Finland) 1985 Seminar, P. Lahti and P. Mittelstaedt editors, World Scientific Pub. Co., Singapore (1985) pp. 481-495.

2. T. Yoneya, Prog. Theoret. Phys. 51 (1974) 1907; J. Scherk and J.H. Schwarz, Nucl. Phys. B81 (1974) 118.

3. A. Casher, Phys. Lett. B195 (1987) 50.

4. T. Appelquist and J. Carazzone, Phys. Rev. D11 (1975) 2856.

5. Y. Ne'eman and Dj. Sijacki, Phys. Lett. B200 (1988) 489

6. C.Y. Lee and Y. Ne'eman, Phys. Lett.(1990) in press.

7. C.Y. Lee, "Renormalization of a Gravity Model with Local $\overline{GL}(4,R)$ Symmetry", U. of Texas Center for Particle Theory preprint (1990), to be published.

8. F. Englert, E. Gunzig, C. Truffin and P. Windey, Phys. Lett. 57B (1975) 73.

9. R. Delbourgo and M.R. Medrano, Nucl. Phys. B110 (1976) 467; J. Dixon, Oxford Thesis, 1975 (unpublished); G. Sterman, P.K. Townsend and P. van Nieuwenhuizen, Phys. Rev. D17 (1978) 1501; Y. Ne'eman, E. Takasugi and J. Thierry-Mieg, Phys. Rev. D22 (1980) 2371.

10.G. Stephenson, Nuovo Cimento, 9 (1958) 263; C.W. Kilmister and D.J. Newman, Proc. Camb. Phil. Soc. 57(1961) 851; C.N. Yang, Phys. Rev. Lett. 33 (1974) 445.

11.Y. Ne'eman and Dj. Sijacki, J. Math. Phys. 26 (1985) 2457; same authors, Internat. J. Mod. Phys. A2 (1987) 1655

12.F.W. Hehl, G.D. Kerlick and P. von der Heyde, Phys. Lett. B63 (1976) 446; F.W. Hehl, E.A. Lord and Y. Ne'eman, Phys. Lett. B71 (1977) 432 and Phys. Rev. D17 (1978) 428; F.W. Hehl, J.D. McCrea, E.W. Mielke and Y. Ne'eman, to be published (1990) in Physics Reports.

13.B.S. DeWitt, Phys. Rev. 162 (1967) 1195; J. Honerkamp, Nucl. Phys. B36 (1971) 130; G. 'tHooft, ibid, B62 (1973) 444; B.S. DeWitt, in Quantum Gravity 2 , C. Isham et al. edts., Clarendon Press, Oxford (1981); L.F. Abbott, Nucl. Phys. B185 (1981) 189; D.G. Boulware, Phys. Rev. D23 (1983) 389.

14.G. 'tHooft and M. Veltman, Ann. Inst. H. Poincare, 20 (1974) 69; S. Deser J. Kay and K. Stelle, Phys. Rev. Lett. 38 (1977) 527.

APPENDIX
Dimensionality, Density-weight and Ghost Number

Field	Dim.	Densi. weight	Ghost number	θ #	Source or f. funct.	Dim.	Densi. weight	Ghost number
$G^{\mu\nu}$	0	1/2	0		J_h	4	1/2	0
$H^{\mu\nu}$	0	1/2	0		J_t	4	1	0
$h^{\mu\nu}$	0	1/2	0	6	J_u	3	1	0
$G_{\mu\nu}$	0	1/2	0		ζ^{μ}	4	1	1
$H_{\mu\nu}$	0	1/2	0					
$h_{\mu\nu}$	0	1/2	0		$\bar{\zeta}_{\nu}$	4	1	-1
$\omega^{a}{}_{b\nu}$	1	0	0		$\zeta^{a}{}_{b}$	3	1/2	1
$U^{a}{}_{b\nu}$	1	0	0					
$u^{a}{}_{b\nu}$	1	0	0	2	$\bar{\zeta}^{b}{}_{a}$	3	1	-1
$e^{a}{}_{\mu}$	0	0	0					
$T^{a}{}_{\mu}$	0	0	0					
$t^{a}{}_{\mu}$	0	0	0	4*				
a^{ν}	1	1/2	0		K_h	3	1/2	-1
b_{λ}	3	1/2	0		K_t	3	1	-1
					K_u	2	1	-1
C^{μ}	0	0	1	4				
					L_{ν}	3	1	-2
$\bar{C}_{\nu}$	0	0	-1		$L^{b}{}_{a}$	2	1	-2
$C^{a}{}_{b}$	1	0	1	2				
					G	4	1/2	0
$\bar{C}^{b}{}_{a}$	1	1/2	-1		X	1	0	1
					P	3	1/2	-1
					F,M,M'	0	0	0
					N,N'	0	0	0

#θ is the power of the momentum in the denominator of the propagator.
*This value envisages the inclusion of terms $\partial S.\partial S$ (where S is the Torsion $\sim(\partial e - \omega e)$).

ON THE NATURE OF THE TRANSITION IN QUANTUM FIELD THEORY FROM FLAT TO CURVED SPACE-TIME

Paul J. Camp
Physics and Astronomy, University of South Carolina
Columbia, SC 29208 U.S.A.

and

John L. Safko
Physics and Astronomy, University of South Carolina
Columbia, SC 29208 U.S.A.

ABSTRACT

A physical argument based on the uncertainty principle is presented which suggests that the standard method of formulating quantum field theory in a curved background space-time is incomplete. It is argued that there should be curvature dependent factors in the canonical commutator relations which vanish in the Minkowski limit. A massive scalar field in a radiation dominated Friedmann-Robertson-Walker universe is studied as an example. On renormalizing the theory, it is seen that this assumption alone is sufficient to provide rapid inflation and avoidance of the initial singularity normally associated with this example.

1. Introduction

Over the course of the past two decades the results obtained from treating gravity classically and everything else quantum mechanically have been alternately startling and satisfying. Parker[1] began the period with an analysis of particle creation in expanding universes followed a few years later by Hawking's[2] elegant demonstration of the production of black body radiation by a static black hole. More recently, Guth[3] and Linde[4] among others have developed the inflationary universe scenario as a means of explaining a number of long standing cosmological puzzles mostly having to do with initial conditions and the observational absence of particles predicted by standard particle theory. All of these results are based on a particular method of making the transition from flat space to curved backgrounds. It is the purpose of this paper to question the completeness of that procedure and to propose an alternative.

It is easily shown that the existence of a commutator bracket between two quantum operators implies an uncertainty relation between the two associated observables. It is not so easily shown, although it has been done, that the reverse is also true. That is, the existence of an uncertainty relation between two observables implies the existence of a commutator bracket between the two associated quantum operators.[5] Yet it is this latter implication which is the most significant since the uncertainty relation is what is physically observed and the commutator is merely the mathematical representation of this observational result. Thus, one should take care when going to a curved background to always think in terms of uncertainty rather than commutator and avoid placing undue stress on the mathematical formalism at the expense of the physical ideas.

The commutator for position and momentum can be expressed in flat space-time in the form:

$$[x^{\mu},p_{\nu}] = i\hbar\delta^{\mu}{}_{\nu} \tag{1.1}$$

Normally, it is argued that one goes to curved geometries by simply preserving this relation. However, in the light of the uncertainty relations we must think of a particle as occupying a volume and this procedure takes no account of how the geometry might be changing across that volume. As an analogy, think of a spring (i.e., wave packet) sitting in the bottom of a bowl (the geometry). Two different bowls are shown in the following figure.

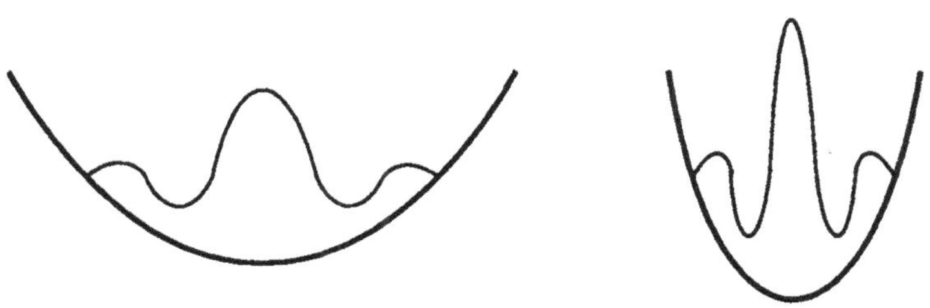

We may certainly choose bowls with different curvatures but which have the same local value of the bowl at the center of the spring. Yet the shape of the springs will be different for the two cases illustrated above due to the difference in curvature of the sides of the two bowls. A purely local relation such as equation 1.1 will take no notice of this. This argument suggests that we should also insert a nonlocal curvature dependent factor in the commutator which goes to one in the limit of zero curvature. The wave function is, after all, a nonlocal entity. We propose modifying equation 1.1 to

$$[x^{\mu},p_{\nu}] = i\hbar f\,\delta^{\mu}{}_{\nu} \tag{1.2}$$

where f may depend on position and momentum as well as curvature. Any non-constant f is equivalent to making Planck's constant a function of space-time. We will investigate only the possibility of an explicit curvature dependence in f.

It is easy to see qualitatively how this idea could neatly solve several problems in the early universe. If we assume that f is an increasing function of curvature (and it is difficult to believe that the universe is less quantum mechanical at high energies) then particles will more and more strongly resist being localized as we go to earlier and earlier times. This is loosely analogous to the electron degeneracy pressure which supports a white dwarf star. The curvature dependence of f thus provides a natural mechanism for inflation and singularity avoidance by an uncertainty induced pressure. Furthermore, this proposal should help to solve the horizon problem (in which causally disconnected portions of the universe nevertheless seem to have been in thermal equilibrium in the past) due to the greatly increased uncertainty in position. These and other cosmological puzzles relevant to inflation are discussed extensively by Blau and Guth. [6]

The inflationary scenario rests on the spontaneous symmetry breaking aspect of unified field theories but this presents a problem. It has been pointed out[6] that the effective potential for the Higgs field is not flat enough at the origin to account for inflation due to radiative corrections from the gauge couplings. Consequently, most workers in the field consider inflation to be driven by a scalar field which may be an axion field, a supersymmetry breaking field, or in many cases a field that serves no other function. None of these fields have been observed. In the scenario we propose the inflation does not care about effective potentials or any other aspects of GUTs; it will happen no matter what.

It should be pointed out that the use of equation 1.2 in black hole radiance calculations will no doubt affect the spectrum of the radiation (barring mathematical accidents of a most unlikely character). However, one should not be too wedded to the idea of black holes as black bodies as this does not seem to be tied to any deep and abiding principles of field theory. It is suggestive that, in a closely analogous problem (the detection of thermal radiation by an accelerated detector) the thermal spectrum has been shown by Pringle[7] to be mathematical accident arising from an EPR-like Minkowski ground state. Since these two cases are so similar, it would not be surprising if this mathematical conclusion should also hold for the black hole problem.

A beneficial effect of our proposal may arise, however, which could counterbalance the loss of such a satisfying picture. It is likely that the singularity inside a black hole is not a true mathematical singularity but some sort of quantum object with a small though definite size l normally taken to be of the order of the Planck length. However, as in the case of inflation, we can now imagine this size to be f dependent. As the black hole shrinks, curvature and f increase and therefore so does l. There may eventually come a time at which l is larger than the radius of the event horizon. If that occurs, the quantum black hole may no longer be a black hole and the ultraviolet catastrophe of Hawking radiation may not occur. What the final state might be is unclear although it seems probable that the black hole would radiate all its remaining mass at once according to whatever the appropriate particle spectrum is. This would not be an infinite-temperature spectrum, however, as the black hole is still of finite size.

This paper is a report on a work in progress. A renormalized energy tensor has not yet been obtained nor back reaction calculated. The question of experimental verification has not yet been addressed. Nevertheless, the results obtained so far are most satisfying and intriguing.

We first address the modified form of the Klein-Gordon equation, its solution in a Friedmann-Robertson-Walker (FRW) metric, and its Feynman propagator. In the third section, we address the complicated problem of renormalization and the interplay between the scalar and gravitational fields. In the final section some problems and possible future avenues are discussed.

We use the Misner, Thorne and Wheeler[18] timelike sign convention with (+ - - -) metric signature.

2. The Klein-Gordon Equation

Equation 1.2 can be implemented by absorbing the factor f into the momentum operator as

$$p^{\mu} = i\hbar f \nabla^{\mu} \tag{2.1}$$

By minimal substitution in the invariant length of the momentum vector, we obtain the Klein-Gordon equation,

$$f \Box \phi + \nabla f \cdot \nabla \phi + \frac{m^2}{f} \phi = 0 \tag{2.2}$$

It is apparent, then, that two conditions on f are that $f \rightarrow 1$ in the limit of zero curvature and that its derivative vanish in the same limit. It will be seen from the Feynman propagator that all derivatives of f must vanish in this limit. A reasonable form for f might then be

$$f = e^{R/R_o} \tag{2.3a}$$

where R_o can be negative or positive to obtain classical or superquantum limits at high curvature. The only natural constant hanging around which might be used for R_o is (Planck length)$^{-2}$ which is a truly huge number. In this case, one would expect significant deviations from standard quantum field theory (SQFT as opposed to our modified quantum field theory MQFT) only at very early times.

Unfortunately, for the simplest realistic models involving a radiation dominated FRW cosmology at early times, R=0 and equation 2.3a fails to produce a modification. The next simplest form is

$$f = e^{-R_{\mu\nu}n^{\mu}n^{\nu}/R_o} \tag{2.3b}$$

where n^{μ} is a unit timelike vector fixed in the metric. The minus sign comes from our choice of metric signature and is needed to make f an increasing function.

The line element for a closed, radiation-dominated FRW cosmology is (see e.g. Misner, Thorne and Wheeler[18])

$$ds^2 = dt^2 - a^2(t)\left[\frac{dr^2}{1-r^2} + r^2\,(d\theta^2 + \sin^2\theta\, d\phi^2)\right], \tag{2.4a}$$

or

$$ds^2 = c(\eta)\left[d\eta^2 - \frac{dr^2}{1-r^2} - r^2\,(d\theta^2 \sin^2\theta\, d\phi^2)\right], \tag{2.4b}$$

where the second expression is in terms of the conformal time parameter, η.

If we introduce the parameter

$$a^{*2} = \frac{8\pi}{3} a_0^4\, \rho_{r0} \tag{2.5}$$

where a_0 and ρ_{r0} are the current values of scale factor and radiation density, then the FRW cosmology line element is given by

$$c(\eta) = a^{*2} \sin^2\eta, \qquad 0 \le \eta \le \pi \tag{2.6}$$

with the curvature

$$R_{00} = -\frac{3}{\sin^2\eta}. \tag{2.7}$$

The Klein-Gordon equation 2.2, expressed in this metric, may be separated into time and space parts as

$$\phi = e^{i\mathbf{k}\cdot\mathbf{x}}\, T(\eta) \qquad \text{with} \qquad |\mathbf{k}|^2 = \omega^2, \tag{2.8}$$

where T satisfies

$$\frac{d^2T}{d\eta^2} + \left(2\,\frac{\cos\eta}{\sin\eta} - \frac{6}{R_0}\,\frac{\cos\eta}{\sin^3\eta}\right)\frac{dT}{d\eta} + \left[m^2 a^{*2} \sin^2\eta\, \exp\left\{-\frac{6}{R_0 \sin^2\eta}\right\} + \omega^2\right] T. \tag{2.9}$$

Define the new parameters and independent variable u by

$$\alpha = \frac{6}{R_0}, \qquad \beta = m\, a^*, \qquad \text{and} \qquad u = \frac{1}{\eta}. \tag{2.10}$$

Then equation 2.9 may be rewritten for early times (small η or large u) as

$$\frac{d^2T}{du^2} + \alpha u \frac{dT}{d\eta} = 0. \tag{2.11}$$

This has the solution

$$T = \frac{\sqrt{\pi}}{2} \operatorname{erf}\left(\frac{\alpha}{2} u\right) \tag{2.12}$$

where erf(x) is variously known as the error function or the probability integral. A useful form valid at early times is obtained by solving equation 2.11 in the original coordinate η

$$T = -\int_0^{\eta} \frac{1}{\xi^2} e^{-\alpha/(2\xi^2)} d\xi . \tag{2.13}$$

This may be integrated by the trapezoid rule since η is small to give

$$T(\eta) \cong -\frac{1}{2\eta} e^{-\alpha/(2\eta^2)} , \tag{2.14}$$

This solution vanishes as η approaches 0, a property which shall shortly prove most interesting.The oscillatory part of T has been neglected at small times in obtaining equation 2.11; over the short period of time in which equation 2.14 is valid, this oscillatory portion would change negligibly.

Bunch and Parker[8] have given a local momentum space expansion of the Feynman propagator and shown that it is equivalent to the proper time formalism developed by Schwinger[9] and DeWitt.[10] We begin with the equation for the Feynman propagator, which must be expressed in self-adjoint form in order for G_F to be a Green function.[11] The propagator must satisfy

$$\nabla(f\nabla)\, G_F(x,x') + \frac{m^2}{f} G_F(x,x') = \sqrt{-g}\, \delta^4(x - x') \tag{2.15}$$

The central idea is to expand equation 2.15 in Riemann normal coordinates[12] around the point x' and then Fourier transform G_F. The equation which one obtains is too frightful for words. This equation is solved by iteration, writing the Fourier transformed propagator as a series of terms, each of which solves a part of the equation involving successively higher derivatives of the metric. After some calculation, one arrives at the expression

$$G(x,x') = -\frac{i\,\Delta^{1/2}(x,x')}{(4\pi)^{n/2}} \int_0^{\infty} i\, ds\, (is)^{-n/2} \exp\left[-\frac{is\, m^2}{f} + \frac{\sigma}{2\, is\, f}\right] \times \left\{a_{-1}\,(is)^{-1} + a_0 + a_1\,(is) + a_2\,(is)^2 + \dots\right\}, \tag{2.16}$$

where σ is half the square of the geodesic distance from x to x', Δ is the Van Vleck-Morette determinant, and the a's are given by

$$a_{-1}(x,x') = -\frac{1}{4}\sigma^{;\alpha} f_{;\alpha}\frac{\sigma}{f^2} - \frac{1}{12} f_{;\alpha\beta}\,\sigma^{;\alpha}\,\sigma^{;\beta}\frac{\sigma}{f^2}, \tag{2.17a}$$

$$a_0(x,x') = 1 - \frac{1}{4}\frac{f_{;\alpha}}{f}\,\sigma^{;\alpha} - \frac{i}{6}\frac{f_{;\alpha\beta}}{f}\,\sigma^{;\alpha}\,\sigma^{;\beta} - \frac{1}{12}\frac{f_{;\alpha\beta}}{f}\,\sigma^{;\alpha}\,\sigma^{;\beta} + \frac{1}{12}\sigma\frac{f_{;\lambda}{}^{\lambda}}{f^2}, \tag{2.17b}$$

$$a_1(x,x') = \frac{f}{6}R - \frac{1}{2}\sigma^{;\alpha}\left(\frac{f}{6}R_{;\alpha} + \frac{m^2 f_{;\alpha}}{f^2}\right) - \frac{1}{3}\,a_{\alpha\beta}\,\sigma^{;\alpha}\,\sigma^{;\beta} + \frac{1}{12}\frac{f_{;\lambda}{}^{\lambda}}{f^2}, \tag{2.17c}$$

and

$$a_2(x,x') = \frac{1}{2}\left(\frac{f}{6}R\right)^2 + \frac{1}{30}\Box R + \frac{m^2 f_{;\lambda}{}^{\lambda}}{6f^2} - \frac{f}{180}R^{\alpha\beta}R_{\alpha\beta} + \frac{f}{180}R^{\alpha\beta\gamma\delta}R_{\alpha\beta\gamma\delta}, \tag{2.17d}$$

where

$$a_{\alpha\beta}(x,x') = \frac{f}{12}R_{;\alpha\beta} + \frac{m^2 f_{;\alpha\beta}}{2f^2} - f\left\{-\frac{1}{30}R_\alpha{}^\lambda R_{\lambda\beta} + \frac{1}{60}R^\kappa{}_\alpha{}^\lambda{}_\beta R_{\kappa\lambda}\right.$$

$$\left. + \frac{1}{60}R^{\lambda\mu\kappa}{}_\alpha R_{\lambda\mu\kappa\beta} - \frac{1}{120}R_{;\alpha\beta} + \frac{1}{40}\Box R_{\alpha\beta}\right\}. \tag{2.17e}$$

The major difference between this and the SQFT result, aside from the appearances of f, is the factor a_{-1} which diverges at the lower limit of integration. Fortunately it, and its relevant derivatives, vanish in the coincidence limit $x \to x'$ and so they may be safely discarded for purposes of renormalization. The series may be carried to higher powers of s but these are all the terms that are essential for isolating divergences in the effective action. It should also be pointed out that this equation is valid for any metrical form with the proviso that f has been taken to be a function of the curvature only.

3. Renormalization in the Effective Action

The primary goal of our research at this point is to renormalize the energy tensor for the field ϕ and to calculate its back reaction on the metric. This project is enormously complicated, even in SQFT. Thus it is beneficial to first approach renormalization through the effective action, introduced, as were so many other things, by DeWitt.[13] We shall proceed by the dimensional regularization scheme due to 't Hooft and Veltman.[14]

In Minkowski space one would simply discard the divergent parts, by normal ordering or some such scheme, but this cannot be done in a curved background since energy is a source of gravity and so may not be fiddled arbitrarily. Furthermore, even discarding a Minkowski-type term leaves a divergent remainder.[15] It is apparent that a more sophisticated procedure will be needed. The trick, ultimately, is to absorb infinities into renormalized gravitational coupling constants. We shall now see briefly how this can be done.

There is, however, a new wrinkle in MQFT which inextricably intertwines the gravitational and matter fields. The action* for the Klein-Gordon equation is

$$S = \int d^4x \sqrt{-g}\, \frac{1}{2} \left(f\, g^{\alpha\beta}\, \phi_{;\alpha}\, \phi_{;\beta} - \frac{m^2}{f}\, \phi^2 \right). \tag{3.1}$$

Variation of this action with respect to the metric yields the energy tensor

$$T^{\alpha\beta} = \frac{1}{2} g^{\alpha\beta} \left(f\, \phi_{;\mu}\, \phi^{;\mu} - \frac{m^2}{f}\, \phi^2 \right) - f\, \phi^{;\alpha}\, \phi^{;\beta} + \left(A^{\alpha\beta\mu\nu}{}_{;\mu\nu} + A^{\mu\nu\alpha\beta}{}_{;\mu\nu} - 2\, A^{\alpha\mu\beta\nu}{}_{;\mu\nu} \right). \tag{3.2a}$$

where

$$A^{\alpha\beta\mu\nu} = -\frac{f}{R_0} \left(\phi_{;\gamma}\phi^{;\gamma} + \frac{m^2}{f^2}\, \phi^2 \right) n^\alpha n^\beta g^{\mu\nu}. \tag{3.2b}$$

This energy tensor is not conserved. However, examination of equation 3.1 or the equation of motion 2.2 reveals that this problem is formally similar to the classical problem of a particle with a time-varying mass such as a rocket. In that case we know the energy of the rocket alone is not conserved since we have neglected that energy carried off by the reaction products. Similarly, the effect of f in our problem is to transfer energy between the gravitational and matter fields. We should therefore not expect the energy of either field to be conserved independently. This transfer is, of course, the physical basis for the inflationary process in the MQFT.

We introduce an auxiliary tensor $\tau_{\alpha\beta}$ into the gravitational field equations. This tensor satisfies the equation

$$\tau^{\alpha\beta}{}_{;\beta} = T^{\alpha\beta}{}_{;\beta}\,, \tag{3.3}$$

and is to be interpreted as geometry rather than matter. Obviously there is a certain amount of gauge ambiguity in $\tau_{\alpha\beta}$, which is arbitrary up to a divergence-free quantity. The full gravitational field equations are then given by

$$E^{\alpha\beta} = G^{\alpha\beta} + 8\pi G\, \tau^{\alpha\beta} + \Lambda\, g^{\alpha\beta} = 8\pi G\, T^{\alpha\beta} \tag{3.4}$$

where $G^{\alpha\beta}$ is the usual Einstein curvature tensor. This additional tensor guarantees that the whole equation has zero divergence although neither side is independently conserved.

For our FRW solution, the divergence of $T_{\alpha\beta}$ is

*It should perhaps be pointed out here that the action for the Schrödinger equation has an additional factor proportional to $f_{;0}/f$. This would introduce an extra curvature dependent phase factor into such phenomena as the Aharonov-Bohm effect.

$$T^{\mu\beta}{}_{;\beta} = \frac{1}{2} f^{;\mu}\left(\phi_{;\alpha}\phi^{;\alpha} + \frac{m^2}{f^2}\phi^2\right) \cong \frac{\alpha k^2}{2\eta^3} e^{\alpha/2\eta^2} \phi^2 \delta^\mu{}_o \, . \tag{3.5}$$

at early times (the contributions of the A's cancel then). Equation 3.3 for $\tau_{\alpha\beta}$ can be solved to the same order of η by

$$\tau^{\mu\nu} = \frac{a^{*2}k^2}{2} \eta^2 e^{\alpha/(2\eta^2)} g^{\mu\nu} \phi^2 , \tag{3.6}$$

which is derivable from variation of the action

$$S_\tau = + \int \sqrt{-g} \left(a^{*2}k^2 \eta^2 \ e^{\alpha/(2\eta^2)} \phi^2 \right) d^4x \tag{3.7}$$

with respect to the metric.

For the background field approach, we use the semiclassical Einstein field equations obtained from 3.4 above by replacing T and τ with their vacuum expectation values. We then define the effective action to be that quantity whose metric variation gives $\langle T_{\alpha\beta}\rangle$; that is,

$$\langle T^{\mu\nu}\rangle = \frac{2}{\sqrt{-g}} \frac{\delta W}{\delta g_{\mu\nu}} \, . \tag{3.8}$$

It is straightforward, though not illuminating, to obtain the relation

$$W = -\frac{i}{2} \mathrm{Tr}[\ln(-G_F)] \tag{3.9}$$

where G_F is interpreted as an operator acting on the space of vectors $|x\rangle$ with delta function normalization. The derivation may be found anywhere in the standard literature and will not be repeated here (we have largely followed Birrell and Davies[15]). In this space, $G_F(x,x') = \langle x|G_F|x'\rangle$. Let H be the differential operator of equation 2.15, with a small negative imaginary part. Then formally we may write the operator equation

$$G_F = -\frac{1}{H} = -i \int_0^\infty e^{-iHs} \, ds \, . \tag{3.10}$$

From this it follows that

$$\ln(-G_F) = -\ln H = \int_0^\infty e^{-iHs} \frac{1}{is} d(is) \, . \tag{3.11}$$

Since the propagator $G_F(x,x') = \langle x|G_F|x'\rangle$, equations 2.16 and 3.10 may be used to evaluate $\langle x|\exp[-iHs]|x'\rangle$. This latter expression is effectively the integrand of equation 2.16. This fact may be combined with equation 3.11 to yield

$$\langle x|\ \ln(-G_F)\ |x'\rangle = -\int_{m^2}^{\infty} \frac{1}{f} G_F(x,x')\, dm^2 \,. \tag{3.12}$$

The trace of this then gives the effective action

$$W = \frac{i}{2}\int_{m^2}^{\infty} dm^2 \int d^n x \sqrt{-g}\, \frac{1}{f(x)} G_F(x,x') \tag{3.13}$$

and the effective Lagrangian,

$$L_{eff} = \frac{i}{2} \lim_{x\to x'} \int_{m^2}^{\infty} \frac{1}{f(x)} G_F(x,x')\, dm^2 \,, \tag{3.14}$$

where the convergence of the integral at the upper end of the integration is guaranteed by the Feynman boundary condition $m^2 \to m^2 - i\epsilon$. Again, nothing significantly different from the standard derivation has occurred.

The effective Lagrangian is explicitly

$$L_{eff} = \frac{1}{2(4\pi)^{n/2}} \lim_{x\to x'} \sqrt{\Delta(x,x')} \sum_{j=-1}^{\infty} \left\{ a_j(x,x') \int_0^{\infty} (is)^{j-1-n/2} \exp\left[\frac{-ism^2}{f} + \frac{\sigma}{2isf}\right] i ds \right\} \tag{3.15}$$

If we allow dimensionality n to be a variable analytically continued into the complex plane then this expression is finite and we may take the coincidence limit $\sigma \to 0$ and $a_j(x,x') \to a_j(x)$. The integral in 3.15 can then be written as a Γ function

$$L_{eff} = \frac{1}{2(4\pi)^{n/2}} \left(\frac{m^2}{f\,\kappa^2}\right)^{-2+n/2} \sum_{j=0}^{\infty} \left\{ a_j(x) \left(\frac{m^2}{f}\right)^{2-j} \Gamma\left(j - \frac{n}{2}\right) \right\} , \tag{3.16}$$

where κ is a mass scaling factor which has been introduced to keep the units consistent as n varies. κ will disappear eventually as the limit $n \to 4$ is taken.

The divergence behavior of L_{eff} is the same as that of Γ which, for $n = 4$, diverges in the range $0 \leq j \leq 2$. Using the expansions

$$\Gamma\left(-\frac{n}{2}\right) = \frac{4}{n(n-2)}\left(\frac{2}{4-n} - \gamma\right) + O(n-4), \tag{3.17a}$$

$$\Gamma\left(1-\frac{n}{2}\right) = \frac{2}{2-n}\left(\frac{2}{4-n} - \gamma\right) + O(n-4), \tag{3.17b}$$

$$\Gamma\left(2-\frac{n}{2}\right) = \frac{2}{4-n} - \gamma + O(n-4), \tag{3.17c}$$

and

$$\left(\frac{m^2}{f\,\kappa^2}\right)^{-2+n/2} = 1 + \frac{1}{2}(n-4)\ln\left(\frac{m^2}{f\,\kappa^2}\right) + O\left((n-4)^2\right). \tag{3.17d}$$

The divergent part of L_{eff} is easily found to be

$$L_{div} = -\frac{\hbar}{(4\pi)^{n/2}}\left[\frac{1}{n-4} + \frac{1}{2}\left\{\gamma + \ln\left(\frac{m^2}{f\,\kappa^2}\right)\right\}\right]$$

$$\times\left[\left(\frac{m^2}{f}\right)^2 \frac{4a_0}{n\,(n-2)} - \left(\frac{m^2}{f}\right)\frac{2a_1}{n-2} + a_2\right]. \tag{3.18}$$

A similar procedure may be followed for the τ Lagrangian by noting that the Feynman propagator is defined by

$$G_F = -i\left\langle 0 \right| T\left[\phi(x)\,\phi(x')\right] \left| 0 \right\rangle, \tag{3.19}$$

where T denotes time ordering. So L_τ may be written as

$$L_\tau = -i\,\frac{a^{*2}k^2}{2}\,\eta^2\,e^{\alpha/(2\eta^2)} \lim_{x\to x'} G_F(x,x')\,. \tag{3.20}$$

This expression is valid for small times, with $K = 3ak^2/a^{*4}$. Properly speaking, this should be a weighted average over field modes k but, as we shall eventually see, all save the $k = 0$ mode are unimportant. The divergent part of this Lagrangian, derived in precisely the same manner as before, is given by

$$L_{\tau,\mathrm{div}} = \frac{1}{(4\pi)^{n/2}} a^{*2} k^2 \eta^2 e^{\alpha/(2\eta^2)} \left[\frac{1}{n-4} + \frac{1}{2}\left\{\gamma + \ln\left(\frac{m^2}{f\,\kappa^2}\right)\right\}\right]$$

$$\times \left[\frac{m^2}{f}\,\frac{2\,a_0}{2-n} + a_1\right]. \tag{3.21}$$

Both equations 3.21 and 3.18 are geometrical in the sense that no matter field operators appear and so they may properly be absorbed into the gravitational Lagrangian through renormalization of the gravitational coupling constants

$$L_g = -\left(A + \frac{\Lambda_b}{8\pi G_b}\right) + \left(B + \frac{1}{16\pi G_b}\right) R + C, \tag{3.22}$$

where the subscript b denotes "bare" quantities and A, B, and C come from the parts of equations 3.21 and 3.18 involving, respectively, a_0, a_1, and a_2. These are

$$A = \left[\frac{4\,m^4}{(4\pi)^{n/2}\, n\,(n-2)\, f^2} - \frac{1}{(4\pi)^{n/2}} a^{*2} k^2 \eta^2 e^{\alpha/(2\eta^2)} \frac{2m^2}{(2-n)\,f}\right]$$

$$\times \left[\frac{1}{n-4} + \frac{1}{2}\left\{\gamma + \ln\left(\frac{m^2}{f\,\kappa^2}\right)\right\}\right], \tag{3.23a}$$

$$B = \left[\frac{m^2}{3(4\pi)^{n/2}\,(n-2)} - \frac{1}{(4\pi)^{n/2}} a^{*2} k^2 \eta^2 e^{\alpha/(2\eta^2)} \left(\frac{-f}{6}\right)\right]$$

$$\times \left[\frac{1}{n-4} + \frac{1}{2}\left\{\gamma + \ln\left(\frac{m^2}{f\,\kappa^2}\right)\right\}\right], \tag{3.23b}$$

and

$$C = \left[\frac{m^2\, f_{;\lambda}{}^{\lambda}}{6(4\pi)^{n/2}\,(n-2)\, f^2} - \frac{a_2(x)}{(4\pi)^{n/2}} - \frac{1}{2(4\pi)^{n/2}} a^{*2} k^2 \eta^2 e^{\alpha/(2\eta^2)} \left(\frac{f_{;\lambda}{}^{\lambda}}{6f^2}\right)\right]$$

$$\times \left[\frac{1}{n-4} + \frac{1}{2}\left\{\gamma + \ln\left(\frac{m^2}{f\,\kappa^2}\right)\right\}\right]. \tag{3.23c}$$

C is quadratic in the curvature tensors, so it involves terms which do not normally appear in the Einstein equations. The message of equation 3.23c is that such terms cannot be a priori excluded. However, the renormalized coupling constants may presumably be set equal to zero as there are rather stringent observational limits on them.[16]

The renormalized cosmological and gravitational constants may be easily written as

$$\Lambda = \Lambda_b + G_b\left[\frac{32\pi\, m^4}{(4\pi)^{n/2}\, n\,(n-2)}\frac{1}{f^2} + \frac{16\pi\, a^{*2}k^2\, m^2}{(4\pi)^{n/2}\,(n-2)}\frac{1}{f}\,\eta^2\, e^{\alpha/(2\eta^2)}\right]$$

$$\times\left[\frac{1}{n-4} + \frac{1}{2}\left\{\gamma + \ln\left(\frac{m^2}{f\,\kappa^2}\right)\right\}\right], \qquad (3.24)$$

and

$$G = \frac{G_b}{1 + 16\pi\, G_b\, B}. \qquad (3.25)$$

There are two possible ways to interpret these equations. First, the bare quantities are infinite valued functions of f which exactly cancel the renormalization factors to yield finite constants. However, we adhere to the Principle of Minimum Mischief, an idea not sufficiently emphasized in the provisional and ad hoc renormalization game. By this principle we mean that after you have calculated according to the best theory available and obtained divergent results, make the minimum adjustments necessary to render the divergent terms finite with the least resulting damage to the physics. For G this means that G_b must be a function of f and G the normal Newtonian constant; there is no consistent way to get a space-time dependent gravitational constant from this approach. A similar conclusion holds for the terms quadratic in the curvature mentioned previously. But the cosmological constant is a different matter; the bare quantity may be taken to be an infinite constant.

From equations 3.24 and 3.25 we may write:

$$\Lambda_b = \Lambda_o - G_b\left[\frac{32\pi\, m^4}{(4\pi)^{n/2}\, n\,(n-2)} + \frac{16\pi\, a^{*2}k^2\, m^2}{(4\pi)^{n/2}\,(n-2)}\frac{1}{f}\,\eta_o^2\, e^{\alpha/(2\eta_o^2)}\right]$$

$$\times\left[\frac{1}{n-4} + \frac{1}{2}\left\{\gamma + \ln\left(\frac{m^2}{\kappa^2}\right)\right\}\right], \qquad (3.26)$$

and

$$G_b = G\left\{1 - 16\pi G\left[\frac{m^2}{3(4\pi)^{n/2}\,(n-2)} + \frac{a^{*2}k^2}{6(4\pi)^{n/2}}\,\eta^2\, e^{\alpha/(2\eta^2)}\right]\right.$$

$$\left.\times\left[\frac{1}{n-4} + \frac{1}{2}\left\{\gamma + \ln\left(\frac{m^2}{f\,\kappa^2}\right)\right\}\right]\right\}^{-1}, \qquad (3.27)$$

where the subscript o denotes today's values. These equations may be substituted in equation 3.24 and, after multiplying numerator and denominator by $n-4$, we discover that there is a finite $n = 4$ limit given by

$$\Lambda = \Lambda_0 + \frac{\frac{3}{2}m^4\left[\left(1 - \frac{1}{f^2}\right) + 2a_*^2k^2\left(\eta_0{}^2\, e^{\alpha/(2\eta_0{}^2)} - \frac{1}{f}\,\eta^2\, e^{\alpha/(2\eta^2)}\right)\right]}{m^2 + a_*^2k^2\, f\, \eta^2\, e^{\alpha/(2\eta^2)}} . \tag{3.28}$$

This expression vanishes at early times except for k = 0. Higher modes do not contribute, as previously indicated. In this case, we obtain the particularly simple expression

$$\Lambda = \Lambda_0 + \frac{3}{2}m^2\left(1 - \frac{1}{f^2}\right) \tag{3.29}$$

which goes to a large (effectively m^2) but finite constant at very early times when $f \to \infty$, and decreases smoothly to today's value.

A possible objection at this point is that we have again arrived at the problem of non-conservation, this time through the cosmological parameter. But since these terms were obtained from the matter Lagrangian, it is reasonable to expect that once again $\langle T_{\alpha\beta} + \tau_{\alpha\beta}\rangle$ has non-vanishing divergence and that it exactly cancels the divergence of the cosmological factor as a square peg fills a square hole. This question can only be answered by a full and explicit renormalization of the energy tensor, a calculation which has not yet been completed.

4. Conclusion

Equation 3.29 offers the possibility of a most entertaining scenario for the early universe. At some early time (after the Planck time) we imagine a highly curved vacuum. In this universe, $\langle T_{\alpha\beta}\rangle = 0$ (recall that ϕ=0 for small η according to equation 2.14), f is very large, the cosmological parameter is very large, and the universe expands rapidly in a quasi-de Sitter fashion. This is an inflationary phase but, in contrast to the standard inflationary model which requires the field to have a large initial value, equation 3.29 requires only that the equations of motion for the scalar field be available for renormalization. As time passes, f decreases and so the inflation smoothly shuts off. However, since the divergence of the field equation 3.4 must vanish, as the cosmological term decreases $\langle T_{\alpha\beta}\rangle$ must increase; so f transfers energy from the inflation into the matter field. Hence, the initially empty universe inflates and creates matter as it goes along.

A similar model was developed by Starobinsky[17] but, as Blau and Guth comment,[6] the initial conditions for his model appear to be implausible. Our model appears to require no specific initial conditions for the matter field; the vanishing of ϕ at η=0 was not imposed at the outset but followed naturally from the equations of motion. Considering the origin of the inflation from an uncertainty-induced pressure, it seems likely that no specific initial conditions are placed on the gravitational field either, other than a time of high curvature. Also, from the same consideration, it is not unreasonable to conjecture that higher spin fields will evolve according to the same general picture. This would eliminate the current necessity of a specifically spin zero field for inflation. This question has not yet been addressed.

This is an amusing idea but to determine its plausibility the renormalized energy tensor, as previously indicated, must be obtained and the back reaction calculated. It might at first be thought that this may be derived from the renormalized effective Lagrangian:

$$L_{ren} = L_{eff} - L_{div} = -\frac{1}{64\pi^2} \int_{s=0}^{\infty} \ln(is) \frac{\partial^3}{\partial(is)^3} \left[F(x,x;is)\, e^{-ism^2/f} \right] d(is) , \tag{4.1}$$

where

$$F(x,x;is) = \sum_{j=0}^{\infty} a_j(x)\, (is)^j . \tag{4.2}$$

This will not work as can easily be seen by reviewing the preceding discussion concerning Feynman propagators. There we obtained only an asymptotic expansion of $F(x,x';is)$, not an exact expression. It is likely that the energy tensor will have to be attacked directly through point splitting or some such formalism, which will in turn have to be redeveloped from the ground up. This problem is currently being investigated.

5. References

1. Parker, L., *Phys. Rev.* **183** (1969) p.1057
2. Hawking, S. W., *Commum. Math. Phys.* **43** (1975) p. 199
3. Guth, A. H., *Phys. Rev.* **D23**, 2 (1981) p. 347
4. Linde, A., *Phys. Lett.* **108B**, 6 (1982) p. 389
5. DeWitt, B. S., *J. Math. Phys.* **3**, 4 (1962) p. 619, 625
6. Blau, S. K. and A. H. Guth in *Three Hundred Years of Gravitation*, ed. S. W. Hawking and W. Israel, (Cambridge University Press, Cambridge, 1987)
7. Pringle, L., Ph.D. dissertation, University of South Carolina, Unpublished.
8. Bunch, T. S. and L. Parker, *Phys. Rev.* **D20**, 10 (1979) p. 2499
9. Schwinger, J., *Phys. Rev.*, **82** (1951) p. 664
10. DeWitt, B. S., in *Relativity, Groups and Topology*, ed. B. S. DeWitt and C. DeWitt (Gordon and Breach, New York, 1965)
    -----, *Phys. Rep.*, **19C** (1975) p. 297
11. See e.g. Arfken, G., *Mathematical Methods for Physicists*, (Academic Press, New York, 1970) on the necessity for this.
12. Petrov, A. Z., *Einstein Spaces*, (Pergamon, Oxford, 1969)
13. DeWitt, B. S., *Phys. Rev.* **160** (1967) p. 1113
    -----, *Phys. Rev.* **162** (1967) p. 1195, 1239
14. 't Hooft, G. and M. Veltman, *Nucl. Phys.* **44B** (1972) p. 189
15. Birrel, N. D. and P. C. W. Davies, *Quantum Fields in Curved Space*, (Cambridge University Press, Cambridge, 1982)

16. Stelle, K. S., *Phys. Rev.* **D16** (1977) p. 953
-----, *Gen. Relativ. Gravit.*, **9** (1978) p. 353
Horowitz, G. T. and Wald, R. M., *Phys. Rev.* **D17** (1978) p. 414
17. Starobinsky, A. A., *Zh. Exp. Teor. Fiz.*, **30** (1979) p. 719
-----, *JETP Lett.* **30** (1979) p. 689
-----, *Phys. Lett.* **91B** (1980) p. 99
18. Misner, C. W., Kip S. Thorne and John A. Wheeler, Gravitation, (W. H. Freeman, San Francisco, 1973)

MUSINGS ON QUANTUM STATISTICS

A. ZEE
Institute for Theoretical Physics, University of California
Santa Barbara, California 93106 U.S.A.

ABSTRACT

The Weisskopf phenomenon, with its connection to the mysteries of quantum statistics, has caused particle physicists much grief.

Surely one of the most amazing features of quantum physics is the appearance of two types of particles: bosons and fermions. While bosons are gregarious, fermions are loners. The wave function acquires a factor of +1 upon the interchange of two bosons, and a factor of −1 upon the interchange of two fermions.

Is it possible to have particles other than bosons and fermions? You may quite naturally guess that if such a generalization is possible the next most interesting number would be i. Such particles are called semions, an etymological hybrid mixing Greek and Latin roots. But semions are apparently impossible: if you interchange them twice you would get $i^2 = -1$. However, they are possible in two dimensional space because in going from our normal three dimensional space to two dimensional space a new physical concept emerges, namely the concept of "going around." One point particle can go around another. The preceding sentence would not be well-defined in three dimensional space. Thus, in two dimensional space, the notion of a phase acquired by the wave function upon the interchange of two particles may be generalized to the notion of a phase acquired by the wavefunction when one particle goes around another. Indeed, not only semions, but anyons, upon whose interchange, the wave function acquires the phase $e^{i\theta}$, are possible. The possibility of this so-called fractional statistics is by now quite well-known.

The interest in semions stems from their possible relevance to high temperature superconductors. This subject is particularly appropriate for this conference, which was organized in part to honor David Bohm: fractional statistics is closely related to the Aharonov-Bohm effect, and the computation to show that the superfluidity of the semion liquid is based on the random phase approximation developed by Bohm and Pines. Since I have already written a detailed review[1] recently, there is no point in my repeating the story here. Instead, I will set down here some of my musings and speculations about quantum statistics.

The nature of quantum statistics lies at the crux of our difficulties in making further progress, since the glory days of the seventies, in understanding fundamental physics. The problem goes back to Weisskopf's discovery in 1934 that while the self-energy of a fermion diverges only logarithmically, the self-energy of a boson diverges quadratically.[2] In classical electrodynamics, the self-energy of a charged particle is just the Coulomb energy contained in the surrounding electric field and this diverges like the inverse of the size of the particle. In other words, the self-energy diverges linearly. Thus, fermi statistics suppresses the self-energy divergence by one power, while bose statistics makes the divergence worse.

As Weisskopf explained in a superb paper published in 1939, the difference in the nature of the divergence can be understood physically in terms of quantum statistics. The "bad" behavior of bosons has to do with their gregariousness. A fermion would push away the fermions contained in the Dirac sea, thus creating a cavity in the vacuum charge distribution surrounding it. Hence its self-energy is less singular than would be the case were quantum statistics not taken into account. A boson does the opposite.

With the advent of grand unified theory, it was realized that the Weisskopf phenomenon means big trouble. In the modern interpretation of renormalization, the self-energy divergence is understood as the difference in the mass (or mass squared) of a particle at one energy scale versus another energy scale. (Another mysterious consequence of statistics is that fermions obey linear equations of motion while bosons obey quadratic equations. Renormalization comes in the mass for fermions, and in the mass squared for bosons.) The mass of a fermion varies logarithmically with energy scale, while the mass squared of a boson varies quadratically. This is a disaster. If we fix the mass of the Higgs field, a bosonic field, to have some generic value at the grand unification scale, typically 10^{15}GeV, then it would have a value of order 10^{15}GeV at the low energy weak interaction scale. We would have to fine tune the mass of the Higgs field at the unification scale in order for it to have at the weak interaction scale a value typical of the weak interaction scale, rather than a value typical of the grand unification scale.

Incidentally, Weisskopf's 1939 paper contains the following prescient statement. "This may indicate that a theory of particles obeying Bose statistics must involve new features at this critical length, or at energies corresponding to this length; whereas a theory of particles obeying the exclusion principle is probably consistent down to much smaller lengths or up to much higher energies." In other words, a theory with an elementary Higgs field cannot be pushed up to grand unification energies. (Notice also that the terms bosons and fermions were unknown in the late thirties.)

Particle physicists have been grappling with this so-called hierarchy problem ever since. Supersymmetry was heralded as providing a solution. Supersymmetry relates bosons to fermions and as such is an extremely attractive notion. It may be said that supersymmetry explains why there are these mysterious fermions in the world. The existence of gauge bosons (and the graviton) follows easily from local symmetries, with all of their intimate connection with geometrical concepts and so on. But who needs fermions? Supersymmetry may provide *raison d'être*. A symmetry connection between fermion and boson solves the hierarchy problem because we know how to give fermions low masses and to keep them low. If we link masses of bosons to masses of their fermionic partners, then bosons could also have low masses.

Unfortunately, instead of linking the observed fermions to the observed bosons, supersymmetry in its actual implementation links the observed particles to as yet unobserved particles. Various schemes have to be devised to hide these unobserved particles. Experimentally, the relevant lower bounds have been improving steadily. Supersymmetry also turns out to be too restrictive in some instances and various convoluted ways to get around these restrictions have to be invented. In short, while low energy supersymmetry has not been ruled out, it has created more problems than it has solved.

Another approach, rather less grand in scope, is to suppose that the Higgs field is not elementary, but merely a composite manifestation of some fermionic interaction at short distances, known as the technicolor interaction. Unfortunately, technicolor also creates problems of its own: in particular, it is difficult to give large enough masses to the observed fermions. A consistent technicolor theory can be constructed but it is exceedingly convoluted.

Thus, in some deep sense, our present difficulties in fundamental physics originate in the nature of quantum statistics and the Weisskopf phenomenon.

Does the story of fractional statistics in two dimensional space shed any light? Apparently not. In quantum field theory, fractional statistics may be incorporated by introducing the Chern-Simons term $\epsilon_{\mu\nu\lambda}A^{\mu}F^{\nu\lambda}$. Suppose we try to generalize it to 3+1 dimensional spacetime. The ϵ symbol has an extra index. To get the indices to match, we may for example introduce a vector n_{μ} and write $\epsilon_{\mu\nu\lambda\sigma}A^{\mu}F^{\nu\lambda}n^{\sigma}$. Thus, for $n_{\mu} = (0,0,0,1)$, each particle would carry a singular tail extending to $z = \pm\infty$. But this breaks Lorentz invariance. We can try to promote n_{μ} to a dynamical field $n_{\mu}(t,\vec{x})$, but then gauge invariance is broken. Perhaps the theory can be hooked up so that the variation under Lorentz transformation may be compensated by a gauge variation. I have not been able to make a sensible theory.

Suppose we promote the abelian gauge potential to a non-abelian gauge potential. Then it can be done. Indeed, it has been known for some time that with non-abelian gauge potentials we can have magnetic monopoles and that amazingly enough, monopoles affect quantum statistics. A composite made of a monopole and a boson in a suitable representation acts as a fermion. Similarly, a composite made of a monopole and a fermion in a suitable representation acts as a boson. Roughly speaking, what happens is that when you interchange two such composites, you pick up an extra phase from the non-abelian gauge potentials. The physics is not entirely dissimilar from the physics responsible for fractional statistics.

It is also possible to construct a fermion out of bose fields, as in the skyrmion. The effective action contains a so-called Wess-Zumino term. Upon interchange of two such lumps of bose fields, the Wess-Zumino term produces an additional phase. This is again not entirely dissimilar from what happens in the fractional statistics case, in which a so-called Hopf term can produce a phase upon the interchange of two lumps of bose fields.

I believe that there remains a great deal to be understood about quantum statistics and that much of it has to do with gauge fields.

Ever since the work of Fermi and Yang, who tried to construct the pion as a nucleon-antinucleon composite, there has been a general prejudice that fermions are somehow more fundamental than bosons, for the obvious reason that if you put two fermions together you get a boson while if you put two bosons together you still get a boson. Thus, in technicolor theory, the bose Higgs field is regarded as a composite of techniquarks. Some people also feel that since fermions are described by Grassmanian variables and since Grassmanian variables are more weird than c-numbers, fermions ought to be more fundamental.

I feel that we should also look at the other possibility, particularly now that we know how to make fermions out of bosons using monopoles and skyrmions. Perhaps quarks and leptons are really composites of some fundamental bose entities and monopoles. Wilczek and I have considered this possibility and have argued that it may be possible to obtain grand unification groups such as $SO(10)$ and $SO(18)$ quite naturally. However, there are also formidable obstacles to be overcome before this sort of idea can be called a model, let alone a theory. In fact, in our speculations, we had to combine fermions with the monopole, rather than bosons.[3]

Recently, I have been playing with another speculative idea, that the half-integral spin carried by fermions may be a delusion of the long distance physicist.[4] The dynamics of a point particle hopping on a lattice in the presence of a suitable arrangement of flux lines may produce an effective fermion field obeying relativistic dynamics in the long distance limit.

These thoughts are exceedingly speculative but they may serve to encourage further work towards a deeper understanding of the nature of quantum statistics.

References

1. A. Zee, "Semionics: a theory of high temperature superconductivity," in *High Temperature Superconductivity*, edited by K. Bedell, D. Pines, and J.R. Schrieffer, (Addison-Wesley, 1990).
2. According to Weisskopf, he had first obtained the erroneous result that the fermion mass diverges linearly. The error was noted by W. Furry.
3. F. Wilczek and A. Zee, *Phys. Rev.* **D25** (1982) 553.
4. A. Zee, "Emergence of spinors from flux and lattice hopping," SBITP preprint (1990).

GRAVITATIONAL CHERN–SIMONS TERM, ANYONS AND AB

S. DESER
Physics Department, Brandeis University
Waltham, MA 02254, USA

ABSTRACT

Solutions of topologically massive gravity generated by a particle source are given. They exhibit some novel effects which we compare to those induced by its electromagnetic counterpart. In particular, a structureless particle source generates a gravi-magnetic field. Conversely, it causes a spinning source to contribute to the static Newtonian potential, thereby violating the equivalence principle; the potential can even become repulsive. The natural presence of Aharonov–Bohm effects in 2+1 dimensional gauge theories is noted, as is the closely related statistics-changing anyon phase.

1. Introduction

Gauge theories in 3 spacetime dimensions provide an exemplary setting for the Aharonov–Bohm effect, so it is a particular pleasure to present here some new results in this field.

Before going into details, we recapitulate how AB comes about here. In gravity, it is well-known that the Einstein and curvature tensors are equivalent in D=3, being just (double) duals of each other. An immediate consequence is that there are no gravitons and so the solutions of the Einstein equations are always locally flat outside the sources. Instead, the exterior is endowed with a global structure that carries the information about the sources, particularly their energy and angular momentum. This carries the usual AB picture to its logical extreme: all matter is a "solenoid," and there is already a classical AB effect in that classical test systems can feel the interior fields even when travelling outside the sources. This has been noted in [1]; closely related four-dimensional examples with cylindrical symmetry were given earlier in [2], where previous D=4 literature is also cited.

The "field-current identity" represented by the Einstein equations (curvature equals stress tensor) is also present in vector gauge theories, where the action of a Chern–Simons term gives rise to field equation expressing proportionality between the field strength $F_{\mu\nu}$, (or rather its dual ${}^*F^\mu$), and the source current j^μ. Again the electromagnetic field vanishes outside the charges, but the global aspects of the vector potentials are nontrivial and carry information in the external field-free regions. In particular, outside a static charge, there exists a transverse (gauge invariant) vector potential whose integral about a closed path measures the interior charge (paradoxically carried here by the magnetic field). The above AB pattern survives even in the more general case of topologically massive theories [3,4] where the field equations are more complex and have their own photon/graviton excitations. Asymptotically, we recover the above salient features of the Einstein and pure vector Chern–Simons models. It should be emphasized that the local flatness/global physics hallmark of our theories is very different from say D=4 Einstein or Maxwell theory, whose exterior Schwarzschild or Coulomb solutions are manifestly not locally field-free, while their asymptotic flatness at spatial infinity is devoid of interesting global properties.

The generation of spin in a structureless charge particle by coupling it to the electromagnetic Chern-Simons term (CST) in 3 spacetime dimensions is by now well-understood. This phenomenon, through the associated change of statistics, may even be physically relevant (see for example [5], where some of the extensive literature is also given). We show here that the gravitational CST has similar effects: A structureless point mass coupled to it generates a gravimagnetic self-field. In each case, a magnetic field can arise because there are transverse circularly symmetric vectors in 2–space. We will also exhibit another common consequence of CST coupling which is the converse of the first: static vectorial sources, namely dipole current and angular momentum density respectively generate Coulomb and Newtonian fields. This represents a violation of the equivalence principle in that inertial and gravitational mass are no longer equal (the latter can even become negative) with a similar implication for charge in the electromagnetic case. Statistics and spin will also be briefly discussed.

There is one important difference between the two theories: the gravitational CST is of third derivative order – one higher than that of the conventional Einstein term, whereas the electromagnetic CST is of first order. Hence in a derivative expansion of the matter-induced effective action, one can neglect the Maxwell term with respect to the CST, while the gravitational expansion will generally be dominated by the Einstein term, the CST representing a (parity violating) correction. [Of the course the dominance of the respective terms in the resulting field equations depends on whether one considers short- or long-range consequences.] One will therefore have to consider the two terms together here; this is topologically massive gravity [3]. Also, because the gravitational CST alone is conformally invariant, it can only couple to a traceless stress tensor such as that of a null particle. To pursue the massive point particle analogy we must in any case include the Einstein term to provide explicit conformal symmetry breaking. For comparison with the Abelian vector model we will consider only the Abelian, linearized, approximation; this should be reasonable in eventual physical applications. The full nonlinear theory, to which we hope to return, will (unlike pure $D = 3$ Einstein theory [3]) be more complicated due to self-energy effects.

2. Vector CST

We first briefly review the electromagnetic case, slightly generalized to include the Maxwell term for later comparison with gravity. A charged point particle is minimally coupled to the vector potential A_μ. The field strength is determined by the field equations [3,4]

$$\partial_\mu F^{\mu\nu} + \mu\, \epsilon^{\nu\alpha\beta}\partial_\alpha A_\beta = -j^\nu \tag{1a}$$

the corresponding action being

$$I = \int d^3x \left[-\frac{1}{4} F^2_{\mu\nu} + \frac{1}{2}\mu\, \epsilon^{\nu\alpha\beta} A_\nu\, \partial_\alpha\, A_\beta + A_\mu\, j^\mu \right] + \frac{1}{2} \int m\dot{x}^2\, dt\,. \tag{1b}$$

Here μ has dimension L^{-1}, while charge has the same dimension as $A_\mu \sim L^{-\frac{1}{2}}$, a different convention from the usual anyon one. Our signature is $(-++)$, $\epsilon^{012} = +1 = \epsilon^{12}$. Consider a circularly symmetric static point current j^ν with

$$j^0 = e\, \delta^2(r) \qquad j^i = g\, \epsilon^{ij}\, \partial_j\, \delta^2(r) \tag{2}$$

where we have included a transverse spatial dipole current of strength g. We use gauges which respect the circular and time invariance symmetries of the source,

$$A_i = \epsilon^{ij}\, \partial_j\, V(r) + \partial_i\, \Lambda(r)\;,\quad A_0 = A_0(r)\;. \tag{3}$$

The longitudinal potential $\partial_i \Lambda$ is irrelevant and remains undetermined by (1). The field equations (1) reduce to

$$\nabla^2(A_0 + \mu V) = e\, \delta^2(r) \tag{4a}$$

$$\epsilon^{ij}\, \partial_j \left[\nabla^2 V + \mu\, A_0 + g\, \delta^2(r) \right] = 0\;. \tag{4b}$$

The Coulomb and Yukawa Green functions

$$-\nabla^2\, C = \delta^2(r)\;,\quad (-\nabla^2 + \mu^2) Y(r) = \delta^2(r) \tag{5a}$$

are

$$2\pi C = -\ln\, r\;,\qquad 2\pi Y = K_0(\mu r)\;. \tag{5b}$$

The Bessel function $K_0(x)$ behaves as $-\ln\, x$ at the origin and as $x^{-1/2}\, e^{-x}$ asymptotically. In terms of these, (4) is solved by

$$\mu V = -e\, C + (e + g\mu) Y,\quad A_0 = -(e + g\mu) Y\;. \tag{6}$$

The pure CST model without a Maxwell term is just the large μ limit of (1), and the corresponding solution there is

$$V = -e/\mu\, C(r)\;,\quad A_0 = -g/\mu\, \delta^2(r)\;. \tag{7}$$

The main effect of dropping the Maxwell term is that A_μ does not propagate, the (dual) field strength becomes proportional to the current and (as discussed in the introduction) is therefore confined to regions containing the sources, just as there is no curvature outside the $T_{\mu\nu}$ sources in $D = 3$ Einstein gravity. Even in this limit, however, the spatial current has given rise to a nonvanishing Coulomb potential A_0 in the interior, $j^i \neq 0$, region. In the extended model's solution (6), this electric field (now of finite range) has strength proportional to $(e+g\mu)$, so that the current contributes to the static electric charge felt by an external test charge; recall that a minimally coupled point charge moves according to the Lorentz force law irrespective of the gauge field's action. The asymptotic field strength, on the other hand, is purely magnetic and is proportional to e only. Thus the "equivalence" between the electric charges defined by the flux from the Gauss law (4a) and by the "Coulomb" force no longer holds, with or without the Maxwell term.

In pure CST theory, the accompanying alteration of statistics, described by the phase change in any circuit about the particle is proportional to $-e\oint A \cdot dl = e^2/\mu$. These results remain valid asymptotically, *i.e.*, outside the particle-field system, when the Maxwell term is included. For example, sufficiently close to the particle, where Coulomb and Yukawa potentials coincide and no self-field is included, a structureless ($g = 0$) particle shows no phase anomaly. Curiously, when the charges are "tuned" to have $e+g\mu = 0$, the two models have the same exterior solution; the "twist" generated by the CST in the extended model has cancelled that present in j^i, and the net result is the same as having a pure charge coupled to CST alone. Asymptotically, the lower derivative term in the gauge field dominates. In this respect (as well as in having a field-source equality) we will see that the Einstein term alone in gravity is like the CST alone in electromagnetism. These results bear out our initial statements that even in the dynamical topologically massive theory the asymptotics have the AB hallmark of a nontrivial potential in a sourcefree region, and that the interior's characteristics can thus be measured in this free region; statistics change is itself one such "measurement."

3. Gravitational CST

We have noted that the appropriate gravitational model incorporating a CST must include the Einstein term as well, both because the latter is of lower derivative order and because the CST alone can only couple to sources with vanishing T^μ_μ, but not to massive particles. [The pure CST model coupled to a lightlike particle will be treated elsewhere. In that theory, spacetime is locally conformally flat except on the null geodesic where the conformal curvature is concentrated, and one has a "partial" AB effect.] The field equations of topologically massive gravity are [3]

$$E^\mu_\nu \equiv \sqrt{-g}\, G^\mu_\nu + \mu^{-1}\, C^\mu_\nu = -\kappa^2\, T^\mu_\nu \; ; \;\; C^\mu_\nu \equiv \epsilon^{\mu\alpha\beta}\, D_\alpha(R_{\beta\nu} - \frac{1}{4}\, g_{\beta\nu}\, R)\,. \tag{8}$$

Unlike pure Einstein gravity in $D = 3$, whose action can have either sign [1], this theory is dynamical, and the overall sign of its action is fixed relative to that of matter in order for it to be non-ghost. This implies that the sign of the Einstein term, i.e., of κ^2, must be opposite to the one in $D = 4$ gravity, which leads to the negative coefficient of $T^{\mu\nu} \equiv \delta I_{\rm matt}/\delta g_{\mu\nu}$ in (8). We shall also see that the asymptotic limit of our solutions is that of Einstein theory essentially because the differences due to C^μ_ν depend on μr, and Einstein theory is the $\mu \to \infty$ limit of our model so that once again there is an automatic AB region at infinity. We will discuss below the significance of the asymptotic metric with the sign required here. The dimension conventions here are that $\mu \sim L^{-1}$, and $\kappa^2 \sim M^{-1}$ is not necessarily related to the $D = 4$ Einstein constant. The Cotton (conformal) tensor density C^μ_ν is identically traceless, symmetric, conserved and invariant under a local rescaling of the metric. The system (8) is also different from Einstein gravity in that, like the electrodynamics of (1), it can have non-vanishing curvature (field strength) outside the sources as well.

We now solve the linearized version of (8) for a time-independent circularly symmetric source. For convenience (linearized curvature being gauge invariant) we gauge fix the potential $\kappa h_{\mu\nu} \equiv g_{\mu\nu} - \eta_{\mu\nu}$ to display these symmetries, drop the (irrelevant) longitudinal part of h_{0i} and choose conformal spatial gauge:

$$h_{0i} = \epsilon_{ij}\,\partial_j\, W(r)\ , \quad h_{ij} = \phi(r)\,\delta_{ij}\ , \quad h_{00} \equiv n(r)\ . \tag{9}$$

All three physical component (W, ϕ, n) will then be determined. Consider a stationary localized particle,

$$\begin{aligned} T^0_0 &= -m\,\delta^2(r)\ , \qquad T_{ij} = 0 \\ T^i_0 &= -\frac{1}{2}\sigma\,\epsilon^{ij}\,\partial_j\delta^2(r) = \sigma\,\delta^2(r)\epsilon^{ij}\partial_j \ln r\ . \end{aligned} \tag{10}$$

The momentum density is that of a stationary ($\int T^i_0\, d^2x = 0$) spinning ($\int \epsilon^{ij}\, x^i\, T^j_0\, d^2r = \sigma$) particle.

With the convention $R^\mu_{\nu\alpha\beta} \sim +\partial_\alpha \Gamma^\mu_{\beta\nu}$, so that $R_{\nu\beta} \sim -\frac{1}{2}\,\Box h_{\nu\beta}$, the linearized Cotton and Einstein tensors for the metric (9) are

$$\begin{aligned} C^0_0 &= \frac{1}{2}\nabla^4 W\ , \quad C^i_0 = -\frac{1}{4}\epsilon^{ij}\,\partial_j\,\nabla^2(\phi + n)\ , \quad C_{ij} = -\frac{1}{2}(\delta_{ij}\nabla^2 - \partial^2_{ij})\nabla^2 W \\ G^0_0 &= \frac{1}{2}\nabla^2\phi\ , \quad G^i_0 = -\frac{1}{2}\epsilon^{ij}\,\partial_j\,\nabla^2 W\ , \qquad G_{ij} = -\frac{1}{2}(\delta_{ij}\nabla^2 - \partial^2_{ij})n\ . \end{aligned} \tag{11}$$

Both tensors are manifestly conserved; C^μ_ν is also traceless and conformal: $C^\mu_\nu(h_{\alpha\beta} + \eta_{\alpha\beta}\xi(x)) = C^\mu_\nu(h_{\alpha\beta})$. The field equations (8) have just 3 independent components,

$$E^0_0 \equiv \frac{1}{2}\nabla^2(\phi + \mu^{-1}\nabla^2 W) = \kappa m\,\delta^2(r) \tag{12a}$$

$$E^i_0 \equiv -\frac{1}{2}\,\epsilon^{ij}\,\partial_j\nabla^2[W + \frac{1}{2\mu}(\phi + n)] = \frac{1}{2}\kappa\sigma\,\epsilon^{ij}\partial_j\delta^2(r) \tag{12b}$$

$$E_{ij} = -\frac{1}{2}(\delta_{ij}\nabla^2 - \partial^2_{ij})[n + \mu^{-1}\nabla^2 W] = 0\,. \tag{12c}$$

The desired solution of (12) is given by

$$\begin{aligned} W &= \mu^{-1}\kappa(m+\mu\sigma)(C-Y) \\ \phi &= \kappa(m+\mu\sigma)Y - 2\kappa m C \\ n &= \kappa(m+\mu\sigma)Y\,. \end{aligned} \tag{13}$$

and was also recently found, for $\sigma = 0$, in [6].

In the absence of any sources, our time-independent field equations have only the trivial solution. In the full nonlinear theory, it has been shown [7,8] that there are no sourcefree time-independent solutions unless $\nabla^2 W \neq 0$; in particular this implies that there are no static ($g_{0i} = 0$) solutions. We conjecture that there are no asymptotically flat stationary solutions either, just as in sourcefree Einstein theory (for any D). There are, however, source-free cosmological solutions; explicit examples are given in [7,8,9], but these are not accessible in our linearization about flat space.

Let us next discuss the special case in which $(m + \mu\sigma)$ vanishes. Then the metric (13) reduces to

$$n = W = 0\,, \qquad \phi = -2\kappa\, mC \tag{14}$$

which has the same form as that generated by a structureless spinless point mass in pure (linearized) Einstein gravity, since the field equations essentially reduce to "$(1 + \mu^{-1}\nabla\times)$"$(G + \kappa^2 T) = 0$: The "twist" effects of the CST and of T^i_0 have cancelled each other, and there is no net angular momentum. If the combination $(m + \mu\sigma)$ does not vanish (in particular if $\sigma = 0$), then neither can W, just as for V in the vector case.

The spacetime represented by the generic solution (13) is asymptotically locally flat. Indeed, near spatial infinity, it has the same form as (the linearization of) the exterior (hence flat) Kerr metric generated by spinning massive sources in pure $D = 3$ Einstein theory [1]. There, m was the inertial mass, the gravitational mass vanished (because the Newtonian potential n did) and the "inertial" spin was just the σ part of the coefficient of C in W; the "gravitational" spin vanished as well because its contribution to the linearized geodesic equation is through $\Gamma^i_{0j}\,\dot{x}^j$ and $\Gamma^i_{0j} \sim h_{0i,j} - h_{0j,i} \sim \epsilon_{ij}\nabla^2 W \sim \epsilon_{ij}\nabla^2 C \sim 0$ (incidentally, Γ^i_{0j} is obviously independent of the longitudinal, gauge, part of h_{0i}). We show next that while the inertial quantities are the same in our theory, the gravitational ones do not vanish. The Newtonian potential is now $(m+\mu\sigma)Y(r)$; since the gravitational mass m_g of a source is defined in terms of the static force it generates on a test particle, m_g is now $(m + \mu\sigma)$. [The fact that the force is of finite, rather than infinite, range in massive gravity is irrelevant.] Thus the equivalence principle is violated (but for quite different reasons than in $D = 3$ Einstein gravity) when there is a nonvanishing spinning T^i_0, just as "charge equivalence" was violated in the electromagnetic system. As noted above, gravitational angular momentum is obtained from the geodesic equation through the term $\Gamma^i_{0j}\,\dot{x}^j \sim \epsilon^i_j\,\nabla^2 W\,\dot{x}^j \sim Y\,\epsilon^i_j\,\dot{x}^j$, which exhibits a torque (again at finite range) imparted to the test particle; by (13) this quantity is the coefficient $(\sigma+\mu^{-1}m)$ of Y in W. The σ contribution is due to the source's intrinsic spin while $\mu^{-1}m$ measures the amount of gravimagnetic field generated by m. In the linearized approximation, the inertial quantities are just the source values, namely m and σ for mass and spin, according to the usual flux integral definition using the left side of the field equations (12), namely

$$\begin{aligned} m_I &= \kappa^{-2}\int E^0_0\,d^2x = \frac{\kappa^{-1}}{2}\oint \nabla(\phi+\mu^{-1}\nabla^2 W)\cdot dS = m \\ S_I &= \kappa^{-2}\int \epsilon^{ij}x^i E^j_0\,d^2x = \kappa^{-1}\oint \nabla[W+\frac{1}{2\mu}(\phi+n)]\cdot dS = \sigma\,, \end{aligned} \tag{15}$$

so that

$$m_g = m_I + \mu\,S_I\,, \qquad S_g = S_I + \mu^{-1}\,m = \mu^{-1}\,m_g\,. \tag{16}$$

In the full nonlinear theory, the results will be more complicated, because of the gravitational field's contributions. However, the relations between the dressed gravitational and inertial quantities should remain similar. It is also an interesting question whether the special properties of the $m+\mu\sigma=0$ source carry over to the full theory, (this conjecture was recently confirmed [10]), and whether any static many-particle solutions are permitted there.

We have seen that (as expected) the asymptotic form of the general solution (13) is that of the pure Einstein solution (and the two agree everywhere in form for the special case $m+\mu\sigma=0$). In particular, we found that the spatial metric behaves near infinity as

$$\phi \sim -2\,\kappa^2\,mC = +\kappa^2\,m/\pi\ln r \tag{17a}$$

whereas the corresponding Einstein result [1] was, in our units,

$$\ln(1+\phi) = -\kappa^2_E\,m/\pi\ln r\,. \tag{17b}$$

The left side of (17b) of course linearizes to ϕ. The sign difference between (17a) and (17b) is due to the fact that in [1] the (*a priori* arbitrary) sign of κ^2_E was so chosen as to give a normal conical 2–geometry with positive angular defect. However, we have no such freedom here and are therefore obliged to accept a negative (linearized) angular defect for a positive mass source. This represents a conical space whose angular range exceeds 2π, a result which (being asymptotic) presumably persists in the full nonlinear solution. Consequently, the source mass can now be arbitrarily large (in contrast with the pure Einstein sign above where there was a maximal allowed value of m) as is physically reasonable for the present, dynamical, system. On the other hand, it is precisely because of our κ^2 sign that the theory describes an attractive Newtonian force (since, $K_0'(x)<0$), at least for $(m+\mu\sigma)>0$ – and so in particular for normal $m > 0$ sources when σ is absent. "Anti-gravity" is, however, possible for sufficiently negative $\mu\sigma$.

3. Conclusions

A structureless massive particle coupled to linearized topologically massive gravity exhibits a gravimagnetic field (proportional to m/μ) to an asymptotic observer, just as a charged particle has a magnetic field $\sim e/\mu$ in the corresponding electromagnetic model.

If the particle has "bare" spin as well then we saw that this spin contributes additively to both the gravitational angular momentum and mass, so that the two contributions can even cancel. We noted further that the equivalence of inertial and gravitational quantities is in general violated, precisely because the gravitational mass and spin as felt by a test particle were "twisted" away from the inertial ones, a phenomenon which permits "anti-gravity" repulsive Newtonian forces.

We come now to the "anyon" aspects of our solutions, namely the statistics change due to the presence of the h_{0i} potential: The latter enters on the same footing as A_i in its coupling to the particle. Hence (for $\sigma = 0$) the anomalous phase here is just obtained by replacing e by κm and so is proportional to $(\kappa m)^2/\mu$. The whole discussion here is parallel to that at the end of Sec. 2, also in its AB aspects. Insofar as the linearized model treated here is Poincaré invariant, the statistics thus induced should agree with the spin defined as the value of the total system's rotation generator including of course the field contribution (beyond the particle's orbital, $r \times p$ part) for our solution. This calculation is in progress.

Finally, one may speculate that to the extent that material stresses are relevant in planar systems, these would induce an effective $D = 3$ gravitational contribution whose lowest terms in a derivative expansion would generate both Einstein and Chern-Simons terms to realize the mechanism presented here. The work presented here is to appear shortly in Physical Review Letters.

4. Acknowledgements

I thank L. Alvarez-Gaumé, L. Susskind and especially Jim McCarthy for enlightening discussions, and A. Shimony for providing ref. [2]. This work was supported by NSF grant PHY88-04561.

5. References

1. S. Deser, R. Jackiw, and G. 't Hooft, Ann. Phys. **152**, 220 (1984).

2. J. Stachel, Phys. Rev. **D26**, 1281 (1982); J. Math. Phys. **25**, 338 (1984).

3. S. Deser, R. Jackiw, and S. Templeton, Ann. Phys. **140**, 372 (1982).

4. W. Siegel, Nucl. Phys. **B156**, 135 (1979); J. Schonfeld, Nucl. Phys. **B185**, 157 (1981), R. Jackiw and S. Templeton, Phys. Rev. **D23**, 2291 (1981).

5. Y-H. Chen, F. Wilczek, E. Witten, and B.I. Halperin, Int. J. Mod. Phys. **B3**, 1001 (1989).

6. B. Linet, Paris preprint.

7. I. Vuorio, Phys. Lett. **163B**, 91 (1985).

8. R. Percacci, P. Sodano, and I. Vuorio, Ann. Phys. **176**, 344 (1987).

9. Y. Nutku and P. Baekler, Ann. Phys. **195**, 16 (1989).

10. M.E. Ortiz, DAMPT preprint.

Topology of Configuration Space and Quantizations

E. C. G. Sudarshan
Center for Particle Theory
University of Texas, Austin, Texas

ABSTRACT

Quantum theory on topologically nontrivial manifolds is studied with particular reference to the fundamental groups of the manifolds and the generic vector quantum theories. Quantum statistics are studied as a special case of a quantum theory on a nontrivial manifold; and the structure of the associated braid groups is examined. Several realizations are explicitly worked out. Inequivalent field quantizations are considered as an application.

Quantum Mechanics generically is a vector space theory where the state vector is a map from a configuration space Q to complex numbers:

$$Q \xrightarrow{\psi} \psi(Q); \quad \int |\psi(Q)|^2 \, d\mu(Q) = 1,$$

where the configuration space Q is taken as a manifold with suitable coordinate charts and $d\mu(Q)$ is a measure. The question naturally arises: Is ψ is a single-valued map? That is to say, when Q describes a loop in the manifold, does $\psi(Q)$ also describe a loop in the vector space? Since the probability density and probability flow are measurable physical quantities, they should be uniquely defined at each point; consequently the ambiguity in $\psi(Q)$ should be by a constant unitary transformation. Since $\psi(Q)$ has only one component, this corresponds to a phase ambiguity which is a topological property; continuous deformations of the loop in configuration space do not alter this "constant" unitary transformation. We can define a natural multiplication law on the loops and thus obtain a group $\pi_1(Q)$ the fundamental (first homotopy) group of the space. The unitary transformations on the state vectors constitute a unitary representation of the fundamental group.

As simple examples of configuration spaces we may consider Euclidean spaces $\mathbb{R}^1, \mathbb{R}^2, \mathbb{R}^3$, etc. For all these, all loops are contractible. Thefundamental group is therefore trivial and the quantization is unique.

For configuration space $\mathbb{R}^2 - \{\text{disc}\}$ or $\mathbb{R}^3 - \{\text{cylinder}\}$ the loop which winds around the disc or cylinder is noncontractible and all powers (and inverse powers) of this basic loop constitute the infinite cyclic group $\mathbb{Z}$. The generating element may be realized as $e^{i\theta}$, $o \leq \theta < 2\pi$ and hence distinct quantizations are represented by this angle. This is of course familiar to us as the Aharanov-Bohm effect, thirty years of which we are celebrating this year.

For the spheres S^2 and S^3, all loops are contractible but for S^1 we have the infinite cyclic group $\mathbb{Z}$ as the fundamental group. For $S^2/\mathbb{Z}_2$ or $S^3/\mathbb{Z}_2$, (spheres with opposite points identified) we have two classes of loops and hence the fundamental group is $\mathbb{Z}_2$, the cyclic group of order 2. Since $S^3/\mathbb{Z}_2$ is the manifold of $O(3) = SU(2)/\mathbb{Z}_2$, this is the familiar two-valuedness of the half-integer spin realizations.

In all these cases we have abelian fundamental groups. The fundamental group is always discrete, finite or infinite. A manifold with an infinite nonabelian group is obtained by $\mathbb{R}^2 - \{\text{two discs}\}$, which is topologically equivalent to the surface of a pair of pants. Hence the fundamental group is generated by loops around the two discs which may not commute: $\pi_1 = \mathbb{Z} * \mathbb{Z}$. For a manifold with discrete nonabelian fundamental group we can consider the orientation of the moment of inertia ellipsoid. These consist of $SU(2)/Q_8$ where Q_8 is the quarternion group generated by reflections in the principal planes. The fundamental group is then the quarternion group.

Similar considerations apply to many particle configuration spaces (Braid groups) and to the configuration space of quantum field theories.

Fundamentally Perfect Spaces Have Unique Quantizations

Do nontrivial fundamental groups and the underlying nontrivial topology of the configuration space automatically imply inequivalent quantizations? As long as we are restricted to ordinary (scalar) quantum theories in which we can only tolerate phase ambiguities, the fundamental group is to have nontrivial one-dimensional representations. But the one-dimensional representations are necessarily the representations of the abelianization of the fundamental group: $\pi_1 \to \pi_{1ab} = \pi_1/[\pi_1, \pi_1]$ which is the quotient of the fundamental group by its first commutator subgroup. It follows that quantization ambiguities depend on whether or not π_{1ab} is trivial or nontrivial.

For cases in which the fundamental group is already abelian, there is no need to abelianize it! Thus the ambiguity of the Aharanov-Bohm manifolds is given by a single angle. But there are other cases with nonabelian fundamental group the situation can be dramatic. For the simplest nontrivial case consider the manifold S^3/I^* where I^* is the 120-element icosahedral group which has a trivial abelianization. Such groups are called perfect; a space which has a fundamental group which is perfect is called fundamentally perfect. Such a space has a unique quantization.

Scalar and Vector Quantizations

We can take the more general view that we choose the state vectors as multivalued maps from Q into $\mathbb{C}^N, N \geq 1$, that is, as sections of a $\mathbb{C}^N$-bundle over Q which we take to be flat[3]. The classical limit will contain an external flat connection. On each such bundle the holonomy of the flat connection provides a representation ρ of the fundamental group $\pi_1(Q)$ by $N \times N$ matrices. Conversely, each N-dimensional representation of $\pi_1(Q)$ is "realized" by some flat $\mathbb{C}^N$-bundle over Q. Without loss of generality we restrict ourselves to irreducible bundles, that is, we assume ρ is irreducible. The corresponding theories are called vector quantum theories if $N > 1$. Since there are additional difficulties when N becomes infinite, we shall assume that N is finite. So there are as many inequivalent quantizations of a system with configuration space Q as there are finite-dimensional irreducible unitary representations of $\pi_1(Q)$.

There are groups with no nontrivial finite dimensional unitary representation. These groups are called U-inert. It follows that if you have a U-inert fundamental group $\pi_1(Q)$ (which must be infinite if nontrivial), then a system with configuration space Q has a unique quantization. Similarly there are groups all whose finite dimensional irreducible representations are 1-dimensional; these are called U-scalar groups. A configuration space with a U-scalar fundamental group has no vector quantizations. We note that there are many nontrivial *finitely presented* U-inert and U-scalar groups, and for every $n \geq 4$ there exists a closed orientable n-manifold M with $\pi_1(M)$ any finitely presented group.

Quantum Statistics and Braid Groups

In particle mechanics we have finite dimensional configuration spaces. If the motion is on an underlying manifold M the topology of M could induce inequivalent quantizations. The simplest such example is the Aharanov-Bohm effect[4)] in which a certain region of Euclidean space is excluded. This leads to $\pi_1(M) = Z$ as is well known. But an entirely new context in which multiply connected configuration spaces arise is when we consider the motion of n identical particles on an underlying manifold. In this case we must identify any two points in M^n which are obtained by a permutation of the particle labels. However the action of the permutation group S_n on M^n has fixed points and hence M^n/S_n is not a smooth manifold. We remove this difficulty by removing from M^n the subcomplex Δ with two or more particles having identical coordinates. $M^n - \Delta$ has no fixed points under S_n and so we may choose the smooth manifold

$$Q \equiv Q_n(M) = (M^n - \Delta)/S_n$$

for the configuration space[(5)]. This space is multiply connected and has the n-string braid group $B_n(M)$ as its fundamental group

$$\pi_1(Q_n(M)) \equiv B_n(M).$$

An element of $B_n(M)$ can be considered as a homotopy class of paths in $M^n - \Delta$ whose initial and final points differ by a permutation of the particle coordinates. Fix the initial point $(q_1, \ldots, q_n)$. Then one set of such paths contains those in which q_1 is taken through a homotopically nontrivial loop with the other coordinates being kept fixed, the set of all such paths being denoted by $\mathcal{L}$. Next, let D be an open disk in M containing $q_1, \ldots, q_n$ and for each $j < n$ consider a path σ_j which interchanges the jth and $(j+1)^{th}$ particles in D, not enclosing any of the other m_j's which all remain fixed. Denote the set of all $\sigma_j, 1 \leq j \leq n-1$, by $\mathcal{P}$. Then $\mathcal{L}$ and $\mathcal{P}$ together generate the braid group $B_n(M)$. The relations among these generators may be rather complicated. We shall discuss some examples later. In the generic context let $\Sigma_n(M)$ denote the group generated by $\mathcal{P}$, called the "statistical subgroup" of the braid group $B_n(M)$. In general the irreducible representations of $B_n(M)$ will be reducible when restricted to $\Sigma_n(M)$. We call two irreducible unitary representations ρ_1 and ρ_2 of $B_n(M)$ *statistically equivalent* if we have

$$(\rho_1 \downarrow \Sigma_n(M)) \otimes \mathbb{1}_s \simeq (\rho_2 \downarrow \Sigma_n(M)) \otimes \mathbb{1}_t$$

with s, t positive integers. The presence of the trivial representations $\mathbb{1}_s$ and $\mathbb{1}_t$, of dimension s and t respectively, in the above equality accounts for differences which only pertain to the distinct dimensionalities of ρ_1 and ρ_2. Statistical equivalence is an equivalence relation on the set of all irreducible unitary representations of $B_n(M)$. Therefore this set breaks up into equivalence classes, each containing only representations whose corresponding quantizations yield the same statistics for the n identical particles on M. If $\pi_1(M) = \{e\}$ then $\Sigma_n(M) = B_n(M)$ and any two distinct irreducible unitary representations of $B_n(M)$ yield distinct statistics for the n particles. (It is possible to demonstrate that changing the disk which we used to define the group $\Sigma_n(M)$ does not affect the notion of statistical equivalence.)

Since the action of S_n on $M^n - \Delta$ is free we obtain the short exact sequence

$$\{e\} \rightarrow \pi_1(M^n - \Delta) \xrightarrow{\alpha} B_n(M) \xrightarrow{\beta} S_n \rightarrow \{e\}.$$

The generators of $\mathcal{P}$ map on to the corresponding transpositions in S_n under β, the elements of $\mathcal{L}$ being in the kernel of β. So $\beta \downarrow \Sigma_n$ is an epimorphism from $\Sigma_n(M)$ onto S_n. Any irreducible unitary representation of S_n can be "lifted" into an irreducible unitary representation of the braid group $B_n(M)$. These correspond to the parastatistics suitably made irreducible. Their number is equal to $p(n)$, the number of partitions of the integer n. For $n \geq 2$ there are only two irreducible unitary representations of S_n of degree one, namely those giving Bose and Fermi statistics. If dim $M \geq 3$ and $\pi_1(M)$ is U-inert, then parastatistics are the only statistics possible for n identical particles on M. However, in all other cases, one obtains other new kinds of exotic statistics.[6]

Exotic statistics

For $d = \dim M \geq 3, \pi_1(M^n - \Delta) \cong \pi_1(M)^n$ and the braid group $B_n(M)$ is a semidirect product of $\pi_1(M)^n$ with S_n:

$$B_n(M) = \pi_1(M)^n \rtimes_\mu S_n,$$

where $\mu : S_n \rightarrow \mathrm{Aut}(\pi_1(M)^n)$ is given by

$$\mu(p)(\ell_1, \ell_2, \ldots, \ell_n) = (\ell_{p(1)} \ldots \ell_{p(n)}); \quad \ell_j \in \pi_1(M), p \in S_n.$$

$B_n(M)$ is called the wreath product[7] of $\pi_1(M)$ with S_n, denoted by

$$B_n(M) = \pi_1(M) \wr S_n, \quad d \geq 3.$$

In this case the statistical subgroup $\Sigma_n(M)$ is S_n, but if ρ is an irreducible unitary representation of $B_n(M)$ then $\rho \downarrow \Sigma_n$ may contain inequivalent irreducible representation of S_n. The statistics so obtained may be called generalized parastatistics. Naturally if dim $\rho = 1$ (i.e., for scalar quantum theory) the only possible statistics are Bose and Fermi.

The wreath product $\pi_1(M) \wr S_n$ has a natural presentation in terms of n copies of the generators ℓ_α of $\pi_1(M)$ and the transpositions σ_k:

$$\ell_{j\alpha}, \sigma_k, \quad 1 \leq j \leq n, \quad 1 \leq k \leq n-1$$

with the relations

$$\begin{aligned}
\ell_{i\alpha}\ell_{j\beta} &= \ell_{j\beta}\ell_{i\alpha}, \quad i \neq j, \quad 1 \leq i, j \leq n \\
\sigma_j \ell_{j\alpha} \sigma_j &= \ell_{j+1,\alpha}, 1 \leq j \leq n-1 \\
\ell_{j\alpha}\sigma_k &= \sigma_k \ell_{j\alpha}, \quad j \neq k, k+1 \\
\sigma_k\sigma_{k+1}\sigma_k &= \sigma_{k+1}\sigma_k\sigma_{k+1}, 1 \leq k \leq n-2 \\
\sigma_k\sigma_\ell &= \sigma_\ell\sigma_k, \quad | k - \ell | \geq 2 \\
\sigma_k^2 &= e, \quad 1 \leq k \leq n-1 \,.
\end{aligned}$$

For $n = 2$ the braid group has a very simple structure. Take

$$M = \mathbb{R}P^3 = S^3/\mathbb{Z}_2$$

which is the group space of SO(3). Since

$$\pi(\mathbb{R}P^3) = \mathbb{Z}_2$$

the braid group is the dihedral group of order 8:

$$B_2(\mathbb{R}P^3) = \mathbb{Z}_2 \wr \mathbb{Z}_2 = D_8.$$

It is generated by just two elements ℓ, σ:

$$B_2(\mathbb{R}P^3) = < \ell, \sigma \mid \ell^2 = \sigma^2 = e;\ (\ell\sigma)^2 = (\sigma\ell)^2 > .$$

Also, in this case

$$\Sigma_2(\mathbb{R}P^3) = \mathbb{Z}_2 = < \sigma > \ .$$

There are five irreducible unitary representations of D_8 of which four are one-dimensional and one has two dimensions.

$$\rho_1(\ell) = 1, \quad \rho_1(\sigma) = 1$$

$$\rho_2(\ell) = 1, \quad \rho_2(\sigma) = -1$$

$$\rho_3(\ell) = -1, \quad \rho_3(\sigma) = 1$$

$$\rho_4(\ell) = -1, \quad \rho_4(\sigma) = -1$$

$$\rho_5(\ell) = \begin{pmatrix} 0 & 1 \\ 1 & 0 \end{pmatrix}, \ \rho_5(\sigma) = \begin{pmatrix} 1 & 0 \\ 0 & -1 \end{pmatrix}.$$

Two of the representations ρ_1, ρ_3 yield Bose statistics and two ρ_2, ρ_4 yield Fermi statistics. But the two dimensional representation ρ_5 provide a new kind of half Bose-half Fermi exotic statistics which may be called *ambistatistics*[(8)].

When M has dimension 2 there are other possibilities[(9)]; now, in general,

$$\pi_1(M^n - \Delta) \neq \pi_1(M)^n.$$

Consequently the generators $\ell_{j\alpha}$ for fixed j no longer generate a subgroup isomorphic to $\pi_1(M)$. The presence of the second particle is tantamount to altering the topology of the configuration space for the first particle, a feature which obtains in the more complex cases of two geons or two closed strings. This has the consequence that new statistics obtain. We consider, in particular $M = \mathbb{R}^2$ and $M = S^2$.

For particles moving in a plane, $M = \mathbb{R}^2$ and $B_n(\mathbb{R}^2)$ has the presentation[(10)]

$$B_n(\mathbb{R}^2) = < \sigma_1, \ldots, \sigma_{n-1} \mid \sigma_j\sigma_{j+1}\sigma_j = \sigma_{j+1}\sigma_j\sigma_{j+1};\ \sigma_i\sigma_j = \sigma_j\sigma_i,\ \mid i-j \mid \geq 2 > .$$

Of course $\Sigma_n(\mathbb{R}^2) = B_n(\mathbb{R}^2)$. For two identical particles moving on a plane

$$B_2(\mathbb{R}^2) = \mathbb{Z}.$$

Therefore the quantizations (which are all scalar) are labelled by an angle θ, yielding the so-called fractional statistics or θ-statistics (or anyon statistics) with Bose ($\theta = 0$) and Fermi ($\theta = \pi$) statistics as special cases. For $n > 2$ we also obtain these since

$$B_n(\mathbb{R}^2)_{ab} = \mathbb{Z}\ .$$

However for $n \geq 3$ we have non-scalar statistics possible. In particular for $n = 3$ we have a more convenient presentation in terms of $a = \sigma_1\sigma_2$ and $b = \sigma_1\sigma_2\sigma_1$:

$$B_3(\mathbb{R}^2) = < a, b \mid a^3 = b^2 > \ .$$

The degrees of the finite dimensional irreducible unitary representations of $B_3(\mathbb{R}^2)$ are unbounded. The representations of dimension 2 are labelled by two angles ϕ and θ;

$$a = e^{2i\phi}\begin{pmatrix} 1 & 0 \\ 0 & \omega \end{pmatrix}, \quad b = e^{3i\phi}\begin{pmatrix} \cos\theta & \sin\theta \\ \sin\theta & -\cos\theta \end{pmatrix}$$

$$0 < \phi < 2\pi; \ \ 0 < \theta \leq \pi/2; \ \ \omega = e^{2\pi i/3} .$$

(For $\theta = 0$ one must restrict the range of ϕ to $0 \leq \phi < \pi$, in order not to overcount.) The representations of dimension 3 are given by

$$a = e^{2i\phi}\begin{pmatrix} 1 & 0 & 0 \\ 0 & \omega & 0 \\ 0 & 0 & \omega^2 \end{pmatrix}, \quad b_{jk} = e^{3i\phi}(\delta_{jk} + 2n_j n_k)$$

where n_1, n_2, and n_3 are the components of a unit vector which are all strictly positive. (For the special case $n_1 = n_2 = n_3$ the range of ϕ is restricted to $0 \leq \phi < 2\pi/3$.) In the general case, $b^2 = a^3$ is a scalar matrix $e^{6i\phi}$ since it commutes with both the generators. So $e^{-3i\phi}b$ should be a reflection matrix:

$$e^{-3i\phi}b_{jk} = \sum_{\alpha=1}^{r} n_j^\alpha n_k^\alpha - \sum_{\beta=r+1}^{n} n_j^\beta n_k^\beta$$

where n^α, n^β are unit vectors so chosen that b does not commute with a.

For 3 identical particles moving on a 2-sphere, $B_3(S^2) = \Sigma_3(S^2)$ has the presentation

$$B_3(S^2) = < a, b \mid a^3 = b^2; \ b = aba >$$

and is the metacyclic group of order 12. Out of the six irreducible unitary representations, four are scalar ($B_3(S^2)_{ab} \cong \mathbb{Z}_4$) and two are two-dimensional:

$$a = \begin{pmatrix} \omega & 0 \\ 0 & \omega^2 \end{pmatrix} \quad b = \begin{pmatrix} 0 & 1 \\ 1 & 0 \end{pmatrix}$$

and

$$a = \begin{pmatrix} -\omega & 0 \\ 0 & -\omega^2 \end{pmatrix} \quad b = \begin{pmatrix} 0 & -i \\ -i & 0 \end{pmatrix}.$$

For $n \geq 4$ the groups are not easily handled. Both $B_n(\mathbb{R}^2)$ and $B_n(S^2)$ show us that in two dimensions $B_n(M)$ is complicated even when $\pi_1(M)$ is trivial.

The case of $M = \mathbb{R}P^2$ is of particular interest:

$$\begin{aligned} B_2(\mathbb{R}P^2) &= < \ell, \sigma \mid \ell^2 = \sigma^2 = (\sigma\ell^{-1})^4 > , \\ \Sigma_2(\mathbb{R}P^2) &= \mathbb{Z}_4 = < \sigma > \ . \end{aligned}$$

$B_2(\mathbb{R}P^2)$ is the dicyclic group of order 16 and has 7 irreducible representations of which 4 are scalar and 3 of dimension 2. They are "identical" with those of $B_2(\mathbb{R}P^3)$ given above except there are two additional ones

$$\rho_6(\ell) = \frac{i}{\sqrt{2}}\begin{pmatrix} 1 & 1 \\ 1 & -1 \end{pmatrix}, \quad \rho_6(\sigma) = i\begin{pmatrix} 1 & 0 \\ 0 & -1 \end{pmatrix}$$

$$\rho_7(\ell) = \frac{-i}{\sqrt{2}}\begin{pmatrix} 1 & 1 \\ 1 & -1 \end{pmatrix}, \quad \rho_7(\sigma) = i\begin{pmatrix} 1 & 0 \\ 0 & -1 \end{pmatrix}.$$

ρ_6 is statistically equivalent to ρ_7, yielding a new type of "fractional ambistatistics".

Inequivalent Field Quantizations

From the work of Friedrichs and others we know that quantum field theories allow for inequivalent realizations by virtue of the infinitely many degrees of freedom. Over and above these well known ambiguities we have other sources of inquivalence stemming from the nontrivial topology of the field configuration manifold. The base manifold Σ is one which is mapped into a target space by maps which are the configurations of the field theory. For a pure gauge theory with compact space manifold Σ of dimension d and structure group G we may take the configuration space to be the orbit space $\mathcal{A}/\mathcal{G}_*$. Here $\mathcal{A}$ is the affine space of all maps from Σ to the Lie algebra of G, and $\mathcal{G}_*$ is the group of all basepoint preserving maps from Σ to G. $\mathcal{G}_*$ is called the group of gauge transformations and has the usual action on $\mathcal{A}$. $Q \equiv \mathcal{A}/\mathcal{G}_*$ is an infinite dimensional manifold. Then for $M = S^3$,

$$\pi_1(Q) \equiv \pi_1(\mathcal{A}/\mathcal{G}_*) = \pi_o(\mathcal{G}_*) = \pi_3(G) = \mathbb{Z}$$

leading to the familiar inequivalent realizations labelled by an angle (θ-vacua).

Similarly for gravity theories defined over a connected compact 3-manifold M, the configuration space is

$$Q = \text{Riem}\,(M)/\text{Diff}_F(M) \equiv R/D$$

and

$$\pi_1(R/D) = \pi_o(\text{Diff}_F(M)),$$

where Riem (Σ) is the affine space of all Riemannian metrics on M and $\text{Diff}_F(M)$ is the group of diffeomorphisms keeping a point in M and a frame at that point fixed. Of particular interest is the choice of the Poincaré sphere

$$M = S^3/I^*$$

with I^* the binary icosahedral group of order 120 which acts freely on S^3. Here $\pi_o(\text{Diff}_F(M)) \cong I^*$ and since I^* is perfect there is a unique scalar quantum theory.

Recently it has been shown that the statistics of geons (localized excitations of the gravity field) exhibit exotic possibilities[11)]. The presence of one geon acts as if the other geon moves on a manifold with nontrivial topology. This leads to ambistatistics for two identical geons. Similar comments apply to closed strings.[(14,15)]

When gravity and gauge theories are simultaneously present further modifications take place. For example the θ-parameter of the gauge theory may get quantized.[(16)]

At the present time many of these exotic possibilities are of interest in theory architecture. But condensed matter phenomena already seem to make use of fractional statistics. The study of vector quantum theories, especially their physical interpretation, is yet to be pursued.

Phases enter in quantum mechanics as observable entities, in spite of the fact that physical states are represented by rays; in particular the Berry phase for closed loops and the phase difference between the two paths in the two-slit experiment stem from a common origin.[12]

The space of states of a quantum mechanical system forms a complex metric projective space, $\tilde{\mathcal{H}} = \boldsymbol{CP}(\infty)$ where $\mathcal{H}$ is the Hilbert space of vectors; the ray of the ket vector $|\psi\rangle$ is denoted by ψ. As rays cannot be added, the superposition principle as conventionally stated needs reformulation, and in a recent paper by one of us [1] it is called the decomposition principle, because the metric (scalar product) has the consequence that a given state (ray) ψ has well-defined projections on other rays.

However, one cannot dispense altogether with the vectors $|\psi\rangle$ of the overlying Hilbert space; for example, the very concept of eigenvalue, crucial to the axions of measurements, is usually stated through a spectral equation involving vectors.

In this paper we set up some simple considerations as to make the totality vector space $\mathcal{H}$ projective space $\tilde{\mathcal{H}}$ the appropriate frame for the formulation of quantum mechanics. In mathematical terms it is the **line bundle** structure which matters:

$$\begin{array}{l} \eta \;:\; \boldsymbol{C}^* \to \mathcal{H} - \{0\} \to \boldsymbol{CP}(\infty) \\ \downarrow \\ \ell \;:\; \boldsymbol{C} \to E \approx \mathcal{H} \to \boldsymbol{CP}(\infty) \end{array}$$

where $C^* = C - \{0\}$. η is the principal bundle, and ℓ the associated line bundle. The upper line can be replaced by

$$U(1) \to S \to \boldsymbol{CP}(\infty)$$

where $S = S^\infty$ is the injective limit of odd spheres; (2) implies the limitation to vectors of constant norm, in particular norm 1.

The motion occurs in $\boldsymbol{CP}(\infty)$ but we need the bundle ℓ t o precisely state the measurement axioms.

The following are some of the reasons why vectors in $\mathcal{H}$ are needed:

1) conventional **observables** are linear hermitian operators acting on $\mathcal{H}$; the spectrum refers to eigenrays or eigensubspaces of $\mathcal{H}$, with a well-defined projection in $\tilde{\mathcal{H}}$, there is a number attached to these fixed objects, the corresponding eigenvalue. Notice the invariant character of the spectrum:

$$\text{if } O|\psi\rangle = o|\psi\rangle, \qquad O(\lambda|\psi\rangle) = o(\lambda|\psi\rangle).$$

The evolution is also defined in the vector (Hilbert) space: that is, the hamiltonian H, as generator of the motion, is given as a vector field in $\mathcal{H}$; of course it defines also a vector field $\overline{H}$ in $\overline{\mathcal{H}}$, which sets the rays (states) in motion. Notice that evolution is also projective invariant, that is, if $|\psi(t_0)\rangle$ evolves to $|\psi(t_1)\rangle$, then $\alpha|\psi(t_0)\rangle$ does to $\alpha|\psi(t_1)\rangle$.

For the interesting case of **closed** evolution in state space, there is possibly a phase mismatch between the initial and the final vector representatives, which is in principle measurable: if $\psi(t_0 = \psi(t_1)$ indicates the loop in state space, then

$$|\psi(t_1)\rangle = e^{i\alpha}|\psi(t_0)\rangle$$

with a well-defined (total phase) α, which is again projective invariant.

The partition $\alpha = \beta + \gamma$ into a dynamical phase $\beta = \int\langle\psi|H|\psi\rangle\, dt$ and a geometrical (Berry) phase $\gamma = \int\langle\tilde{\psi}|\, id\, |\tilde{\psi}\rangle$, where $\psi(t) \rightarrow |\tilde{\psi}(t)\rangle$ is a **section** of the line bundle ($Cfr.$[3]) is very often done these days. The uniqueness of the splitting is related to a hermitian and holomorphic connection in the line bundle ([2], [3]); notice also that α, β and γ are all projective quantitites.

More physically speaking, all interference experiments from Young double slit (ca. 1800) to neutral kaon decay (1955) seem to make appeal to vectors and angles and *a priori* are difficult to understand in terms of rays. Nevertheless, by considering the interference pattern as due to an optical phase difference, we can compare the process with the loop evolution of 2), and relate the angle to the eigenvalue of a (unitary) operator. That the common interpretation of the two-slit experiment is misleading has been also recently shown by the other of us; the possibility of paths encircling the two slits was considered a long time ago by Biedenharn.

The central issue we want to emphasize is the *line-bundle* structure; then one represents states as vectors by means of a **section** $\psi : \tilde{\mathcal{H}} \rightarrow \mathcal{H}$, which is the general concept of wavefunction; these sections have an equivalent relation

$$\psi \sim \psi' \qquad \text{if} \qquad \psi'(x) = \lambda(x)\psi(x)$$

with $\lambda(x)\varepsilon\mathbf{C}^*$. This makes the physics section-independent, and it is a kind of "gauge invariance", first advocated by Weyl (1927) to "incorporate" electromagnetism in wave mechanics. Of course, nothing is lost if $\lambda(x)$ is just a phase, i.e., $|\lambda(x)| = 1$, namely if we take the principal bundle as being (2); this is so because the evolution, that is to say the flow of a vector field on $E \approx \mathcal{H}$, is unitary.

Let us now see how this extended framework of line bundles helps us to understand the physics; we consider the three previous situations, namely observables and measure theory, dynamical evolution and the interference phenomena.

1) conventional observables do exist and act on vectors; for a state ψ (a ray) and an observable A (an hermitian operator) the statistical result of a measurements

$$\exp A[\psi] = Tr\, A\pi(\psi).$$

where $\pi(\psi) = |\psi\rangle\langle\psi|/\langle\psi|\psi\rangle$ is the ray ψ represented by a trace-one projector. If $A = \Sigma\lambda_a P_a$ is the spectral decomposition of A, and if ψ is eigenstate of A, then $\pi(\psi) \subset P_a$ for some a.

2) The evolution is given by the vector field H acting on $E \approx \mathcal{H}$; therefore

$$|\psi(t)\rangle = U(t, t_0)|\psi(t_0)\rangle$$

with

$$U(t,t_0) = P\ \exp(-i/\hbar) \int_{t_0}^{t} H(\tau)\, d\tau.$$

As we said, this is a projective-invariant evolution; of course, this defines also an evolution in $\mathcal{H} = \mathbf{C}P(\infty)$, which is the real evolution of states.

Now for closed loops in state space, the unitary evolution produces an eigenvalue equation

$$|\psi(t_1)\rangle = U(t_1,t_0)|\psi(t_0)\rangle = e^{i\alpha}\,\psi(t_0)\rangle$$

when $\psi(t_0) = \psi(t_0)$. ***The eigenvalues of unitary operators are as measurable as those of hermitian operators***, because both are equally projective. In fact the (total) phase α can be and has been measured (e.g. in optical fibers); we already remarked on the question of invariant splitting of α in a geometrical part γ and a dynamical part β.

We come now to consider the two-slit experiment. This crucial experiment (the *only* mystery of quantum mechanics, according to Feynman) takes a quantum particle (s), e.g. electrons or photons through two paths, say C_1 and C_2 and they recombine at a point P on the screen.

We want to compare it with the closed loop evolution mentioned before; namely the loop $C_2^{-1} \circ C_1$ in which the system goes to the screen at point P through C_1 and returns through C_2; the phase for the closed loop is

$$\begin{aligned} \alpha &= \int_{C_1} n(x)\, ds - \int_{C_2} n(x)\, ds, \\ \exp\, i\alpha &= U(C_2^{-1} \circ C_1) \end{aligned}$$

(where **nds** is the infinitesimal optical path) is clearly what determines the intensity at point P at the screen.

References and Footnotes

1. For a review and extensive references see, for example, P. A. Horvathy, G. Morandi and E. C. G. Sudarshan, Nuovo Cimento **11D**, 201 (1989).
2. T.D. Imbo and E. C. G. Sudarshan, Phys. Rev. Lett. **60**, 481 (1988).
3. E. C. G. Sudarshan, T. D. Imbo and T. R. Govindarajan, Phys. Lett. **B213**, 471 (1988).
4. Y. Aharanov and D. Bohm, Phys. Rev. **115**, 485 (1959).
5. L. S. Schulman, J. Math. Phys. **12**, 304 (1971); MGG Laidlaw and C. Morette-deWitt, Phys. Rev. **D3**, 1375 (1971); J. S. Dowker, J. Phys. **A5**, 936 (1972); J. Phys. **A18**, 352 (1985).
6. For a review, see "A User's Guide to Exotic Statistics" by T. D. Imbo, C. S. Imbo, R. S. Mahajan and E. C. G. Sudarshan (paper under preparation).
7. A. Kerber, "Representations of Permutation Groups I and II" (Springer-Verlag, Berlin 1971 and 1975).
8. T. D. Imbo, Chandni Shah Imbo and E. C. G. Sudarshan, Phys. Rev. A (in press).
9. Y-S. Wu, Phys. Rev. Lett. **52**, 2103 (1984).
10. E. C. G. Sudarshan, T. D. Imbo and Chandni Shah Imbo, Ann. Inst. Henri Poincare **40**, 387 (1988).
11. C. Anezeris, A. P. Balachandran, M. Bourdeau, S. Jo, T. R. Ramdas and R. D. Sorkin, Mod. Phys. Lett. **A4**, 331 (1989).
12. Luis J. Boya and E. C. G. Sudarshan, "Rays and Phases in Quantum Mechanics" (paper under preparation).

QUANTUM DAMPING AND ITS "PARADOXES"

Leo Stodolsky
Max-Planck- Institut für Physik und Astrophysik
Föhringer Ring 6, 8000 Munich 40

ABSTRACT

We review the understanding and applications of Quantum Damping arrived at in studying two state systems, particularly optical isomers and neutrino oscillations, using a density matrix formalism. The damping parameter D, derived from the underlying scattering amplitudes gives a quantitative measure of the "reduction of the wavefunction." In the limit of large D the equations give the "Turing" (or "Xeno" or "Watched-Pot") effect. To go beyond two levels, a continuum model is presented and solved.

1. Introduction

One of the most intriguing questions in Quantum Mechanics has to do with the questions of how particular directions in the quantum mechanical vector space get singled out. We describe things, after all, in terms of a linear vector space where any direction or any linear combination of vectors is as good as any other. So how the devil do some directions or some vectors seem to get a preferred meaning? Why do we tend to have things in more or less definite positions and not in eigenstates of $x + p$ or some other funny operator? Or why are you and I not in a symmetrized state with half of us here and half of us there? Or, as we shall shortly discuss, why in a two-state system like an optical isomer, do we seem to have a preference for two particular states and not some others as a basis? This question is all the more acute in view of our growing technical understanding of the early universe, which allows us to wonder about the absolute basics, like where space-time and geometry come from.

In the following I would like to describe some understanding and insights into this question which we have been able to reach in the study of two-level systems in various contexts. This system is simple enough so that much can be understood in quite explicit detail, yet being a full quantum mechanical system it incorporates the essence of the problem. It is to be hoped that the generalization to more complicated systems involves only "technical problems", and at the end we attempt one such extension.

We shall see that in the process we have been able to understand and clarify some closely related "paradoxes": Hund's, Turing's (also called Xeno's or the Watched Pot); and to give explicit physical formulas for the calculation of the associated effects. In particular the formula for the "damping parameter" D, to be introduced below, may be viewed as a scattering theoretic expression for quantitatively measuring the "reduction of the wavefunction". In work with R. A. Harris and with others, we have been able to apply these results to the calculations of various phenomena such as tunnelling in optical isomer molecules or the damping of neutrino oscillations and to suggest some experimental consequences and tests.

2. The two level system

To begin chronologically, in the mid-seventies we were looking for manifestations of "neutral currents", at the time the key new aspect of the now well-established "Electro-weak standard model". Harris and I wondered if in looking at left and right-handed molecular isomers we might be able to see some manifestations of the parity-violating — that is handed— forces produced by "neutral currents"[1]:

Imagine the good old double potential well problem. This serves as an idealization of the two configurations possible for an optical isomer. If we think of something like an ammonia molecule (the three hydrogen atoms would actually have to be different, say H, T and D, for the two configurations to have distinct handedness), we have the state $|L>$, when the system is in the left well (nitrogen above the plane), and the state $|R>$ when it is in the right well (nitrogen below the plane). Since parity violation will be a very small effect we anticipate that we will have two almost degenerate states for a given configuration of the molecule. The Hamiltonian for the resulting two level system:

$$H = const + \begin{pmatrix} \epsilon & \delta \\ \delta & -\epsilon \end{pmatrix} \tag{1}$$

has a tunnelling term δ, and the two diagonal elements representing the energy of each state. Now if parity is conserved the two diagonal elements must be equal, $\epsilon = 0$, since the energy of a left and a right isomer of the same molecule must be the same. In this case this eigenstates are

$$\frac{|L> \pm |R>}{\sqrt{2}} \tag{2}$$

with an energy splitting 2δ.

We can now state our first "paradox": a question raised by Hund in the late twenties[2]. If the true eigenstates are not $|L>$ and $|R>$ but rather this linear combination, how come what we seem to see in nature (say for handed sugars), are not these eigenstates but in fact just $|L>$ or $|R>$? Here we have the question of the selection of a direction in the vector space.

Hund answered himself with what must be in effect one of the earliest tunnelling calculations. He found that if we start with a big $|L>$ molecule the tunnelling energy is so tiny (a whole huge molecule must be uncoiled and recoiled) that the time to go to $|R>$ is so enormously long that $|L>$ is effectively stable. This is of course true. However as we shall see, it is not the full answer.

3. Parity and the Isomer

Let us now come back to the seventies, when weak neutral currents came along and raised the possibility $\epsilon \neq 0$ to first order in weak interactions. Now there can be a small but perhaps not hopelessly small difference ϵ in the energies for the diagonal elements, say 10^{-15} eV[1]. This is still a very tiny energy. But the tunnelling energy usually will be even much smaller, since 10^{-15} eV correspond to only one inverse second.

This brings us to the our first new answer to Hund's question. Let $\epsilon \gg \delta$, that is let the tunnelling time be much more than seconds or minutes, as is most often the case. Then the above H is effectively diagonal in the $|L>$,$|R>$ basis and these are indeed the eigenstates, not (2). So given a big enough molecule – which actually can be rather small – the weak interaction makes Hund's paradox go away. This answer is also correct and would in fact be the real answer for a totally isolated molecule. A third answer, however, as we shall see, is probably the *most* real answer. In any event, we certainly can't fault Hund for not knowing about parity violation in 1927.

Now the idea that attracted us was a possible way to see the small parity violation energy ϵ by creating an apparently big effect out of it[1]. The idea is a molecular analogy to the famous K^0 phenomena in particle physics. We will let P violation play the role for the isomer that CP violation does for the K^0's. Due to the tunnelling energy a molecule, started at $t = 0$ in the pure left state $|L>$, is not in an eigenstate of H and will oscillate into the right state $|R>$. If parity is conserved, that is if the diagonal elements are equal, we will get symmetric oscillations between left and right. When the diagonal elements are not equal we get asymmetric oscillations, where the molecule can be more often $|L>$ than $|R>$. For this asymmetry to be a big effect we merely need that the tunnelling energy be smaller than the parity violating difference in the diagonal elements. As just mentioned, this would not seem to be hard to arrange, 10^{-15} eV corresponding to a tunnelling time of about 1 sec. If we take even a small molecule, say the ammonia-like structure but with phosphorous substituted for nitrogen, the tunnelling time is down to minutes, and with bigger molecules the tunnelling times get enormous. Optimally, for the most striking effect, ϵ and δ should be about the same size.

So this sounds easy. All we need is an obliging chemist who will make us a beam of the appropriate handed molecules. If they start all $|L>$ at $t = 0$, then we look to see if the oscillations to $|R>$ and back are asymmetric. If so, then $\epsilon \neq 0$ and parity violation in the molecule has been seen.

Well, this sounds too good to be true. It can't be that easy to see 10^{-15} eV. Otherwise the Nobel prize wouldn't have just been awarded for the development of techniques to approach this sensitivity. But where's the difficulty? This is just the most elementary quantum mechanics and the two-by-two Hamiltonian is as simple as can be.

The difficulty, clearly, is in the neglect of the environment. The assumption of a molecule undisturbed with perfect phase coherence for a whole second is, under most conditions, totally unrealistic. In any even very good vacuum or in any container with walls, there will be a great number of collisions in a second. These effects certainly make the assumption of the perfect quantum coherence implicit in the treatment of the oscillations dubious. Or put another way, we cannot neglect energies which might be comparable to the tiny energy we are trying to see.

It may be that the advent of single atom traps and the like, as recognized by the aforementioned Nobel prizes, might make such experiments less impossible than they sounded a while ago. But their containment fields are also " walls" and in any event it is clear we must think about how to treat the problem of the loss of coherence. For the optical isomer this means studying the relaxation of optical activity induced by the surroundings, raceimazation, as it is called. But it is obviously a more general issue with many implications, so let us consider the general two-state system. Instead of $|L>$ and $|R>$ let us discuss two general basis states $|+>$ and $|->$.

4. The density matrix

To consider such questions of coherence and incoherence of the wavefunction we must go beyond consideration of the wavefunction – beyond pure states – and use the density matrix.

Usually, whatever we do, be it field theory, the Aharonov-Bohm effect, molecules, or whatever, is to try to solve the Schroedinger equation in one form or another. Here we are going to take a somewhat different approach and concentrate on studying the evolution of the density matrix. Of course the time dependence of the density matrix is given by the Hamiltonian, so you will say there's no real difference. A difference does arise, however, when we consider "open" systems, systems which don't have a Hamiltonian. This can arise where there is an environment interacting with our system, one whose Hamiltonian we don't know, or perhaps want to ignore because it's too complicated. This is not as unusual or as undefined as it may sound. For example the common thermodynamic system of a body which is not thermally isolated has no Hamiltonian because there are the suppresed variables of the heat bath. Nevertheless we do have the rules of thermodynamics to tell us what to do. Similiarly, even when we are not doing thermodynamics, as is the case here, we can try to make certain more or less obvious models of what the environment does to the density matrix and look at the resulting physics. As we shall see, this can lead to some non-trivial conclusions.

We are interested, then, in the density matrix of a two level system imbedded in an environment. We don't propose to observe the environment, so our object of interest is the two-by-two density matrix for the two level system produced by summing over the variables of the environment.

We can, however, imagine the whole big system, with its typically 10^{23} variables, to be in a pure state described by a wavefunction ψ. Then we write

$$\psi = \phi_+|+> + \phi_-|-> \tag{3}$$

where the phi's are functions of the all the other variables of the system except the two-valued variable describing the subsystem. The two-valued variable is formally identical to and can naturally be thought of as a kind of "spin".

The density matrix of the two-level system can then be written as

$$\rho = \begin{pmatrix} (\phi_+|\phi_+) & (\phi_+|\phi_-) \\ (\phi_-|\phi_+) & (\phi_-|\phi_-) \end{pmatrix} \tag{4}$$

where the normalization of (3) means that the trace is equal to one. In this representation it is clear that the " purity " of the state is characterized by the degree to which ϕ_+ and ϕ_- are different. If the two ϕ's are the same,"parallel", then they can be factored out in (3), or in terms of ρ, we have a matrix which can be reduced to a single element, signifying a pure state. If, on the contrary, the two are orthogonal, we have (1/2) the identity matrix, the maximally impure state.

If we imagine our isomer or general two level system to be in a pure state at $t = 0$ and then to be exposed to the environment, the growing incoherence of the two-level sub- system will be represented by the decreasing overlap of the two phi's. If the environment is so big that the state of the subsystem essentially has no influence on it, we expect the system to see an essentially stationary environment and so a kind of stationary decrease in the overlap.

5. Time evolution of ρ

Since the two level system can be thought of as a kind of spin, where "up" and "down" are identified with the two basis vectors, we have a convenient visualization. Any linear combination corresponds to some direction for the "spin". Our two-by-two density matrix can be expanded in terms of the three Pauli matrices σ as

$$\rho = \frac{1}{2}(1 + \mathbf{P} \cdot \sigma) \tag{5}$$

where the $\mathbf{P}$ can be thought of as the polarization vector. When it has length one we have a pure state and complete polarization, when it has length zero, we have complete depolarization or incoherence. Similarly, a Hamiltonian like (1) can be expanded in terms of the Pauli matrices as $H = const. + \mathbf{V} \cdot \sigma$. Pursuing the spin analogy, $\mathbf{V}$ is a kind of "magnetic field" which will make the "polarization" precess around it. In the case of (1) we have $V_z = \epsilon$, $V_x = \delta$, $V_y = 0$. Neglecting for the moment the incoherence question, i.e. considering a purely isolated system, the equation of motion of $\mathbf{P}$ is simply that for precession around $\mathbf{V}$, namely

$$\dot{\mathbf{P}} = \mathbf{V} \times \mathbf{P}. \tag{6}$$

In the general case $\mathbf{V}$ may be time dependent, which will lead to a complicated behavior for $\mathbf{P}$. Observe, however, that with this equation $\mathbf{P}$ cannot change its length; the degree of coherence is preserved in time.

This brings us to our main problem, the description of the loss of coherence, that is the description of how P shrinks in length. We envision situations where our probe – the two level system – does not have a significant back-reaction on the big environment, so we can treat the properties of the environment as unchanging. This assumption leads, as we shall see in a simple model in the next section, to a constant relaxation rate, that is, to an exponential decay in time. Thus we anticipate that (6) becomes

$$\dot{\mathbf{P}} = \mathbf{V} \times \mathbf{P} - D_T \mathbf{P_T} - D_L P_z \tag{7}$$

where the D's are certain "damping parameters". These lead to an exponential damping of the corresponding component of $\mathbf{P}$. We distinguish between a longitudinal damping D_L which acts on P_z and a transverse damping D_T, which acts on $\mathbf{P_T} = (P_x,\ P_y)$.

This division into transverse and longitudinal damping is often used in the spin relaxation problem of nuclear magnetic resonance. Here, the longitudinal or z axis corresponds to the "quality" or property in question. For the optical isomer it corresponds to the amount of $(L - R)$, for K mesons or oscillating neutrinos it corresponds to the "flavor" difference and so forth. The longitudinal relaxation is a rather trivial matter: it corresponds to an averaging out of the "quality" due to the environment simply directly flipping the system from one state into the opposite state. For the isomer it corresponds to the atoms of the ambient medium knocking the system over the potential hump of the double potential well. It is no surprise that this kind of thing leads to relaxation, indeed it is the standard classical relaxation mechanism. For simplicity of presentation, and to be able to concentrate on the really subtle part of the problem, we shall therefore leave this rather pedestrian effect out in the following. We set D_L to zero and simply refer to D_T as D. This can also be achieved physically by having a low enough temperature so that the ambient atoms are below the barrier energy for the molecule. For the K's or neutrinos it corresponds to the neglect of flavor nonconserving scattering , which is the common assumption anyway. The longitudinal damping could easily be put back in, but the really interesting point is to see how a combination of the incoherent but "quality" diagonal environment and the coherent "quality" rotating Hamiltonian (1), neither of which would lead to relaxation themselves, combine to give a relaxation, one with rather unusual properties.

6. The Damping Parameter

We now turn to the question of how to evaluate D.[3] If we wish, D can be thought about in measurement- theoretic terms. It describes how fast the surroundings are destroying the off-diagonal elements of ρ, which represent the coherence or the possiblity of inteference between the two basis states. When the off-diagonal elements of ρ are zero, the system is totally " reduced"; there can be no interference effects between the two basis states. This is as if the environment has "measured " the subsystem, establishing a correlation between its state and that of the subsystem. Thus the calculation of D amounts to a quantitative statement on how much "reduction" a "measurement" accomplishes. We will now explain how to arrive at a concrete scattering theoretic formula for D on the basis of a simple model.

What we want to see is how successive interactions – "measurements" – by the environment reduce the value of the off-diagonal elements of ρ. We imagine the environment to consist of a very large, if not infinite number of distinguishable units, "atoms". These are taken to scatter successively on our system. Thus at a given moment a certain number of the atoms, say up to atom number N, have interacted with our system, and the rest have not. Thus, approximating the environment as a simple product wavefunction we write

$$\begin{aligned} \phi_+ &= \chi_+(1)\chi_+(2)\ldots\chi(N)\chi(N+1)\ldots \\ \phi_- &= \chi_-(1)\chi_-(2)\ldots\chi(N)\chi(N+1)\ldots \end{aligned} \tag{8}$$

where the χ's are the wavefunctions of the individual atoms of the medium. The χ up to N have a subscript because these atoms have interacted with the system and may be correlated with it. The atoms N and beyond have not yet interacted and so have no way of knowing about the system.

Up to number N the atoms have interacted, gone away and will never come back. It is with this assumption that we sneak in irreversibility, something that we couldn't do if we didn't have an "open system". Now let atom N scatter on the subsystem and inquire as to the change in the overlap of the two phi's. If there is a sufficiently long time between scatterings so that the particles may be regarded as free, we can convenienently describe the process by the S matrix, which is what turns an incoming state into an outgoing state. In one such interaction then,

$$\begin{aligned} S\ \chi(N) &\rightarrow S_+\chi(N) \equiv \chi_+(N) \quad \text{for } \phi_+ \\ S\ \chi(N) &\rightarrow S_-\chi(N) \equiv \chi_-(N) \quad \text{for } \phi_- \end{aligned} \tag{9}$$

The subscripts on the S refer to the fact that in one case the medium atom interacts on a (+) object and in the other with a (−) object so we have to use different S's.

The new overlap is then the old one times the overlap of the S's, or in one collision

$$(\phi_+|\phi_-) \rightarrow (\phi_+|\phi_-)\langle\chi(N)|S_+^\dagger\ S_-|\chi(N)\rangle. \tag{10}$$

Note that since S is unitary, the modulus of this factor must be one or less (by the Schwartz inequality). The overlap is always decreasing, hence irreversibility.

Thus the "reduction per collision", by which we mean the change in $\rho_{+-} = (\phi_+|\phi_-)$, is

$$(\phi_+|\phi_-)\langle\chi|1 - S_+^\dagger S_-|\chi\rangle \qquad (11).$$

This change is proportional to ρ_{+-} itself, so that given a uniform collision rate we have a constant proportional rate of decrease, corresponding to the exponential damping assumed in the previous section.

We now assume that the incoming wavepackets of the medium are uncorrelated with the target, so that as in the usual treatment of scattering we may treat the χ's as a series of nearly plane -wave like wave packets.

If we work out the flux factors we get the following: Let Λ be the following expression:

$$\Lambda = Nv\frac{\pi}{k^2}i\Sigma_l(2l+1) < \chi|1 - S_+^{l\dagger}S_-^l|\chi > \qquad (12),$$

where S has the usual form $e^{2i\delta}$. (This is for the simple case of spinless collision partners, for the more general case see ref.(3), or for neutrinos ref.(4). D is then the imaginary part of Λ:

$$D = Im\Lambda \qquad (13)$$

It may be that $\langle\chi|1 - S_+^\dagger S_-|\chi\rangle$ turns out, as will usually be the case, to also give a phase, corresponding to the real part of Λ. What is the meaning of this? Since ρ_{+-} and ρ_{-+} are complex conjugates the real part of Λ gives equal and opposite phases to the off-diagonal elements of ρ. This corresponds to a rotation of the "spin" around the z axis, so the real part gives an effective "magnetic field " along the z axis, a contribution to V induced by the medium, an addition or renormalization of the internal Hamiltonian (1):

$$V_z^{Med} = Re\Lambda \qquad (14)$$

A number of simplifications and assumptions are involved in this set of formulas. First there is the assumption involving the S matrix and the treatment in terms of a series of distinct collisions. This a kind of dilute gas approximation for the medium, and does not seem terribly dangerous. Even if we had condensed matter, we could often talk about a dilute gas of phonons or other excitations. Or it may be that the collisions happen rapidly so that the particles are not free between collisions and so something more complicated than the S matrix is needed; but this does not seem to imply any fundamental problems.

A very important simplification, however, is that we have an equation for the reduced density matrix (4) in terms of itself alone. In general, although the evolution of the *total* density matrix is given in terms of itself at an earlier time, there is no reason for this to be true of a reduced density matrix where we have thrown information away. In a completely general situation we can imagine that in (4) there are various phi's which give the same reduced density matrix to begin with, but which will lead to a different evolution of the system and of the reduced density matrix at latter times. Our simple one-and- only-one interaction model, however, gives an essentially unique prescription for the development of the reduced matrix. A related assumption is that their is no correlation between the medium and the subsytem before the collision. Our results depend on this, as we can see by taking an extreme counter example: Say the interaction medium-subsystem was only in some particular configuration, in the s-wave , for example. Let the medium know about this somehow so that the incoming χ wavepackets are all in the s-wave. Then even if $|+>$ and $|->$ have different s-wave scattering the successive scatterings or "measurements" will simply create a phase shift that is an energy V_z^{Med} but not any incoherence or damping D. Thus it makes a difference if our incoming wavepackets are plane waves, that is s-,p...-waves coherently or first s-,then p-,...

waves seperately. That this kind of correlation does not exist is also important for our assumption that each collision gives only an infinitesimal change in the overlap; most of the packet should remain unscattered. I believe these points are relevant in understanding how our argument seems to get irreversibility practically from nothing, just from the unitarity of S and the Schwarz inequality. The situation is like that (as suggested to me by R. Haag) in the Boltzman H argument, where one also "derives" irreversiblity, but there is a loophole in the neglect of the correlations before the collisions[4].

Incidentally, the reader should not be mislead by this discussion to think that anything so far has anything to do with thermodynamics. All we have assumed is a kind of steady state randomness and we have not mentioned temperature , thermodynamic equilibrium or anything of the sort. This is to be distinguished from other attempts to understand this kind of problem where damping is introduced by coupling the system to a heat bath of harmonic oscillators and making use of thermodynamics in the path integral form. I believe one of the advantages of the present treatment is that it shows how most of the results follow without thermodynamics, just from a more primitive randomness assumption[5].

7. Properties of D

Our formula for D, the "reduction rate" seems quite plausible. We note some of its features:

- If the two components of the probe scatter the medium the same way, that is if $S_+ = S_-$, then there is no damping. This is because the collisions establish no correlation between the state of the subsystem and the state of the medium, the subsystem is not "measured" by the medium. This can have some striking consequences in practice. For example[1], if we have our optical isomer in a very neutral gas like Helium, rather high partial waves are necessary for S to be different for the $|R>$ and the $|L>$ isomer. Thus at low temperature , where most of the scattering is s-wave, there is little damping, even though the total scattering rate may be high. The isomer can bounce around a lot without having its coherence disturbed. Hence it would not be correct to think that every scattering "stops the clock"(or "watches the pot", "Turing's the system" according to your favorite expression) the scattering must distinguish the components of the subsystem. Formula (13) tells how strong the measurement is.

- Another special case is $S = 1$ for one component. In terms of the scattering amplitude T, where $S - 1 = iT$, the damping is proportional to the diagonal T or forward scattering amplitude for the active component. In particular if we take the imaginary part and use the optical theorem to relate the forward scattering amplitude to the total cross section we obtain

$$D = (1/2)Nv\sigma, \tag{15}$$

where σ is the total cross section and N is the number density in the medium so that we have (1/2) v (mean-free-path) for the component which can scatter.

This is in turn (1/2) the scattering rate and so the result here corresponds to the notion that each scattering accomplishes a "reduction ".

- The real part of Λ, through the equal and opposite phases it gives to the density matrix signifies a rotation around the z axis. This supplementary "magnetic field" along the z axis adds to that due to the internal Hamiltonian. This extra energy, given in terms of the *real* part of the forward scattering amplitudes, reminds us of the fact that the index of refraction, which can be viewed as giving the change in the energy of the particle due to the medium, is determined by the real part of the forward scattering amplitude. Indeed one finds[3,6] (to linear order in T)that the formula for V_z^{Med} corresponds to

$$n_+ - n_- = \frac{2\pi}{k^2}N(f_+^0 - f_-^0), \tag{16}$$

where f^0 is the forward scattering amplitude. This is just the familiar Lorenz-Lorentz formula, which we get here as a bonus from our method.

In general the expression (12) can be evaluated if the underlying scattering amplitudes are known. In Flaig's diploma thesis[7], for example, D is evaluated for the mixing of two neutrino flavors in terms of Feynman graphs for neutrino scattering.

Formula (12) makes clear that D and therefore the "reduction" involves a definte energy -the potential that makes $S-$ and that the reduction goes away as the interaction gets weaker. You can't collapse the wavepacket just by thinking about it.

8. Character of the Solutions

Now that we have both the equation for $\mathbf{P}$, namely

$$\dot{\mathbf{P}} = \mathbf{V} \times \mathbf{P} - D_T \mathbf{P}_T \tag{17}$$

and a way of calculating D, we can inquire as to the nature of the solutions for $\mathbf{P}$. With a time independent environment, so that D and $\mathbf{V}$ are constant, this is simply a set of three linear equations describing damped oscillations. In fact the equations can be cast in the form of the damped harmonic oscillator[8]. Let the transverse component of V be in the x direction. Then taking a second time derivative, inserting the equation in itself and using $V_x P_y = \dot{P}_z$ one obtains

$$\ddot{P}_z + D\dot{P}_z + V_x^2 P_z = 0 \tag{18},$$

which is the equation for an oscillator with friction D. This is a funny kind of friction however, one that uses up coherence or entropy, not energy .

Different regimes of behavior will be characterized by the relationship between the damping parameter D and the energy V. As we increase the strength of the damping we expect a continuous change from slightly damped oscillations to strong damping and finally to overdamping, where there are no oscillations at all. As long as a transverse V is present and D is non-zero and non- singular, the final fate of the polarization vector will always be that it shrinks to zero[9], but the different time scales may correspond to different kinds of physics. In particular for the case of strong overdamping, to be discussed in the next section, the time scale may get so long that it implies a qualitatively new physical situation.

Another class of problems occurs when V and D might be time varying. In neutrino astrophysics there has been considerable interest[10] in the "Michaeev-Smirnov effect" where a neutrino of one flavor (P_z up) gets rotated into the other flavor (P_z down) by a varying $\mathbf{V}$. This can occur through a slow inversion of V, since a "spin" will tend, under adiabatic conditions, to follow the "magnetic field". This may happen when the density or other properties of the environment vary with time or space, such as inside the sun, in the early universe or in a supernova. In ref.(7) we studied the effect of damping on such processes , using the methods developed in ref.(6). In general the solutions must be found by following the evolution of (17) numerically. Of particular interest to us was the case where P_z might be totally inverted. We found, for example that a strong enough D could block the inversion of the neutrino flavor. One can say the neutrino does not have time to rotate its flavor before a collision stops it.

A further interesting point, also studied in ref.(7), which arises in such problems is the possiblility of feedback or non-linear effects. The motion of $\mathbf{P}$ implies a change in the neutrino composition of the medium which then can effect $\mathbf{V}$ and D, which then effects the evolution of $\mathbf{P}$ and so forth.

Recently in a Melbourne thesis[11] and in papers by the Helsinki group[12] neutrinos in the early universe were studied in detail with the help of these methods.

These kinds of arguments will have applications to many kinds of systems, Perhaps the most interesting case of all is the extreme of strong damping, to which we now turn.

9. Turing's-Xeno's-Watched Pot Paradox and Anti-intuitive Relaxation

In these "paradoxes" one wonders what happens if we repeatedly measure something. If the measurement collapses the wavefunction to a eigenstate of the measurement operator does repeated measurement fix the system in that state?

Something very much like this actually happens from our equations. Consider the limit of very strong damping, with D so large that we neglect V. In this extreme, it will be seen that $\mathbf{P}$ constant along the z axis is a solution of (17). If a small V is now switched on, one finds, following the appropriate eigenvalue, that the time behavior of this "frozen" solution becomes (for[6] $V_z = 0$)

$$P_z \sim \epsilon^{-t(V_T^2/D)}. \tag{19}$$

This is now overdamping; the polarization relaxes without oscillations. Mathematically this simply happens because the root of a quadratic for the eigenvalues of our set of linear equations becomes purely imaginary. Physically, however, it corresponds to some striking effects.

We may view the coefficient of t as a relaxation rate for the quality in question

$$Relaxation\ Rate = \frac{V_T^2}{D} \tag{20}$$

This rate has the at first sight puzzling and anti-intuitive property that when we increase the damping D–that is the rate of jostling by the surroundings— the relaxation slows down. In fact if we make the damping very big we will approach the condition where the polarization is practically constant and whatever asymmetry we started with stays frozen in.

In view of the possiblity of interpreting D as being induced by measurements this "freezing in" may be understood as just the manifestation of the Turing paradox[3]: repeated "measurement" keeps the state from evolving. The phenomemon arises simply as a solution to our equation (17) for large D and can be turned on and off by varying the underlying scattering amplitudes in (12).

With $1/D$ on the order of the scattering rate, say 10^8/sec or more and $1/V_T$ the tunnelling time, say minutes or more, the effect can be quite enormous, and for any even nearly macroscopic objects the system will become frozen forever.

With small enough systems, however, the "paradox" could be made to manifest itself experimentally; in particular when relexation must take place through tunnelling. In our example of the isomer, imagine that the relaxation rate is studied as a function of temperature (or pressure) to vary the collision rate and thus D. As the temperature goes down the usual type of relaxation, induced by barrier hopping will also go down. At low enough temperature , this will freeze out completely and only tunnelling will remain. Now the anti-intuitive relaxation will take over and lowering the temperature further will lead to a speeding up of the relaxation. Finally, at small enough D the notion of relaxation rate looses its meaning and the system breaks into oscillation.

Observation of this effect, first a slowing down of the relaxation rate with decreasing temperature or pressure and then a speeding up, would provide beautiful evidence both of tunnelling and of the "paradox".

How can we understand these effects which seem so contrary to our usual intuition that shaking something up should make it relax faster? First of all, we should recognize that this usual intuition corresponds to the barrier hopping (longitudinal D). This is present in general but we have been neglecting it in virtue of the assumption that the temperature is so low that no molecules can be knocked over the barrier.

The tunnelling relaxation is something qualitatively different. Mathematically, the damped oscillator form of the equations (18)makes it clear that increased D means increased friction and increasing friction of course will lead to slower relaxation. Nevertheless it is curious how this friction arises because in some sense the interactions with the surroundings tend to interrupt the oscillations or tunnelling, and can become so large as to stabilize an asymmetry like handedness.

It might be helpful to think about the successive "measurements", for strong D, as "stopping the clock ", which then has to start again. If the behavior for short times is like cosine $\sim 1 - t^2/2$, as in our two-level problem, then the repeated stops and starts mean that we will never get away from one. This is because the deviation from one starts quadratically, there is a "flat top" at $t = 0$.

This viewpoint also helps us to understand why you can't stop a state from decaying by continually observing it, a proposal that has been entertained sometimes. Decays are exponential in time and in contrast to cosine, starting and stopping an exponential is the same as letting it run,$e^{-t_1}e^{-t_2} = e^{-(t_1+t_2)}$. To obtain the paradoxical behavior one must have the "flat top" at $t = 0$.

Finally, we have come to our third and most definitive answer to Hund's question. Appropriately enough, the question is answered by the "paradox": the handness is frozen in by the interaction with the environment. The isomer, if it starts out $|L>$, will be frozen there for a time $\frac{D}{V_T^2}$. This time can be very long, typically very much longer than the tunnel time $1/V_T$ by a large factor. Thus for any but the most isolated of isomers this third answer will prevail: the molecule will be fixed in its state by the surroundings.

10. Fixing the Direction

We seem to have accomplished what we set out do, to select a prefered direction in the vector space of our two-state system. After a time on the order of $1/D$ the density matrix of the subsystem will be "reduced " so that the states $|+>$ and $|->$ cannot interfere. If a system starts out in a coherent linear combination of $|+>$ and $|->$, it will be reduced to an incoherent sum of $|+>$ and $|->$ in a short time. We accomplished this through the influence of the environment, in particular by means of its diagonal but different interaction with the two-states.

Furthermore, if the measuring is strong enough, we can get a very strong stabilization of the state, the Turing -Xeno- Watched Pot effect. The measure for how strong the measuring has to be is that the measuring time (inverse of D) must be short compared to the coherent transition (tunnelling) time:

$$1/D << t_{tunnel} \tag{21}$$

Although these two-state systems are simple, one can see how these conclusions might plausibly generalize to more complicated situations. The interaction with either $|+>$ or $|->$ without transitions between them corresponds to locality in the variable in question. That is, let the environment be diagonal in a given variable. Then by repeatedly interacting with the system it will fix it in *that* variable and not in some other linear combinations.

Consider for example the problem of why molecules have definite shapes. Actually a purely isolated molecule, say our ammonia-like example, does not in general have a definite shape. Even though it's like a tetrahedron, the isolated molecule is typically in some rotational state. To have a definite shape it must be fixed in a definite orientation, in an angle eigenstate, which is the contrary of a rotational state. The environment, however interacts with the molecule through interactions roughly local in position. This tends to fix the molecule into some orientation. The countervailing tendency of the molecule to travel from one orientation to another is represented by the rotational energy, the splitting of the rotational levels $L^2/2I$, where I is the moment of inertia. Thus the condition for fixing is $\frac{L^2}{2I} \ll D$, where L must be at least big enough to represent the orientational localization created by D. Hence, big molecules with a big I, will be fixed more easily, i.e. are more "classical". This is a nice example, because we can imagine that taking smaller and bigger molecules or by taking a less or more frequently interacting medium that we can go continuously from one case to another and turn the fixing on and off.

It is presumably this same mechanism which puts things–like you and me— in definite places, against the tendency of our wavepackets to spread. Going a step further, we might speculate that we have space because it is that variable in which things are local.

11. Beyond two levels

It is tempting to consider how these considerations would generalize to multi-state systems. One approach might be to consider an n-level system in terms of SU(n) in analogy to our treatment of the two-state system in terms of SU(2) with the Pauli matrices[13]. It is found that the tendency to "fix" the system by random noise is generic and can be expected in general for strong damping, although there might be some exceptions .

Another approach might be to try to consider an infinity of levels by going straight to the continuum and guessing what the analogy of our equations for the density matrix would be. We could proceed as follows.

Analogously to $|+>$ and $|->$ in our two level problem, we have a system with a basis $|x_1>$, $|x_2> \ldots$, and an "internal" Hamiltonian like (1) causing the system to go from one x_i to another, which in the continuum limit we take as ∇^2, as for the free particle . We will also have a density matrix $\rho(x', x)$. The effect of the environment, that is of the other interactions, we suppose to be represented by a damping parameter $D(x', x)$. This plays the role of "distinguishing" different x's. That is, we suppose that the outside interactions are sensitive to the position of the particle in x, so that correlations are set up between x and the state of the environment. D is then taken to give the rate of an exponential damping of the off-diagonal elements; that is ρ will shrink at a rate given by $\dot{\rho} = -D\rho$. Putting this together with the usual evolution of ρ as given by the commutator with the internal Hamiltonian, we get for the equation of motion for the density matrix

$$\dot{\rho} = -D\rho - i[H, \rho]. \tag{22}$$

A state can't be distinguished from itself, so when x approaches x' this correlation must vanish:

$$D(x, x) = 0. \tag{23}$$

The density matrix is Hermitian $\rho(x',x) = \rho^*(x,x')$ and in order that this property be conserved in time we should also have

$$D(x',x) = D(x,x') \tag{24}$$

If we assume the situation has translational symmetry then furthermore

$$D(x',x) = D(x-x'). \tag{25}$$

Taking these properties together, we have that D is a real, even function with $D(0) = 0$. If it is analytic near the origin we have $D(z) \sim z^2$. An important parameter characterizing D will presumably be the distance $x - x' = l$, in which D rises to a significant value, l having the interpretation of a kind of correlation length.

Taking the non-relativistic free particle Hamiltonian $\nabla^2/2M$ we can then write (22)as

$$\dot{\rho}(x',x) = -D(x'-x)\rho(x',x) - \frac{i}{2M}(\nabla^2_{x'} - \nabla^2_x)\rho(x',x) \tag{26}$$

Note that because of (23)our equation preserves $Tr\rho = \int \rho(x,x)dx = 1$ as it should. This equation plays the role for the present continuous or infinite level problem that (17) does for the two level system. It is noteworthy that we can find explicit solutions of the equation, at least in one dimension, where (26) becomes an equation in two variables. This is possible because if we introduce the variables $x_\pm = x \pm x'$, we can write $(\nabla^2_{x'} - \nabla^2_x) = -4\nabla_{x_+}\nabla_{x_-}$ and then use the fact that D is a function only of x_-. Then the x_+ dependence is trivial and Fourier transform gives a first order equation in x_-. The analogy to a plane wave solution is $\rho = e^{-i\epsilon t}\rho_\epsilon$, where

$$\rho_\epsilon = e^{iP_+x_+/2 - iP_-x_-/2 - \frac{M}{P_+}\int_0^{x_-} D(x_-)dx_-} \tag{27}$$

with $\epsilon = P_+P_-/2M$. Note that this is not Hermitian, so it must appear together with its Hermitian conjugate.

Going further, we can even find the explicit solution which starts from a given ρ at $t = 0$. Let a given Fourier component in x_+ be called ρ_{P_+}. Then (26) gives the following equation:

$$(\partial_t + \frac{2p_+}{M}\partial_{x_-})\rho_{P_+}(x_-,t) = -D(x_-)\rho_{P_+}(x_-,t) \tag{28}$$

Let $\rho^0_{P_+}(x_-)$ be the desired boundary value of ρ_{P_+} at $t = 0$. Introducing the combination of variables y which is orthogonal to the differential operator on the left, $y = x_- - \frac{2P_+}{M}t$, we have that the differential operator gives zero on any function of this variable, and in particular on $\rho^0_{P_+}(y)$. Thus

$$\rho_{P_+}(x_-,t) = \rho^0_{P_+}(y)e^{-\frac{M}{2P_+}\int_y^{x_-} D(z)dz} \tag{29}$$

is a solution of (28), and gives the desired boundary value at $t = 0$.

In addition observe that since $\rho^0_{P_+}(y)$ is a solution for $D = 0$, it is the answer to the corresponding problem without damping.

It now remains to Fourier transform over P_+ to get back to x_+ space. Note however that (29) is simply a product of two functions of P_+. Hence by the *Faltung* theorem we have for the answer in x space

$$\rho(x_+, x_-, t) = \int dx'_+ \rho^0(x'_+, x_-, t)\rho^D(x_+ - x'_+, x_-, t), \tag{30}$$

where ρ^0 is the Fourier transform of the first term in (29) and ρ^D that of the second. Since, as just mentioned, ρ^0 is a solution for $D = 0$, and by construction it has the given boundary value at $t = 0$, it is what we would get for the situation without damping. On the other hand, since it becomes one in P_+ space for $t = 0$, ρ^D is the delta function in x_+ for $t = 0$.

This rather simple result says that to get the effects of D one just takes the density matrix that one would get for no damping at a given time and smears it in x_+. We smear it with

$$\rho^D(x_+, x_-, t) = \int \frac{dP_+}{2\pi} e^{iP_+x_+} e^{-\frac{M}{2P_+}\int_{(x_- - \frac{P_+}{2M}t)}^{x_-} D(z)dz} \tag{31}$$

One may verify that this object is Hermitian and that as a consequence of $D(0) = 0$ its trace, that is its integral over x_+ for $x_- = 0$ is one.

With a large D, the effect in ρ^D on the off-diagonal elements of the density matrix will be to suppress them exponentially, representing a rapid loss of coherence. For example, the average behavior with x_-, as given by the integral of (31) over x_+ is $e^{-tD(x_-)}$. Thus an initial wavepacket representing a pure state will become an incoherent mixture in a time $1/D$.

To investigate the effect on the probability density of the particle we set $x_- = 0$ to get the diagonal elements of ρ. In this case (31) is simple and we can put in various models for D, such as $D(z) \sim z^2$ or a symmetric zero box around zero. The resulting $\rho^D(x_+, 0, t)$, as seems clear from looking at (31), has the usual feature of giving a larger spread in x_+ for a smaller l, that is for a faster rise of D. The width of $\rho^D(x_+, 0, t)$ will increase with time, in a way which depends on the model chosen for $D(z)$. An interesting question here, which could be studied with different models is whether the spreading is dominated by D and or by the initial localization as given by ρ^0.

Our result that the effect of D is a simple smearing wouldn't seem to give any "freezing " effect and corresponds to the usual intuition that the repeated localizations represented by D just lead to a further spread in position[14]. Why don't we get a Turing-Xeno-Watched Pot effect in x for large D? I believe the explanation can be sought in the fact that our continuum model (26) cannot represent the strong damping condition (21). This is an interesting question that I hope to able to return to later.

REFERENCES

1. R.A. Harris and L. Stodolsky *Phys. Lett.* **78 B** (1978) 313 . The idea that that the state of the isomer might be fixed by the environment was discussed in a general way by M. Simonius *Phys. Rev. Lett.* **40** (1978) 980.
2. F. Hund Z. Phys.**43** 805 (1927).
3. R.A. Harris and L. Stodolsky J. Chem. Phys. **74** (4), 2145 (1981); R.A. Harris and L. Stodolsky *Phys. Lett.* **116 B** (1982) 464.
4. See *Elements of Statistical Mechanics* by D. ter Haar, Rinehart and Co., N.Y. (1954).
5. For a discussion of the connection of the present approach with other methods see R. A. Harris and R. Silbey, J. Chem. Phys.**83** 1069 (1985) and earlier work cited there .
6. L. Stodolsky *Phys. Rev.* **D 36** (1987) 2273. For other treatments of the problem see A. Dolgov Sov. J. Nucl. Phys. **33**, 700(1981), A. Manohar *Phys. Lett.* **186 B** (1987) 370.
7. J. Flaig, *Neutrino - Oszillationen in thermischer Umgebung*, Diploma thesis submitted to the University of Munich (1989).
8. R. A. Harris and R. Silbey, J. Chem. Phys. **78** 7330 (1983).
9. This assumes that the the energy splitting of the eigenstates, is much less than the other energies in the problem. For example in a thermal situation we should relax to a non-zero P_z as given by the Boltzman factor.
10. S.P. Michaeev and A.Yu. Smirnoff, NC C9 (1986) 17; for a simple introduction see L. Stodolsky , *Neutrino Oscillations and Their Damping,* lectures at the Schladming School, Feb. 1988, published in *Particle Physics and Astrophysics, Current Viewpoints*, Eds. H. Mitter and F. Widder, Springer Verlag (1989).
11. *Neutrino Oscillations in the Early Universe* by M. J. Thomson, thesis submitted to the University of Melbourne, (1990).
12. K. Enquist, K. Kainulainen, and J. Maalampi, University of Helsinki preprints HU-TFT-90-22, HU-TFT-90-13, HU-TFT-90-29.
13. I. B. Khriplovich and V. V. Sokolov, Physica **A141** 73 (1987).
14. I would like to thank Y. Aharonov for discussions on these points.

QUANTUM-MECHANICS AND VOLITION

DAVID Z. ALBERT
Department of Philosophy
Columbia University
New York City

ABSTRACT

This paper consists of a few very preliminary remarks on the influence of the mathematical form of the fundamental physical laws on the possible mental lives of sentient physical objects.

Professor Aharonov used to tell me that he thought that there were things about how human beings behave (things like their _freedom of will_, for example) that no physical theory yet imagined could account for.

I used to say that I didn't entirely understand what it _meant_ to hold that people's wills are "free"; and that to whatever extent I _did_ understand what it meant to hold that, I suspected that it just wasn't _true_, that it was some sort of a _delusion_; and I used to point out that it wasn't implausible _at all_ that the existence of a _delusion_ to the effect that people's wills are "free" (_whatever_ that might turn out to mean) could be accounted for in terms of _very familiar_ sorts of physical laws.

And of course we weren't the first two people in the world to have a conversation like that. And we never made a lot of progress with it.

But lately I've begun to think that the whole business is a good deal more complicated and more interesting than I understood back then. Now I think that there are very familiar sorts of physical laws which allow for the existence of physical systems whose mental lives are fantastically richer than I would previously have imagined possible. And what I want to do here is to explicitly lay out a particularly simple special case of that.

* * *

Let me start out by telling you a science-fiction story in the tradition of famous stories of quantum-mechanics like the one about Schrodinger's cat and the one about Wigner's friend. Those stories both begin with the assumption that every physical system in the world (not merely subatomic particles, but measuring instruments and tables and chairs and cats and people and oceans and stars too) is a quantum-mechanical system, and that all such systems evolve entirely in accordance with the linear quantum-mechanical equations of motion, and that every self-adjoint local operator of such systems can, at least in principle, be measured. Those are the rules of the game we're going to play here; and what I want to tell you about is a move which is possible in this game, but which hasn't been considered before.

The old stories of Schrodinger's cat and Wigner's friend end at a point where (in the first case) the cat is in a superposition of states, one in which it is alive and the other in which it is dead; or where (in the second case) the friend is in a superposition of states that entail various mutually exclusive beliefs about the result of some given experiment.
Suppose, for example, that Wigner's friend carries out a measurement of the y-spin of a spin-1/2 particle p that is initially prepared in the state $|\sigma_x = +1/2\rangle_p$. He carries out the measurement by means of a measuring device that interacts with p and that he subsequently looks at in order to ascertain the result of the measurement. The end of that story looks like this:

$$|\alpha\rangle \equiv 1/\sqrt{2}\,\Big[|\text{Believes that } \sigma_y = +\tfrac{1}{2}\rangle_{\text{Friend}} \cdot |\text{Shows that } \sigma_y = +\tfrac{1}{2}\rangle_{\substack{\text{Measuring}\\ \text{Device}}} \cdot |\sigma_y = +\tfrac{1}{2}\rangle_p$$
$$+ |\text{Believes that } \sigma_y = -\tfrac{1}{2}\rangle_{\text{Friend}} \cdot |\text{Shows that } \sigma_y = -\tfrac{1}{2}\rangle_{\substack{\text{Measuring}\\ \text{Device}}} \cdot |\sigma_y = -1/2\rangle\Big]$$

(The phrase "Believes that σ_y = 1/2," of course, doesn't completely specify the quantum state of Wigner's friend's very complicated brain. But the many other degrees of freedom of that system -- those, for example, that specify what sort of ice cream Wigner's friend prefers, simply don't concern us here, and so, for the moment, we'll ignore them). Now, such endings as this are usually judged to be so bizarre, and so blatantly to contradict our daily experience, as to invalidate the assumption that gives rise to these stories. That is, these stories are usually judged to imply that there must be physical processes in the world that <u>cannot</u> be described by linear equations of motion, processes like the collapse of the wave function.

There are, on the other hand, as everybody knows, a number of ways of attempting to _deny_ this judgment; there are a number of ways of attempting to suppose that _this_ is genuinely the way things _are_ at the end of a measuring process, and that this state somehow manages to _appear_ to Wigner's friend or to _count_ for Wigner's friend either as a case of believing that σ_y = +1/2 or as a case of believing that σ_y = -1/2.

One of these attempts goes back to Hugh Everett, and has come to be called the Many Worlds interpretation of quantum mechanics. I think it's too bad that Everett's very simple thesis (which is just that the linear quantal equations of motion are always exactly right) has come to be called that at all, because that name has sometimes encouraged a false impression that there are supposed to be more physical universes around after a measurement than there were before it. It might have been better to call what Everett came up with a "many-points-of-view" interpretation of quantum mechanics, or something like that, because it is surely true of Everett's picture (as it is in all other pictures of quantum theory that I know about) that there is always exactly one physical universe. However, the rules of Everett's game, which he insists we play to the very end, require that _every_ one of the physical systems of which that universe is composed -- including cats and measuring instruments and my friend's brain and my _own_ brain -- can be, and often _are_, in those bizarre superpositions. The various elements of such a superposition, in the case of brains, correspond to a variety of mutually exclusive points of _view_ about the world, as it were, all of which are simultaneously associated with one and the same physical observer.

Needless to say, in some given physical situation, different observers may be associated with different numbers of such points of view (they may, that is, inhabit different numbers of the so-called "Everett worlds"). Suppose, for example, that we add a second friend (Friend #2) to the little story we just told. Suppose at the end of that story, when the state of the composite system consisting of p and of the measuring device for σ_y and of Friend #1 is $|\alpha\rangle$, that Friend #2 measures A, where A is a maximal observable of that composite system such that $A|\alpha\rangle = \alpha|\alpha\rangle$. Friend #2 carries out that measurement by means of an A-measuring device (which, according to the rules of the game, can always be constructed) which interacts with that composite system and which Friend #2 subsequently locks at to ascertain the result of the measurement. When that's all done (since the result of this measurement will with certainty be $A = \alpha$), things will look like this:

$$|\beta\rangle \equiv |\text{Believes that } A=\alpha\rangle_{\substack{\text{Friend}\\ \#2}} \cdot |\text{Shows that } A=\alpha\rangle_{\substack{\text{A-measuring}\\ \text{Device}}} \cdot |\alpha\rangle$$

In this state, Friend #1 inhabits two Everett worlds (the world in which $\sigma_y = +1/2$ and the world in which $\sigma_y = -1/2$, whereas Friend #2 inhabits only one (the world in which $A = \alpha$), which by itself encompasses the entire state $|\beta\rangle$. Moreover, in his single world, Friend #2 possesses something like a photograph of the two worlds which Friend #1 simultaneously inhabits (he possesses, that is, a recording in his measuring device of the fact that $A = \alpha$). By means of his measurement of A, Friend #2 directly sees the full superposition of Friend #1's brain states; and indeed, he can even specify the relative sign between those states.

Nothing ought to be very surprising in this, and indeed, it was all very well known to Everett and his readers. So far as Friend #2 is concerned, after all, Friend #1, whatever else he may be, is a physical system out there in the external world; and consequently, according to the rules of our game, Friend #1 ought to be no less susceptible to being measured in superpositions than a single subatomic particle. But this need not be the very end of the game. One more move, which is fully in accordance with the rules of the game, is possible; a move that Everett never mentions. Here it is: Suppose, at the end of the slightly longer story we just told, when the state of things is $|\beta\rangle$, that Friend #2 shows his photograph of the two Everett worlds that Friend #1 simultaneously inhabits to Friend #1. Suppose, that is, that Friend #1 now looks at the measuring apparatus for A. Well, it's quite trivial to show that the result of such a move will be this:

$$|\gamma\rangle \equiv |\text{Believes that } A=\alpha\rangle_{\substack{\text{Friend}\\ \#2}} \cdot |\text{Shows that } A=\alpha\rangle_{\substack{\text{A-measuring}\\ \text{Device}}}$$

$$\cdot \frac{1}{\sqrt{2}}\Big[|\text{B.t. } A=\alpha, \text{B.t. } \sigma_y=+\tfrac{1}{2}\rangle_{\substack{\text{Friend}\\ \#1}} \cdot |\text{S.t. } \sigma_y=+\tfrac{1}{2}\rangle_{\substack{\sigma_y\\ \text{M.D.}}} \cdot |\sigma_y=+\tfrac{1}{2}\rangle_P$$

$$+ |\text{B.t. } A=\alpha, \text{B.t. } \sigma_y=-\tfrac{1}{2}\rangle_{\substack{\text{Friend}\\ \#1}} \cdot |\text{S.t. } \sigma_y=-\tfrac{1}{2}\rangle_{\substack{\sigma_y\\ \text{M.D.}}} \cdot |\sigma_y=-\tfrac{1}{2}\rangle_P\Big]$$

This is in a number of respects a somewhat extraordinary state of affairs. Let's look at it carefully. To begin with, note that we have here brought an additional degree of freedom of Friend #1's brain explicitly into consideration (I've called this degree of freedom the A-memory of Friend #1). This is the degree of freedom wherein Friend #1 remembers the information in the photograph that Friend #2 has just shown him. Now, what's going on in this state is that Friend #1 still simultaneously inhabits two different and mutually exclusive Everett worlds, one in which $\sigma_y = +1/2$ and the other in which $\sigma_y = -1/2$; but now, in each of those two worlds separately, Friend #1 knows that $A = \alpha$; he knows, that is, that another world exists; indeed, he has literally seen what amounts to a photograph of that other world!

Maybe I can make this a little clearer by saying it in a somewhat different language.

Let's imagine telling this story from a purely <u>external</u> perspective (from the perspective of some <u>third</u> observer, say, who isn't involved in the action at all), without speculating about what it might be like to <u>be</u> either Friend #1 or Friend #2, as if they both were merely, say, automata. If we tell it that way, then we'll be able to say what it is that's interesting about this peculiar state at the end without having to indulge in any talk about multiple worlds.

Here's how that works: Suppose that the state $|\gamma\rangle$ obtains, and that the automaton called Friend #1 is ordered to predict the outcome of an upcoming measurement of σ_y, and suppose that that σ_y measurement is carried out, and that the automaton's prediction of that outcome is measured as well; then invariably it will be the case that the outcomes of <u>those</u> two measurements (the final measurement of σ_y and the measurement of the automaton's <u>prediction</u> about the outcome of that measurement) will coincide. And precisely the same thing applies (when $|\gamma\rangle$ obtains) in an upcoming measurement of A as well (since, when $|\gamma\rangle$ obtains, there is an element of the memory bank of this automaton whose state is correlated to σ_y, and there is also <u>another</u> element of the memory bank of this same automaton whose state is, at the same time, correlated to A); and as a matter of fact, precisely the same thing applies even to upcoming measurements of <u>both</u> A <u>and</u> σ_y (but note that in this case the <u>order</u> in which those two measurements are carried out will be important).

This automaton, then, when $|\gamma\rangle$ obtains, <u>knows</u> (in whatever sense it may be appropriate to speak of automata <u>knowing</u> things), accurately and simultaneously, the values of both A and σ_y, even though those two observables don't commute. What this means (<u>leaving aside</u>, as I said, all of the foggy questions about what it might be like from the perspective of the _automaton_, which is what originally drove us to the talk about multiple worlds) is that this automaton, in this state, is in a position to predict, correctly, without peeking, the outcomes of upcoming measurements of either A or σ_y or both, even though A and σ_y are, according to the familiar dogma about measurement theory, incompatible.

Moreover, no automaton in the world other than this one (no _observer_ in the world other than Friend #1, in science fiction talk) can _ever_ even in _principle_, be in a position to simultaneously predict the outcomes of upcoming measurements of precisely those two observables (even though they can, of course, know either one). The possibility of Friend #1's being able to make such predictions hinges on the fact that A is an observable of (among other things) _Friend #1 himself_. There is a well-defined sense here in which this automaton, this friend, has privileged epistemic access to _itself_.

Perhaps it deserves to be emphasized that there are no paradoxes here, and no violations of quantum theory from which, after all, it was all derived. We have simply discovered a new move here, a move that entirely accords with the rules of quantum mechanics (if the quantum-mechanical rules are all the rules there are) whereby quantum-mechanical observers can sometimes effectively carry out certain measurements on _themselves_. This move just wasn't anticipated by the _founders_ of quantum mechanics, and it happens that when you make a move like this, things begin to look very odd, and the uncertainty relations cease to apply in the long familiar ways.

O.K. Now let me say all this just one _more_ time, quickly, in yet _another_ language; a language in which something interesting will come up that's related to the question of volition.

There is _another_ famous attempt to suppose that the linear quantum-mechanical equations of motion are invariably the true equations of motion of the wave-function of the entire physical world. This attempt goes back to Bohm, and has recently been championed and further developed by John Bell.

Bohm's theory is completely deterministic, and it entails that particles invariably have perfectly determinate positions, and that they have perfectly continuous trajectories, and moreover (by means of a perfectly classical sort of averaging over the various possible values of positions of the particles) Bohm's theory reproduces every last one of the statistical predictions of non-relativistic quantum mechanics. And since this theory makes all the same predictions as quantum mechanics does, every one of those predictions, including the ones in our story about quantum-mechanical automata, will necessarily arise in this theory too.

That produces an odd situation. Remember the two automata in the story (Friend #1 and Friend #2). Suppose that [γ> obtains, and suppose that things are set up so that some future act of #1 is to be determined by the results of upcoming measurements of σ_y and A. On Bohm and Bell's theory, there is, right now, a matter of fact about what that act is going to be, and it follows from what we discovered above that automaton #1 can correctly _predict_ what that act is going to be, but _not_ so for automaton #2, nor for any other one, anywhere in the world. So it turns out that it can arise, in a _completely deterministic_ physical theory, that an automaton can in principle be constructed that can ascertain certain of its own acts in advance, even though no other automaton, and no external observer whatever can ascertain them, even if that external observer can measure with infinite delicacy and infinite precision, and calculate with infinite speed.

You might say that an automaton like that has a _absolutely private_ will. I suppose that that isn't quite like having a _free_ will (whatever it _is_ like, exactly, to have a free will); but it strikes me as something of a surprise that any completely deterministic physical theory allows any completely physical system even to have a will as subtle and as curious as _this_ one; and I want to offer it to Yakir here, in the spirit of our old conversations, aevidence that the _familiar_ sorts of physical theories, the ones that have _already_ _been_ imagined, may yet have room in them for mental lives of heretofore unimagined richness.

QUANTUM TIME MACHINE

YAKIR AHARONOV*, Jeeva ANANDAN, and Lev VAIDMAN

Physics Department, University of South Carolina
Columbia, South Carolina 29208, U.S.A.

ABSTRACT

A method to obtain a superposition of time evolutions of a quantum system which correspond to different periods of time is considered[1]. Such superpositions may lead to unusual consequences as demonstrated by two examples presented in this letter. In the first example we show that a suitable superposition of time evolutions due to the action of weak forces is equivalent to a time evolution resulting from an action of a strong force. This corresponds to a new method of amplification. In the second example we show that a particular superposition of time evolutions with the same Hamiltonian but different periods of time corresponds to a time evolution for a very different (possibly even negative) period of time. This provides an example of a new type of "time machine" which is peculiar to quantum systems and has no classical analog. As a purely gedanken experiment this "time machine" can move you to the future to meet your grand-grand-grandchildren. We believe, however, that the amplification effect can be achieved in a real laboratory[2].

---

*Also at School of Physics and Astronomy, Tel-Aviv University, Ramat Aviv 69978, Israel

[1]Y. Aharonov, J. Anandan, S. Popescu and L. Vaidman, "Superpositions of time evolutions of a quantum system and a quantum time machine", Phys. Rev. Lett., 64 (1990) 2965.

[2]J.M. Knight and L. Vaidman, "Weak Measurement of Photon Polarization", *Phys. Lett.* **143 A** (1990) 357.

QUANTUM MAGNETIC FIELD FLUCTUATIONS AND ELECTRON COHERENCE

L. H. Ford
Department of Physics and Astronomy, Tufts University
Medford, Massachusetts 02155, USA

and

Chandra Pathinayake
Institute for Theoretical Physics, University of Bern
Sidlerstrasse 5, CH-3012 Bern, Switzerland

Abstract

The vacuum fluctuations of the electromagnetic field are altered by the presence of a conducting boundary, so that the root-mean-squared magnetic field is nonzero. Quantum mechanical particles, such as electrons, are shown to be capable of detecting this fluctuating magnetic field through their coherence properties. It is found that an electron interference pattern is distorted by a factor of $\exp(-\frac{1}{2}e^2 F_R)$. Here $F_R = \int da^i \int da'^j < B_i(\mathbf{x}) B_j(\mathbf{x}') >_R$ is the double surface integral of the renormalized magnetic field two-point function over the surface enclosed by the electron path. This effect is analogous to the Aharonov-Bohm effect in that it is nonlocal; the electrons are sensitive to changes in the quantum fluctuations in regions from which they are excluded. However, in contrast to the Aharonov-Bohm effect, the positions of the interference minima are not shifted. Rather, the relative heights of the maxima are altered in a way which depends upon the conductor's geometry. It may be feasible to detect this effect experimentally.

1. Introduction

As is well known, especially to the participants of this conference, the Aharonov-Bohm effect[1] arises from the coupling of a quantum mechanical particle to a classical electromagnetic field. The subtle and surprising aspect of this effect is the dependence on the total magnetic flux enclosed by a path, regardless of whether the particles move in a region where the magnetic field is nonzero. Nonetheless, it is clearly a real effect and has been verified experimentally.[2] Another subtle and surprising effect in quantum physics is the Casimir effect[3], the attraction of a pair of electrically neutral plates due to quantum fluctuations of the electromagnetic field. The Casimir force may be regarded as due to the presence of a nonzero vacuum energy density in the region around a conducting boundary. More generally, the retarded van der Waals force between dielectrics is a manifestation of the electromagnetic vacuum energy. These forces have also been confirmed experimentally.[4–6] At first sight, it would seem that the Aharonov-Bohm and Casimir effects have little in common. However, we will see that this is not the case.

A conducting boundary alters the vacuum fluctuations of the electromagnetic field in such a way that although the expectation of the magnetic field is zero in the vacuum state, the expectation value of the *squared* magnetic field is nonzero. This raises an interesting question: Is there any way that coherent electrons can detect this fluctuating magnetic field? It is our purpose to answer this question in the affirmative and to propose a new effect. This effect is analogous to the Aharonov-Bohm effect, but arises from quantum fluctuations of the electromagnetic field and can occur even in the absence of any classical fields.

2. Formalism and the Key Result

The physical situation which we wish to analyze is a nonrelativistic electron coupled to the quantized electromagnetic field. This situation can be obtained as the limit of quantum electrodynamics in which the Dirac field is in a single-particle state describing nonrelativistic motion. In this limit, we may describe the electron by a "wave functional", ψ, which is a solution of a Schrödinger equation containing second-quantized electromagnetic field operators. Here we will assume that the electron's magnetic moment may be neglected. Let the quantized electromagnetic field be a free field in the Coulomb gauge. Then it is described by a divergence-free vector potential operator, $\mathbf{A}(\mathbf{x})$. Here it will be convenient for us to use the Schrödinger picture to describe the quantum dynamics of both the electrons and the electromagnetic field. Consequently, $\mathbf{A}$ is time-independent. The electron may also be coupled to a classical electrostatic potential, V. In units where $\hbar = c = 1$, the wave functional, $\psi = \psi(\mathbf{x}, t; \mathbf{A})$, satisfies the Schrödinger equation:

$$-\frac{1}{2m}(\nabla - ie\mathbf{A})^2\psi + eV\psi = i\frac{\partial\psi}{\partial t}. \tag{1}$$

Note that ψ, a functional of $\mathbf{A}$, is an operator on the Hilbert space of photon states. Observable quantities are formed by taking the expectation value of bilinear functionals of ψ in the state of the photon field. For example, the number density of electrons at the spacetime point $(\mathbf{x}, t)$ is

$$n(\mathbf{x},t) = <\psi^{\dagger}(\mathbf{x},t;\mathbf{A})\psi(\mathbf{x},t;\mathbf{A})>, \tag{2}$$

where $<>$ denotes the expectation value in the photon state.

Suppose that χ is a (c-number) solution of Eq. (1) with $\mathbf{A}$ set equal to zero; that is, the wave function of an electron in the potential V. Then a solution for the wave functional is

$$\psi(\mathbf{x},t;\mathbf{A}) = \chi(\mathbf{x},t)\exp(ie\int^{\mathbf{x}}\mathbf{A}\cdot d\mathbf{l}). \tag{3}$$

Here the line integral is along any path with an endpoint at $\mathbf{x}$. Let us now consider an experiment in which electrons from a coherent source are split into two beams and then later recombined and allowed to form an interference pattern, as illustrated in Fig. 1.

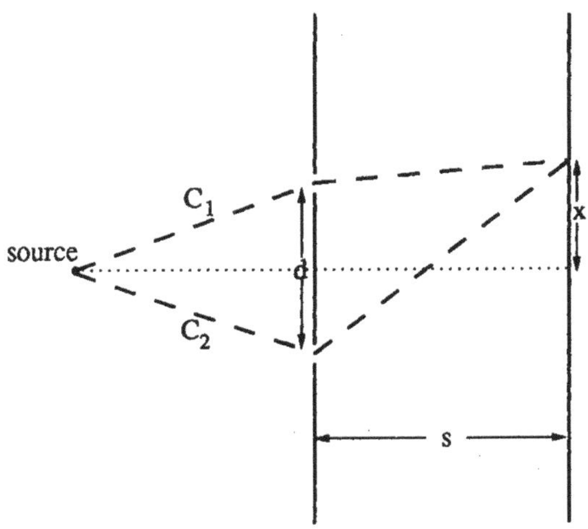

Fig 1. The geometry for electron interference. The electrons from the source on the left travel by classical paths, C_1 and C_2, and interfere at the screen on the right.

There are two classically distinct paths, C_1 and C_2, by which the electrons can get from the source to the detector. Following the procedure used in the absence of the quantized electromagnetic field, we write $\psi = \psi_1 + \psi_2$, where ψ_1 and ψ_2 are both of the form of Eq.(3) with the line integral taken along C_1 and C_2, respectively. That is,

$$\psi_i = \chi_i e^{ie\alpha_i}, \qquad \alpha_i = \int_{C_i} \mathbf{A} \cdot d\mathbf{l}, \tag{4}$$

where χ_i is the c-number wavefunction of an electron which reaches the detector along path C_i, ignoring the coupling to the quantized photon field.

We now wish to assume that the electromagnetic field is in its vacuum state, but in the presence of nontrivial boundary conditions. An example is the presence of perfectly reflecting boundaries of the sort that give rise to the Casimir effect. In such a vacuum state, the expectation value of a quantity linear in the field operator vanishes (e.g., the magnetic field), but quadratic expectation values, such as of the squared field, $< \mathbf{B}^2 >$, do not. As discussed above, we may suspect that there is an Aharonov-Bohm-type effect produced by this fluctuating magnetic field. To verify that this is the case, we must calculate the number density, $n(\mathbf{x})$, in the electromagnetic field vacuum state, $|0>$. Note that the field operator $\mathbf{A}$ and hence the operators α_i may be split into positive and negative frequency components: $\alpha_i = \alpha_i^+ + \alpha_i^-$, where $\alpha_i^-|0>= 0$ and $<0|\alpha_i^+ = 0$. We now wish to calculate the number density using Eqs. (2) and (4). The result is

$$n(\mathbf{x}) = |\chi_1|^2 + |\chi_2|^2 + (\chi_1\chi_2^* + \chi_2\chi_1^*)\exp(-\frac{1}{2}e^2F), \tag{5}$$

where

$$F = \oint dx^i \oint dx'^j < A_i(\mathbf{x})A_j(\mathbf{x}') > = \int da^i \int da'^j < B_i(\mathbf{x})B_j(\mathbf{x}') >. \tag{6}$$

This quantity is expressible either as a double line integral around the closed contour, $C = C_1 - C_2$, of the two-point function of $\mathbf{A}$, or as a double surface integral over the area enclosed by C of the two-point function of the magnetic field, $\mathbf{B}$. Here we have used the Campbell-Baker-Hausdorf formula, $e^{(A+B)} = e^A e^{-\frac{1}{2}[A,B]} e^B$, for any two operators, A and B, which commute with their commutator, $[A, B]$. We have also used the fact that

$$[\alpha_i^-(\mathbf{x}), \alpha_j^+(\mathbf{x}')] =< \alpha_i(\mathbf{x})\alpha_j(\mathbf{x}') > . \tag{7}$$

A very similar calculation, although with a rather different physical interpretation, arises in the theory of global symmetry breaking.[7]

The two-point function for the magnetic field diverges in the coincidence limit, $\mathbf{x} \to \mathbf{x}'$, as $|\mathbf{x} - \mathbf{x}'|^{-4}$. Thus the integral in Eq. (6) is divergent. This is just the usual ultraviolet divergence coming from the part of the two-point function that is independent of the boundary conditions. Physically, we are interested in comparing the behavior of the electrons in the presence of boundaries with that in the absence of boundaries. Hence, we should use not the complete two-point function, but rather the renormalized two-point function defined by

$$< \mathbf{B}(\mathbf{x})\mathbf{B}(\mathbf{x}') >_R = < \mathbf{B}(\mathbf{x})\mathbf{B}(\mathbf{x}') > - < \mathbf{B}(\mathbf{x})\mathbf{B}(\mathbf{x}') >_0, \tag{8}$$

where $< \mathbf{B}(\mathbf{x})\mathbf{B}(\mathbf{x}') >_0$ denotes the empty-space two-point function (that in the absence of boundaries). Thus, we replace Eq. (5) for the number density by

$$n(\mathbf{x}) = |\chi_1|^2 + |\chi_2|^2 + N(\chi_1\chi_2{}^* + \chi_2\chi_1{}^*)\exp(-\frac{1}{2}e^2 F_R), \tag{9}$$

where F_R denotes F as defined in Eq. (6), but using the renormalized two-point function, and N is a normalization constant to be determined by the details of the geometry.

We can think of this constant as an additional arbitrariness beyond the overall normalization of χ_1 and χ_2 introduced by the need to perform an infinite subtraction from F. It is admittedly not obvious that the contribution of the empty-space two-point function to F is a constant when suitably regularized. It will be of interest to elaborate the details of the formal renormalization in this theory, and its relation to the order-by-order renormalization of perturbative quantum electrodynamics. (Note that here we are working to all orders in the coupling constant, e.) However, the requirement that $n(\mathbf{x})$ depend only on the renormalized two-point function seems to dictate that the answer must be of the form of Eq. (9). Recall that $|\chi_i|^2$ is the number density when only one of the slits (slit i) is open. Thus we can form an observable quantity which is clearly independent of the empty-space two-point function. This quantity is

$$\frac{n(\mathbf{x}) - (|\chi_1|^2 + |\chi_2|^2)}{n_0(\mathbf{x}) - (|\chi_1|^2 + |\chi_2|^2)}.$$

The experiment which determines this quantity involves measuring the number density; first, in the absence of the boundary, for both slits open and for each slit open alone, and then, repeating these measurements in the presence of the boundary. Because this observable is a ratio of a quantity proportional to $\exp(-\frac{1}{2}e^2 F)$ and one proportional to $\exp(-\frac{1}{2}e^2F_0)$, it depends only on F_R. Here F_0 is the contribution to F from the empty-space two-point function. This shows that the interference term in the presence of the boundary is proportional to $\exp(-\frac{1}{2}e^2F_R)$.

Let n_0 be the number density that we would have without the effects of the quantized electromagnetic field; i.e., n_0 is given by Eq. (9) with $N = 1$ and $F_R = 0$ (Eq. (17) below). Because the electromagnetic vacuum fluctuations do not change the number of electrons present, the integrals over all space of n and of n_0 for the same χ_1 and χ_2 must be equal. This condition determines N. Note that in the usual Aharonov-Bohm effect, the dependence on the magnetic field appears in a complex exponential whose absolute value is always unity; here the magnetic field appears in a real exponential. In both cases, the electrons are sensitive to the electromagnetic field in regions from which they may be excluded. If $< \mathbf{B}(\mathbf{x})\mathbf{B}(\mathbf{x}') >_R$ is zero everywhere that the electrons are allowed to travel, but the double flux integral, F_R, is nonzero, there will still be an effect on the interference pattern.

We have obtained our result, Eq. (9), working from a viewpoint in which the crucial factor, $\exp(-\frac{1}{2}e^2F_R)$, is regarded as the effect of vacuum fluctuations. Y. Aharonov and L. Susskind have suggested to us that it might also be possible to regard it as coming from decoherence due to soft photon emission. In this latter view, the photon emission modifies the electron wavefunctions, χ_1 and χ_2, and the presence of the conductor in turn alters the emission and introduces the factor in question. Whether this is an equivalent viewpoint or a physically distinct effect is not yet clear to us. If they are indeed equivalent, then it must be possible to use photon emission considerations to derive Eq. (9). This is a problem requiring further work.

3. A Simple Geometry: A Single Conducting Plate

To better understand the nature of the effect of quantum magnetic field fluctuations, let us consider a geometry where the two-point functions may be explicitly computed. The simplest such geometry is the case of an infinite perfectly-conducting plane, where the two-point function for the vector potential in the Coulomb gauge may be found by the method of images. The full two-point function in the presence of the plane is the sum of the empty-space two-point function and an image term:

$$< A_i(\mathbf{x})A_j(\mathbf{x}') >=< A_i(\mathbf{x})A_j(\mathbf{x}') >_0 -(1-2\delta_{jz}) < A_i(\mathbf{x})A_j(\tilde{\mathbf{x}}') >_0 . \qquad (10)$$

Here the conductor lies in the xy plane and $\tilde{\mathbf{x}}' = (x', y', -z')$ is the image of the point $\mathbf{x}'$. Recall that we are working in the Schrödinger picture; however, the two-point function in this picture is just the Heisenberg picture two-point function evaluated at coincident times, $t = t'$. The renormalized two-point function is just the second term on the right hand side of Eq. (10), which in the Coulomb gauge is

$$< A_i(\mathbf{x})A_j(\mathbf{x}') >_R = -\frac{(1-2\delta_{jz})(x_i - \tilde{x}'_i)(x_j - \tilde{x}'_j)}{2\pi^2\Sigma^2}, \tag{11}$$

where $\Sigma = (x-x')^2 + (y-y')^2 + (z+z')^2$. The renormalized magnetic field two-point function is obtained by taking the curl of this expression in both $\mathbf{x}$ and $\mathbf{x}'$; the result is

$$< B_i(\mathbf{x})B_j(\mathbf{x}') >_R = (1-2\delta_{jz})\frac{2(x_i - \tilde{x}'_i)(x_j - \tilde{x}'_j) - \delta_{ij}\Sigma}{\pi^2\Sigma^3}. \tag{12}$$

It is instructive to give here a measure of the scale of the magnetic field fluctuations due to a conductor. For the case of a single plate, the expectation value of the squared magnetic field at a height z above the plate is $< \mathbf{B}^2 >_R = -\frac{3}{16\pi^2 z^4}$. This result arises from the coincidence limit of Eq. (12). The negative sign represents a decrease in magnetic field fluctuations compared to empty space. (Recall that only changes in the vacuum fluctuations are observable.) The associated rms magnetic field is

$$B_{rms} = \frac{\sqrt{3}}{4\pi z} = (7.7 \times 10^{-2} G)(\frac{1\mu}{z})^2. \tag{13}$$

Thus we see that the magnetic field fluctuations may be of a macroscopic scale. This supports the hope that their effects are detectable.

Let the electrons be constrained to move in a plane parallel to and a height z above the conductor. Then the flux integral, F_R, may be calculated as the double line integral of the renormalized two-point function, Eq. (11), around the closed contour, C. Let L be the circumference of this contour and assume that $L \gg z$. In this limit, we may obtain an expression for F_R valid for an arbitrary contour (so long as the radius of curvature is always large compared to z). In this case, we may divide C into N straight segments, each of length ℓ, so that $C \gg \ell \gg z$, and write

$$F_R = N \int_0^\ell dx^i \int_0^\ell dx'^j < A_i(\mathbf{x})A_j(\mathbf{x}') >_R. \tag{14}$$

If we choose the x-axis to be the direction of integration and let $\xi = x - x'$, then we have

$$F_R = -N \int_0^\ell dx' \int_{-x'}^{\ell - x'} d\xi \frac{\xi^2}{(\xi^2 + 4z^2)^2}. \tag{15}$$

The dominant contribution to the above integral comes from values of ξ of the order of z. However, because $\ell \gg z$, for most values of x', $\ell - x' \gg z$ and $x' \gg z$. Consequently, we may replace the upper and lower limits of the ξ integration by $+\infty$ and $-\infty$, respectively. Evaluation of this integral, and use of the fact that $L = N\ell$, yields

$$F_R = -\frac{L}{8\pi z}. \tag{16}$$

As noted above, this result is independent of the shape of the contour, C. We have also directly evaluated F_R for the cases of rectangular and circular loops parallel to the conductor. If the dimensions of the loop are large compared to z, the results agree with Eq. (16).

4. Electron Interference near a Conducting Plate

Let us now consider a double-slit interference experiment with the geometry illustrated in Fig. 1. Again we assume that the electrons are constrained to move in a plane which lies a distance z above a perfectly conducting plate. In the absence of the plate, the interference pattern produced at the screen is given by

$$n_0(\mathbf{x}) = |\chi_1|^2 + |\chi_2|^2 + (\chi_1\chi_2{}^* + \chi_2\chi_1{}^*), \tag{17}$$

where χ_1 and χ_2 are the amplitudes for electrons to go through the upper and lower slits, respectively. If the slit separation, d, is small compared to the distance to the screen, s, then we have, for monoenergetic electrons with wavenumber k,

$$n_0 = K[1 + \cos(\frac{kdx}{s})], \qquad d \ll s, \tag{18}$$

where K is a constant.

Now include the effect of the coupling to the vacuum fluctuations. As discussed above, the number of electrons/unit time arriving at point x will be modified by the insertion of a factor of $N \exp(-\frac{1}{2}e^2 F_R)$ in the interference term. In the geometry of

Fig. 1, the total path length, L (the *sum* of the lengths of C_1 and C_2), is expressible as

$$L(x) \approx L_0 + \frac{x^2}{s}, \qquad x \ll s, \tag{19}$$

where L_0 is the total path length to the point $x = 0$. Using the result, Eq. (16), for F_R, we find that the modified interference pattern is, for $x \ll s$,

$$n(x) = K[1 + N' \cos(\frac{kdx}{s}) \exp(\frac{e^2 x^2}{16\pi s z})]. \tag{20}$$

Here the L_0 dependence has been absorbed into the constant N'. The effect of the vacuum fluctuations has been to distort the interference pattern. The minima are still in the same locations, but the relative heights of the maxima are altered. For small x, these heights are enhanced by a factor of $\exp(\frac{e^2 x^2}{16\pi s z})$. Thus, observing this effect requires measuring the shape of the interference pattern to sufficient accuracy to detect the distortion. If the argument of this exponential is small compared to one, then the fractional change in the maxima heights is $\frac{e^2 x^2}{16\pi s z} \approx 10^{-4}\frac{x^2}{sz}$. Because $x < s$, this requires measuring these heights to a precision of the order of $10^{-4}x/z$. The shifting of intensity away from the center of the interference pattern is characteristic of a negative F_R. If this quantity were positive, then the shift would be in the other direction.

5. Discussion

We have seen that coherent electrons are not only capable of detecting a magnetic field in a region from which they are excluded, but that they are even capable of detecting the zero-point fluctuations of the magnetic field in such a region. Thus the Aharonov-Bohm and the Casimir effects are united.

It seems possible that the effect which we are proposing may indeed be experimentally detectable. At this point it is in order to note what other effects may need to be taken into account in an actual experiment. Of course, we ourselves are theoreticians, not experimentalists, so our real purpose here is to pique the interest of our experimental colleagues, rather than to give an exhaustive discussion of experimental issues. As noted above, our analysis ignores the effect of the electron's magnetic moment. However, it is expected to be small and can probably safely be neglected. We have also not included the classical image-charge effects. Electrons moving in the vicinity of a conductor necessarily feel an electrostatic potential due to their images. It may be regarded as part of the classical potential, V, which appears in the Schrödinger equation, Eq. (1). Such an electrostatic potential can alter the c-number wavefunction χ. For free electrons moving parallel to a single conducting plate, the image charge will attract the electrons toward the plate. This imposes a limitation on how large s could be for a given value of z. For example, for $z = 1\mu$, electrons with an energy of 100keV can travel a distance of the order of 1cm parallel to the plate. For electrons in a thin film, this limitation does not arise. In this case, the image potential is effectively cancelled by an internal potential in the film. The electrostatic image effect can also be cancelled for free electrons in more elaborate geometries, such as a pair of parallel plates. A second image effect is the magnetic field of the image charges. This classical magnetic field generates an Aharonov-Bohm shift in the interference pattern. Such a shift is proportional to the electron's velocity and so should be distinguishable from the effects of electromagnetic vacuum fluctuations, which are velocity-independent.

Gunther[8] has recently suggested that effects of magnetic field fluctuations might be observable in a SQUID. If the SQUID is deposited on a thin insulator substrate, which is over a plane conductor, the quantum magnetic field fluctuations should alter the voltage versus current relation of the SQUID. The critical current of the SQUID increases by a factor of $\exp(-\frac{1}{2}e^2 F_R)$. This seems to be a promising avenue for experimental investigation.

6. Acknowledgements

We would like to thank B. Allen, L. Gunther, and A. Vilenkin for helpful discussions. This work was supported in part by the National Science Foundation under Grant PHY-8905400.

7. References

1. Y. Aharonov and D. Bohm, Phys. Rev. **115** (1959) 485.
2. For a recent review, see S. Olariu and I.I. Popescu, Rev. Mod. Phys. **57** (1985) 339. See also the contributions of Tonamura and of Webb in this volume.
3. H.B.G. Casimir, Proc. K. Ned. Akad. Wet. **51** (1948) 793.
4. M.J. Sparnaay, Physica **24** (1956) 751.
5. B.V. Deriagin and I.I. Abrikosova, Sov. Phys. JETP **3** (1957) 819.

6. J.N. Israelachvili and D. Tabor, Proc. Roy. Soc. London, **A331** (1972) 19.

7. L.H. Ford and A. Vilenkin, Phys. Rev. D **33** (1986) 2833.

8. L. Gunther, "On the Increase of the Critical Current of a DC SQUID by Electromagnetic Vacuum Flucuations", to be published.

Down-conversion Photon Pairs: A New Chapter In the History of Quantum Mechanical Entanglement

Michael HORNE, Abner SHIMONY, and Anton ZEILINGER

Department of Physics, Stonehill College
North Easton, MA 02356

Departments of Philosophy and Physics
Boston University, Boston, MA 02215

Atominstitut der Österreichischen Universitäten
Schüttelstrasse 115, A-1020 Vienna, Austria

ABSTRACT

An entangled state of a many-particle system is one which cannot be expressed as a product of single-particle states. We summarize the major steps in the history of this concept, notably Schrödinger's original recognition of entanglement, its role in the Gedankenexperiment of Einstein-Podolsky-Rosen, and the Bohm-Aharonov demonstration that photon pairs from positronium annihilation are entangled. We then show that interferometric phenomena of photon pairs generated in parametric down-conversion are manifestations of entanglement, and we analyze two experiments in which different types of entanglement appear: that of Ghosh and Mandel, in which there is entanglement of *directions* of wave-vectors, and that of Kwiat et al., in which there is entanglement of *magnitudes* of wave-vectors (or wave-numbers).

1. Introduction

This paper is two years late. It should have been presented in 1987 at the Conference on the Foundations of Quantum Mechanics to Celebrate the Thirtieth Anniversary of the Demonstration by Bohm and Aharonov of the Existence of Entangled States of Spatially Separated Systems.[1] That conference did not take place – or, in the language of quantum mechanics, it was a virtual conference – and therefore we wish to use the occasion of the present conference, honoring Yakir Aharonov and David Bohm for their illumination of the effect that bears their name, to recognize an earlier contribution of theirs, which in our opinion is of the same order of magnitude.

But first, since we are evidently preoccupied with history, we wish to ensure that due recognition be given to two other physicists who made a pioneering contribution to the subject of the present conference. In 1949 W. Ehrenberg and R.W. Siday noticed the effect on an electron's wave function due to a magnetic field which is present only in the region where the wave function is zero.[2] The formula which they wrote was essentially that which Aharonov and Bohm[3] investigated. Although Ehrenberg and Siday were struck by the peculiarity of the effect, they made no attempt to analyze its conceptual significance, whereas Aharonov and Bohm pointed out that the effect implies either that quantum mechanics endows the electromagnetic potentials with a kind of reality over and above that of the fields, or that it implies a kind of nonlocality, or both. Science is not a zero-sum game, and therefore a large element of glory is due to each of these two pairs of pioneers, without detraction from the other.

2. Some History of Entanglement

We now turn to the concept of entanglement and to its history. Schrödinger recognized in his pioneering series of papers of 1926 that the quantum state of an n-particle system can be entangled, even though at the time he did not give a name to the concept.

> We have repeatedly called attention to the fact that the ψ-function itself cannot and may not be interpreted directly in terms of three-dimensional space – however much the one-electron problem tends to mislead us on this point – because it is in general a function in configuration space, not real space.[4]

Since in general a function of 3n variables (configuration space) does not factor into a product $u_1(\mathbf{r}_1)\ \ldots\ u_n(\mathbf{r}_n)$, $\mathbf{r}_i$ being the position variable of the ith particle, Schrödinger has effectively recognized the kinematical possibility of entangled states. To our knowledge, it was not until 1935 that Schrödinger used the

English word "entangled"[5] and the German word "verschränkt"[6] to characterize a wave function that cannot be expressed in product form, and also to characterize the knowledge encapsulated in such a wave function; and he also uses the nouns "entanglement" and "Verschränkung" to refer to the property which such wave functions share. At that time Schrödinger could hardly have been more emphatic about the significance of entanglement:

> When two systems, of which we know the states by their respective representation, enter into a temporary physical interaction due to known forces between them and when after a time of mutual influence the systems separate again, then they can no longer be described as before, viz., by endowing each of them with a representative of its own. I would not call that *one* but rather *the* characteristic trait of quantum mechanics.[5]

For clarity in our subsequent discussion, it will be useful to give a more abstract definition of "entangled" than Schrödinger's, by using the terminology of Hilbert space theory. Suppose that $H_i (i = 1, 2)$ is a Hilbert space in which the states of system i are represented. Each state of system i is associated with a unique ray (or one-dimensional subspace) of H_i, but one can also speak of the state being represented by any normalized vector of the ray. (In fact, for short, we shall often use the locution "the state $|u\rangle$," where $|u\rangle$ is a vector in H_i). The states of the composite system 1+2 are represented in the tensor product Hilbert space $H_1 \otimes H_2$. If a state of 1+2 is represented by a vector $|\Psi\rangle$ of $H_1 \otimes H_2$ such that

$$|\Psi\rangle = |w\rangle_1 |z\rangle_2 \quad , \quad \text{where} \quad |w\rangle_1 \in H_1 \text{ and } |z\rangle_2 \in H_2 \, , \tag{1}$$

the state is called "a product state." On the other hand, if the state is represented by a vector that cannot be expressed in the form of Eq. (1) for any choice of vectors $|w\rangle_1$ and $|z\rangle_2$ then the state is called "entangled," and it is then sometimes also convenient to apply the adjective "entangled" to the system itself. Finally, all that has been said about a composite system with two parts can be naturally generalized to a system with more than two parts.

That entanglement is not merely kinematically possible but is indispensable for a realistic quantum mechanical description of many-particle systems was shown by early analyses of systems with several electrons. Heisenberg[7] explained the splitting of the helium term scheme into orthohelium and parahelium terms by postulating that in the former the spins combine into triplet states and in the latter they combine into a singlet state (the spatial wave function being anti-symmetrized or symmetrized in the respective cases in order to yield over-all anti-symmetrization). The singlet spin state and one of the three triplet basis states are entangled, as is the spatial wave function associated with the triplet spin state. Dirac[8] showed more generally that the Pauli exclusion principle can be recovered quantum mechanically by anti-symmetrizing the total many-electron wave function, and the Bose-Einstein statistics can be recovered quantum mechanically by symmetrizing the many-boson wave function, which automatically entails entanglement if more than one single-boson state is present in the many-boson state.

There was, of course, a wealth of applications of these ideas to many-body systems in the years immediately following the discovery of quantum mechanics. In none of these, however, was entanglement exhibited for a pair of particles which are spatially well separated over macroscopic distances. To be sure, the entanglement of the valence electrons in a metallic crystal, or of the photons in a black-body enclosure, is manifested over the macroscopic *extent* of these systems, but neither case is an instance of spatial *separation.*

A very important abstract discussion of entanglement appears in von Neumann's book in 1932 on the foundations of quantum mechanics.[9]. In chapters 5 and 6 von Neumann analyzes the measuring process quantum mechanically, taking system 1 to be a microscopic object and system 2 to be a macroscopic apparatus. The initial quantum state of 1+2 evolves under the time-dependent Schrödinger equation into a final state which in general is not a product state (see especially section 2 of chapter 6). If von Neumann's abstract formulation were realized in a concrete physical situation, then in principle the object and the apparatus could be spatially separated after ceasing to interact. However, we find no mention in von Neumann's book of the oddity of the resulting entanglement of spatially separated systems, for his preoccupation seems to be with aspects of the measurement problem, especially with understanding in what sense definite measurement results are finally achieved in spite of entanglement (see especially sections 1 and 3 of chapter 6).

It is only in the 1935 paper of Einstein, Podolsky, and Rosen[10] (hereafter abbreviated by EPR) that explicit attention is paid to the entanglement of a spatially separated pair of systems. EPR consider a pair of spinless particles constrained to move in one dimension, in the quantum state

$$|\Psi\rangle = \int\!\!\int dx_1\, dx_2\, \delta(x_1 - x_2 - 2a)\, |x_1\rangle|x_2\rangle\,. \tag{2}$$

If the momentum rather than the position basis is used, then $|\Psi\rangle$ can be re-written as

$$|\Psi\rangle = \int\!\!\int dk_1\, dk_2\, e^{ia(k_1+k_2)}\delta\,(k_1 + k_2)\, |k_1\rangle|k_2\rangle\,. \tag{3}$$

In the state $|\Psi\rangle$ both the positions of 1 and 2 and their momenta are strictly correlated, so that if the position of one is determined by measurement, that of the other can be inferred with certainty, and likewise with the momenta. Schrödinger[5,6] pointed out in comments on EPR that $|\Psi\rangle$ fails to be in product form not only in the position and momentum bases but in any basis – i.e., it is an entangled state.

Furthermore, it is easy to see that any state that would serve EPR's critical purposes – by implying strict correlation both of A_1 with A_2 and of B_1 with B_2, where A_i and B_i ($i = 1, 2$) are observables of particle i represented by non-commuting operators, hence not simultaneously measurable – must be entangled. Of course, it is also essential for EPR's critical purposes that particles 1 and 2 be spatially separated, in order that considerations of no action-at-a-distance can be brought to bear on the measurements performed upon the particles.

EPR's analysis suggested to Schrödinger the question of whether nature ever permits a pair of spatially separated particles to be in an entangled state. Either answer would be highly significant. A positive answer would accentuate the peculiarity of the quantum mechanical phenomenon of entanglement, while a negative answer would entail a limitation upon the scope of quantum mechanics. In the following remarkably penetrating passage Schrödinger says that he regards the question to be (as of 1935) an open one.

> It seems worth noticing that the paradox [of EPR] could be avoided by a very simple assumption, namely if the situation after separating were described by the expansion (12) [essentially an expression of correlations], but with the additional statement that the knowledge of the *phase relations* between the complex constants a_k has been entirely lost in consequence of the process of separation. That would mean that not only the parts, but the whole system, would be in the situation of a mixture, not of a pure state. It would not preclude the possibility of determining the state of the first system by *suitable* measurements in the second or *vice versa*. But it would utterly eliminate the experimenter's influence on the state of that system which he does not touch.
>
> This is a very incomplete description and I would not stand for its adequateness. But I would call it a possible one, until I am told, either why it is devoid of meaning or with which experiments it disagrees. My point is, that in a domain which the present theory does not cover, there is room for new assumptions without necessarily contradicting the theory in that region where it is backed by experiment.[11]

Schrödinger's profound question was not answered until 1957, when Bohm and Aharonov[1] analyzed the scattering data obtained by Wu and Shaknov[12] in 1950 on coincident scattering of pairs of photons produced by positronium annihilation. Bohm and Aharonov showed that the correlations of scattering could be explained accurately by ascribing an entangled state (of zero angular momentum and negative parity) to the photon pair, but could not be explained by any statistical mixture of quantum mechanical product states. Neither Bohm nor Aharonov was aware (as we have determined by asking them) that Schrödinger had raised the question that they answered, and indeed they attributed the question to Furry, even though the latter did not present the question as an open one, as Schrödinger had done. Consequently, although Schrödinger was still alive in 1957, there is no evidence that any one informed him of the remarkable answer to the question that he had posed more than two decades earlier. It is also noteworthy that Bohm and Aharonov avoided the need to perform a new experiment by brilliantly exploiting the results of an experiment which had been performed for an entirely different purpose – an exemplary case of quantum archaeology!

To our knowledge, no further example of the entanglement of spatially separated systems was exhibited until 1972, when the first test of a Bell's Inequality was completed by Freedman and Clauser.[13] Bell[14] and his followers[15] had derived inequalities for the purpose of testing the family of local hidden variables theories against quantum mechanics, but when one examines the various derivations one sees that Bell-type Inequalities also follow if the two-particle system under examination is described by a quantum mechanical product state. The reason is that such a product state ensures that the probability of joint behavior of the two particles factors into the product of single probabilities, which is the crucial premiss for deriving Bell-type Inequalities. We believe that most of the specialists in Bell's Theorem were aware of the fact that a quantum mechanical product state implied this factorizability of the joint probability, but there seems to have been no notice of this fact in print before a 1989 paper of Werner[16] (brought to our attention at this conference by L. Ballentine). In almost all of the dozen or so tests of Bell's Inequalities, the experimental data conflicted with the Inequality, and hence by the argument just summarized no quantum mechanical product state could account for the data. This conclusion is not as dramatic as the one that is usually cited, namely, the refutation of the family of local hidden variables theories, but it is nevertheless a conclusion of considerable conceptual significance. It is worth remarking that the so-called "detection loophole," which determined defenders of local hidden variables theories may invoke to ward off experimental refutation,[17] does not protect the hypothesis of a quantum mechanical product state from refutation, because there is no information encoded in a quantum mechanical product state (as there might conceivably be in a putative hidden variables state) that could account for a biased detection of photon pairs.

3. Down-Conversion Pairs of Photons

A new field for striking exhibitions of entanglement was opened up since 1987 by the application of interferometric techniques to pairs of photons generated in parametric down-conversion.[18] In retrospect, one sees that entanglement was implicit in all down-conversion phenomena since the pioneering experiment of Burnham and Weinberg[19] in 1970. They showed that the illumination of certain crystals by a pumping laser beam gives rise to pairs of photons which are generated simultaneously (as shown by coincidence counting) and have frequencies which sum to that of the pumping photons. To see that entanglement is implicit, consider the down-conversion radiation coming through two pinholes in a screen some distance downstream from the crystal, so that two well-defined beams A and B are formed, if the luminous region in the crystal is small. With suitable filters there approximately is only one mode in each beam, characterized by the wave-vectors $\mathbf{k}_A$ and $\mathbf{k}_B$. Since the two-photon state must be symmetrized, the only possible quantum mechanical description of the radiation into the beams A and B is

$$|\Psi\rangle = \frac{1}{\sqrt{2}}\left[|\mathbf{k}_A\rangle_1|\mathbf{k}_B\rangle_2 + |\mathbf{k}_B\rangle_1|\mathbf{k}_A\rangle_2\right] , \qquad (4a)$$

or equivalently one can write $|\Psi\rangle$ as a state in Fock space, which automatically guarantees symmetrization:[20]

$$|\Psi\rangle = |\mathbf{k}_A \, , \, \mathbf{k}_B\rangle \, . \qquad (4b)$$

Until interferometric experiments were performed using down-conversion pairs, however, there were no data in which the entanglement of the two-photon state was immediately exhibited, and physicists could (and sometimes did) fall into the non-quantum mechanical habit of thinking of one photon in beam A and the other in beam B, which would be formally expressed as

$$|\Phi\rangle = |\mathbf{k}_A\rangle_1|\mathbf{k}_B\rangle_2 \, . \qquad (5)$$

Furthermore, we think that the notation of Fock space in Eq. (4b), by making symmetrization automatic, hides the essential role of entanglement. For this reason, in our subsequent exposition we shall use the tensor product notation of Eq. (4a) in order to demonstrate the direct connection between entanglement and two-particle interference.

A warning is in order at this point against regarding symmetrization to be the essence of entanglement. To see this, one need merely contemplate the entanglement of non-identical particles, which are not subject to the rules of symmetrization and anti-symmetrization. What is essential to entanglement is the multiplicity of ways in which a source can communicate via two particles with two detectors.

We shall illustrate our thesis about entanglement by commenting on two beautiful experiments – that of Ghosh and Mandel of 1987[21], which was a pioneering interferometric experiment with down-conversion pairs, and that of Kwiat, Vareka, Hong, Nathel, and Chiao of 1990[22], which has been described by Chiao at this conference. Although these two experiments have in common the fundamental feature of exploiting entanglement, they differ radically and instructively in several respects. In the former there is entanglement of wave-vectors (directions), whereas in the latter there is entanglement of wave-numbers (or equivalently, frequencies).

Furthermore, in the latter, but not in the former, the placement of ideally narrow filters in front of the photodetectors, so as to admit only a single frequency, would eliminate the two-photon interference fringes. Finally, in the former, but not in the latter, beams of down-conversion radiation are "mixed" by means of beam-splitters before illuminating the detectors.

A simplified version of the experiment of Ghosh and Mandel is shown in Figs. 1a and 1b.

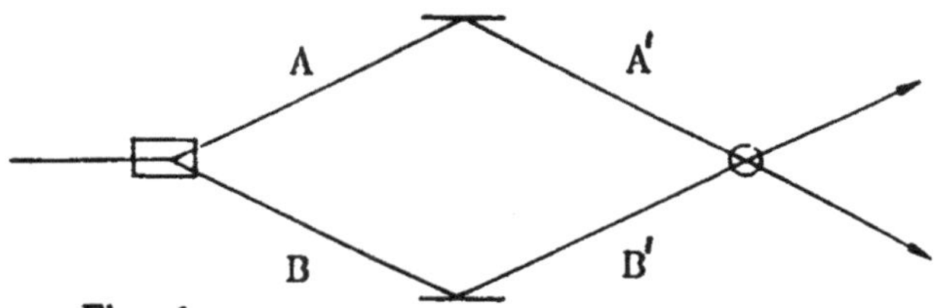

Fig. 1a

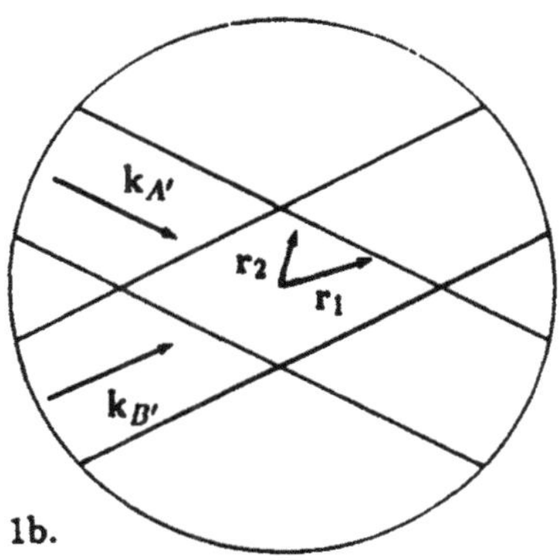

Fig. 1b.

Fig. 1a shows two beams A and B of down-conversion radiation, directed by mirrors through wave-number filters and pinholes (not shown) to a region of overlap, which is shown enlarged in Fig. 1b. Let A′ and B′ label the overlapping beams. If these beams are described by the two-photon state of Eq. (4a) or (4b), then the amplitude for joint detection at $\mathbf{r}_1$ and $\mathbf{r}_2$ (both points in the overlap region) is

$$\Psi(\mathbf{r}_1, \mathbf{r}_2) \sim e^{i(\alpha+\beta)} \left[e^{i\mathbf{k}_{A'}\cdot\mathbf{r}_1} e^{i\mathbf{k}_{B'}\cdot\mathbf{r}_2} + e^{i\mathbf{k}_{B'}\cdot\mathbf{r}_1} e^{i\mathbf{k}_{A'}\cdot\mathbf{r}_2} \right] , \tag{6}$$

where α and β are the over-all phases associated with the A beam and the B beam respectively. Hence the probability of joint detection at $\mathbf{r}_1, \mathbf{r}_2$ is

$$P_{12}^{(\Psi)}(\mathbf{r}_1,\mathbf{r}_2) \sim [1+\cos\{\mathbf{K}\cdot(\mathbf{r}_2-\mathbf{r}_1)\}] \ , \tag{7}$$

where

$$\mathbf{K} = \mathbf{k}_{A'} - \mathbf{k}_{B'} \ . \tag{8}$$

The probability of joint detection given by Eq. (7) shows the genuine two-photon interference fringes obtained by Ghosh and Mandel. The fringes sinusoidally depend on the projection of $\mathbf{r}_2 - \mathbf{r}_1$ upon the fixed vector $\mathbf{K}$ in a manner that cannot be derived from single-photon interference, since the counting rate of single photons at either $\mathbf{r}_1$ or $\mathbf{r}_2$ (obtained by integrating over the other position variable) yields a constant. The two-photon interference fringes result directly from the fact that the entanglement in Ψ provides two different contributions to the probability amplitude. Ghosh and Mandel very emphatically made this point by writing, "In essence there is an interference between two two-photon amplitudes, because the apparatus cannot distinguish between photons from A and B being detected at x_1 and x_2 respectively, or vice versa."

Eq. (7) can be compared with the probability of joint detection implied by the unentangled or product state Φ of Eq. (5):

$$P_{12}^{(\Phi)}(\mathbf{r}_1,\mathbf{r}_2) \sim |e^{i\mathbf{k}_{A'}\cdot\mathbf{r}_1} e^{i\mathbf{k}_{B'}\cdot\mathbf{r}_2}|^2 = const. \tag{9}$$

This result is not changed by allowing the joint detection to be due either to photon 1 being detected at $\mathbf{r}_1$ and photon 2 at $\mathbf{r}_2$ or vice versa, for then even though there are two contributions to the total probability there is no interference. Furthermore, the non-existence of two-particle interference fringes expressed by Eq. (9) is not due to the idealization to a single mode in each of the beams A and B. For suppose that the product form of Eq. (5) is kept, but $|\mathbf{k}_A\rangle$ and $|\mathbf{k}_B\rangle$ are each replaced by superpositions of states with different wave-vectors. The resulting product state, which we label $|\hat{\Phi}\rangle$, would imply a probability of joint detection of $\mathbf{r}_1$, $\mathbf{r}_2$ of the form

$$P_{12}^{(\hat{\Phi})}(\mathbf{r}_1,\mathbf{r}_2) = f(\mathbf{r}_1)g(\mathbf{r}_2) \ , \tag{10}$$

where f and g may each depend sinusoidally on $\mathbf{r}_1$ and $\mathbf{r}_2$ respectively, but their product would not depend sinusoidally on $\mathbf{r}_2 - \mathbf{r}_1$. Returning again to the prediction from the entangled state $|\Psi\rangle$ of Eq. (7), we emphasize that it should not be construed as the interference of two photons with each other. Dirac's famous dictum, "Each photon interferes only with itself. Interference between two different photons never occurs,"[23] is not violated. The dictum must be generalized, however, without loss of the original spirit: "Each entangled system interferes with itself. Interference between subsystems of an entangled system never occurs."

Eq. (7) is the same as the joint probability function which Ghosh and Mandel obtained by a quantum optical analysis of the down-conversion field, which is not surprising, since they restricted their attention to a two-photon subspace of Fock space, with guaranteed symmetrization. We note that they compared their result with the prediction of a classical description of the down-conversion field, which gives a fringe visibility of no more than half of the 100% visibility shown in Eq. (7). We have been interested here in an entirely different comparison. We took for granted that the battle between a classical and a quantum mechanical characterization of the radiation field has been won by the latter. The question of product versus entangled state is a further battle within the victorious quantum point of view. Our foregoing argument, together with the observation of two-photon interference fringes by Ghosh and Mandel, unequivocally settles this battle in favor of entanglement.

The experiment of Ghosh and Mandel was followed by many others[24] that demonstrated the type of entanglement exhibited in Eq. (4a) – i.e., entanglement of wave-vector *directions*. With a suitable arrangement two-photon fringes can also be produced by exploiting the entanglement of wave-vector *magnitudes* (i.e., wave numbers,or, equivalently, frequencies or "colors"). This phenomenon was exhibited in an experiment of Kwiat et al.[22], who carried out a proposal of Franson[25] but adapted it to down-conversion pairs. A diagram of their arrangement is shown in Fig. 2.

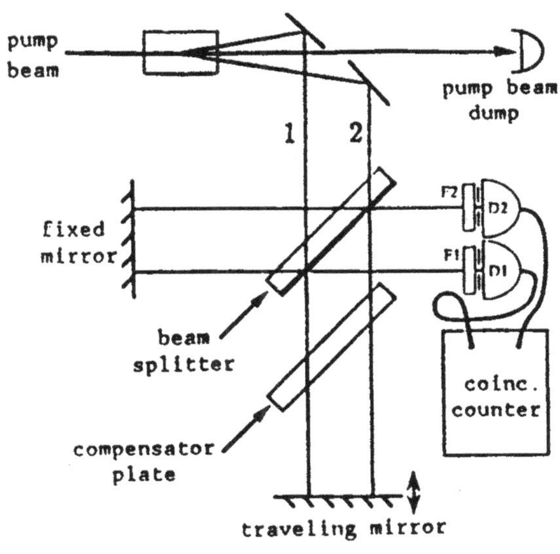

Fig. 2.

Two beams, 1 and 2, selected by pinholes (at the detectors) and made parallel by mirrors, were fed into a Michelson interferometer and then into the filter-detector assemblies, also labeled 1 and 2. Notice that beams 1 and 2 were never "mixed" before reaching their respective detectors, unlike in *all* of the experiments of Ref. 24. Also, because the beams do not meet, the entanglement of Eq. (4a) is irrelevant to the phenomenon of Kwiat et al. Nevertheless, as we shall explain, the coincidence count rate at the detectors exhibited fringes as the Michelson mirror was moved, provided that Δ, the optical path-length difference in the two routes through the Michelson interferometer, exceeded the coherence length associated with the filters. Note that Δ is *twice* the difference of the lengths of the interferometer arms.

In order to show how the fringes arise, we must characterize the state of the two-photon radiation incident upon the Michelson interferometer. Consider a single-photon in the pump beam. Suppose, for simplicity, that this photon is ideally monochromatic, with wave-number $2k_o$, so that its wave function in wave-number space is simply

$$\Phi(k) = \delta(k - 2k_o) \ . \tag{11}$$

If this photon converts in the crystal into a pair of photons, and this pair subsequently enters the beams 1 and 2, then the wave function of the pair in the two-dimensional wave-number space is simply

$$\Phi(k_1, k_2) = \delta(k_1 + k_2 - 2k_o)\,\phi(k_1) \ , \tag{12}$$

where the distribution $\phi(k_1)$ is determined by the details of the down-conversion process and by the beam selection apertures. Note that the wave-numbers k_1 and k_2 of the pair are not fixed, but must satisfy $k_1 + k_2 = 2k_o$ (energy conservation). For conceptual clarity suppose now that the wave-number filters are placed upstream of the Michelson interferometer, instead of just before the detectors. Also suppose, as is commonly done, that within the bandwidth of these filters $\phi(k)$ is essentially a constant. Then downstream from the filters the two-photon wave function of Eq. (12) is "clipped" to become

$$\Phi(k_1, k_2) = \delta(k_1 + k_2 - 2k_o)\, e^{-(k_1-k_o)^2/2\sigma^2} e^{-(k_2-k_o)^2/2\sigma^2} \ , \tag{13}$$

where for simplicity we have assumed that the two filters are gaussian with identical widths σ and centers k_o, as was the case in the experiment of Kwiat et al. (If the filters would have different centers, then among the resulting modifications of Eqs. (19)-(20) below are beating fringes of the sort observed by Ou and Mandel[26], even though the beams 1 and 2 in Fig. 3 never mix.) Since the Michelson interferometer operates in real space, it is convenient to Fourier transform the k_1, k_2-dependent wave function, with the time-dependence inserted, into the configuration space wave function

$$\Psi(x_1, x_2, t_1, t_2) = \int\!\!\int dk_1\, dk_2 \{\Phi(k_1, k_2)\, e^{-ic(k_1 t_1 + k_2 t_2)}\} e^{ik_1 x_1} e^{ik_2 x_2} \ . \tag{14}$$

For the Φ of Eq. (13) this yields

$$\Psi(x_1, x_2, t_1, t_2) = e^{ik_o(x_1 - ct_1)} e^{ik_o(x_2 - ct_2)} e^{-\frac{\sigma^2}{4}[(x_1 - ct_1) - (x_2 - ct_2)]^2} \ . \tag{15}$$

Note that both the general Eq. (14) and the specific Eq. (15) explicitly exhibit entanglement. In Eq. (14) one sees that the entangled state is a superposition of products of one-photon "color" states; and in Eq. (15) the non-factorizability of the two-photon wave function is apparent in the third exponential factor.

The wave function (15) describes the photons of the pair before they enter the Michelson interferometer, that is, $\Psi^*\Psi dx_1\, dx_2$ is the joint probability of finding photon 1 within dx_1 at position x_1 at time t_1 and photon 2 within dx_2 at position x_2 at time t_2, where x_1 (x_2) is a coordinate along beam 1(2). In stating this interpretation of the configuration space wave function of the photon pair we are not oblivious of the well known literature concerning the localization of a photon,[27] but we wish to delay a discussion of this matter to another paper. If detectors are placed in the beams equally distant from the source, so that $x_1 = x_2$, the joint probability density takes the particularly simple form

$$\Psi^*\Psi = e^{-\sigma^2 (c\tau)^2/2} , \tag{16}$$

where $\tau = t_2 - t_1$ is the delay time in the arrival of the photons. Note that the probability of detecting a pair is essentially zero unless the delay time is of the order of $(c\sigma)^{-1}$ or smaller.[28]

Now, with the Michelson interferometer in place, each photon has two routes to its respective detector, the "short" route via the "near" fixed mirror and the "long" route via the "far" movable mirror. Consequently, the total wave function illuminating the detectors at times t_1 and t_2 is a superposition of *four* amplitudes, each of the type of Eq. (15):

$$\begin{aligned}\Psi_{tot}(t_1,t_2) &= \Psi(0,0,t_1,t_2) + \Psi(\Delta,\Delta,t_1,t_2) \\ &\quad + \Psi(\Delta,0,t_1,t_2) + \Psi(0,\Delta,t_1,t_2) ,\end{aligned} \tag{17}$$

where Δ is the path difference between the "long" and "short" routes, and, for simplicity, the detectors have been positioned at $x_1 = x_2 = 0$.[29] From Eq. (17) the joint probability density of detecting photon 1 at t_1 and photon 2 at t_2 is proportional to

$$\begin{aligned}\Psi^*_{tot}\Psi_{tot} &= [2 + 2\cos(2k_o\Delta)]\, f^2(c\tau) \\ &\quad + 4\cos(k_o\Delta)\, f(c\tau)\,[f(\Delta - c\tau) + f(-\Delta - c\tau)] \\ &\quad + [f(\Delta - c\tau) + f(-\Delta - c\tau)]^2 ,\end{aligned} \tag{18}$$

where $\tau = t_2 - t_1$ and $f(u) = e^{-[\sigma^2 u^2/4]}$. Note that this joint probability density, like that of Eq. (16), depends on the times t_1 and t_2 only through their difference. This result is expected, since we modeled the pump photon by an everlasting monochromatic wave. For comparison with experiment the joint probability density of Eq. (18) must be integrated from $\tau = -\frac{T}{2}$ to $\tau = +\frac{T}{2}$, where T is the coincidence window of the experimental arrangement. The result is that the probability for a coincidence count is

$$P_{12}^{(\Psi)} \sim \frac{\sigma c}{\sqrt{2\pi}} \int_{-\frac{T}{2}}^{\frac{T}{2}} \Psi^* \Psi d\tau = A(1+\cos 2k_o\Delta) + B\cos k_o\Delta + C\ , \tag{19}$$

where

$$A = 2\,\mathrm{erf}\left[\frac{cT\sigma}{2\sqrt{2}}\right] \tag{20a}$$

$$B = 4e^{-\frac{\sigma^2\Delta^2}{8}}\{\mathrm{erf}\left[(cT+\Delta)\frac{\sigma}{2\sqrt{2}}\right] + \mathrm{erf}\left[(cT-\Delta)\frac{\sigma}{2\sqrt{2}}\right]\} \tag{20b}$$

$$C = \mathrm{erf}\left[(cT+2\Delta)\frac{\sigma}{2\sqrt{2}}\right] + \mathrm{erf}\left[(cT-2\Delta)\frac{\sigma}{2\sqrt{2}}\right] + 2e^{-\frac{\sigma^2\Delta^2}{2}}\,\mathrm{erf}\left[(cT)\frac{\sigma}{2\sqrt{2}}\right], \tag{20c}$$

and where $\mathrm{erf(u)} = \frac{2}{\sqrt{\pi}}\int_0^{\mathrm{u}} \mathrm{e}^{-\mathrm{u}'^2}\,\mathrm{du}'$ is the error function. In order to make a clear comparison with experiment, it is useful to examine the behavior of Eq. (19) in two cases: (1) $\Delta \ll \frac{1}{\sigma}$, and (2) $\Delta \gg \frac{1}{\sigma}$. Of course, all intermediates between these two cases are also contained in Eqs. (20a)-(20c).

Case 1: This is the case where the path difference Δ is less than the coherence length $\frac{1}{\sigma}$ of the radiation, as determined by the filters, so that the *single* count rate in each detector will exhibit ordinary single-photon Michelson fringes as Δ is varied. To examine the behavior of the *coincidence* count rate, assume for simplicity that $cT \gg \frac{1}{\sigma}$, so that each error function in Eqs. (20a)-(20c) is +1. Then $A = 2$, $B = 8$, $C = 4$, so that

$$P_{12}^{(\Psi)} \sim 2(1+\cos 2k_o\Delta) + 8\cos k_o\Delta + 4 = [2+2\cos k_o\Delta]^2\ . \tag{21}$$

The last expression is proportional to the square of the single-photon probability when there are ordinary single-photon Michelson interference fringes. In other words, in this case all variation of the coincidence count rate as a function of Δ is simply the product of the single rates at the two detectors. No true two-photon fringes are present.

In case 2 the path difference Δ is greater than the coherence length $\frac{1}{\sigma}$. In this case, the ordinary single-photon Michelson fringes disappear, and the fringes which will appear are genuine two-photon interference fringes. Note that, in the limit of ideally narrow filters the coherence length $\frac{1}{\sigma}$ is greater then the path difference Δ, case 1 always holds, and hence the two-photon fringes do not appear. It will be convenient to subdivide case 2 into subcases: (2a) $\frac{cT}{\Delta} \gg 1$, (2b) $\frac{cT}{\Delta} \ll 1$. Of course, all intermediates between these two extreme subcases are also contained in Eqs. (20a)-(20c).

Case 2a: $\Delta \gg \frac{1}{\sigma}$, $\frac{cT}{\Delta} \gg 1$, as in the experiment of Kwiat et al, where $cT \approx 1.5\, m, \Delta \approx 240\mu m$, and $\frac{1}{\sigma} \approx 50\mu m$. Then $A = 2$, $B = 0$, and $C = 2$, so that

$$P_{12}^{(\Psi)} \sim 4 + 2\cos 2k_o\Delta \,, \tag{22}$$

Thus, the two-photon fringes have a visibility of 0.5, as observed by Kwiat et al. Physically, the extra "background" coincidence counts (i.e., the $C = 2$ contribution) arise because with the large coincidence window the electronics responds not only to long-long and short-short amplitudes but also to long-short and short-long amplitudes.

Case 2b: $\Delta \gg \frac{1}{\sigma}$, $\frac{cT}{\Delta} \ll 1$. Then $A = 2$, $B = C = 0$, so that

$$P_{12}^{(\Psi)} \sim 2 + 2\cos 2k_o\Delta \,, \tag{23}$$

That is, in this ideal limit the two-photon fringes have a visibility of unity. The narrowed coincidence window no longer accepts the long-short or short-long amplitudes, hence $C = 0$. We note that our derivation of Eq. (23) via Eqs. (19)-(20) is unnecessarily complicated, for once one realizes physically that there is no contribution from the long-short and short-long routes in Case 2b, one sees that the last two terms in Eq. (17) are negligible, and then Eq. (23) follows immediately[22]. Finally, it should be pointed out that an experimental effort to realize Case 2b is underway by greatly increasing the path difference Δ and modestly decreasing the coincidence window T, thereby improving the two-photon fringe visibility.[30]

We conclude with some remarks about procedure and substance. Our analysis of down-conversion photons has employed only elementary quantum mechanical procedures, and specifically we have refrained from using the field theoretical methods common in quantum optics. We have been able to derive completely by elementary means (largely in unpublished calculations) the results of all the two-photon interferometric experiments to date, including a recent one of Ou et al.,[31] in which a single pair of photons is generated by the superposed operation of two down-conversion crystals. The crucial point is that in all these experiments, including the last, the down-conversion radiation can be adequately treated in the two-particle subspace of Fock space.

Substantively, we call attention to the curious type of entanglement shown in the two-photon wave packet of Eq. (15). Each of the first two factors on the right hand side of this equation is a monochromatic wave ideally present throughout the longitudinal extent of the respective beam and throughout all time, and the third factor is a real Gaussian depending on a space-time point (x_1,t_1) in beam 1 and a space-time point (x_2,t_2) in beam 2. The phase differences that account for interferences are only determined by the first two factors, for there is no phase in the third. But the third factor, which is not a product of a function of x_1 and t_1 and another function of x_2 and t_2, determines a spatio-temporal correlation of the two particles – i.e., it tells at what *pairs of points in space-time* the particles can be found. It is this non-factorizability that accounts for the remarkable feature of the experiment proposed by Franson and carried out by Kwiat et al.: that two-photon interference can be produced without any mixing of beams.

After the completion of this paper we received a preprint from Ou, Zou, Wang, and Mandel,[32] the publication of which was delayed, reporting an independent realization of the experiment proposed by Franson. Their results agree essentially with those of Kwiat et al.

Acknowledgment

We wish to acknowledge very helpful conversations with Raymond Chiao, Z.Y. Ou, and Wolfgang Schleich.

This work was supported in part by the National Science Foundation, Grants No. DIR-8810713 and No. DMR-8713559, and by Fonds zur Förderung der wissenschaftlichen Forschung (Austria), No. 6635.

References

1. D. Bohm and Y. Aharonov, *Phys. Rev.* **108** (1957) 1070.

2. W. Ehrenberg and R.W. Siday, *Proc. Phys. Soc. London.* B **62** (1949) 8.

3. Y. Aharonov and D. Bohm, *Phys. Rev.* **115** (1959) 485.

4. E. Schrödinger, *Collected Papers on Wave Mechanics* (Chelsea Publishing Co., New York, 1978), 130 (translated from *Annal. d. Phys.* **81** (1926) 109.

5. E. Schrödinger, *Proc. Camb. Phil. Soc.* **31** (1935) 555, **32** (1936) 446.

6. E. Schrödinger, *Naturwissenschaften* **23** (1935) 807, 823, 844

7. W. Heisenberg, *Zeits. f. Phys.* **38** (1926) 44.

8. P.A.M. Dirac, *Proc. Roy. Soc.* A **112** (1926) 661.

9. J. v. Neumann, *Mathematical Foundations of Quantum Mechanics* (Princeton University Press, Princeton, 1955) (translated from *Mathematische Grundlagen der Quantenmechanik* (Springer, Berlin, 1932).)

10. A. Einstein, B. Podosky, and N. Rosen, *Phys. Rev.* **47** (1935) 777.

11. See the first item of Ref. 5.

12. C.S. Wu and I. Shaknov, *Phys. Rev.* **77** (1950) 136.

13. S.J. Freedman and J. Clauser, *Phys. Rev. Lett.* **28** (1972) 938.

14. J.S. Bell, *Phys.***1** (1965) 195 (reprinted in J.S. Bell, *Speakable and Unspeakable in Quantum Mechanics* (Cambridge University Press, Cambridge, England, 1987)).

15. Surveys are J.F. Clauser and A. Shimony, *Rep. Progr. in Phys.* **41** (1978) 1881, and M. Redhead, *Incompleteness, Nonlocality, and Realism* (Oxford University Press, Oxford, 1987).

16. R.F. Werner, *Phys. Rev.* A **40** (1989) 4277.

17. N.D. Mermin and G.M. Schwartz, *Found. Phys.* **12** (1982) and references therein.

18. C.O. Alley and Y.H. Shih, in *Proceedings of the Second International Symposium on Foundations of Quantum Mechanics in the Light of New Technology*, eds. M. Namiki et al. (Physical Society of Japan, Tokyo, 1986) 47; R. Ghosh and L. Mandel, *Phys. Rev. Lett.* **59** (1987) 2044, and previous experimental and theoretical papers referred to therein. A proposal for two-photon interferometry was also made by M.A. Horne and A. Zeilinger in *Proceedings of the Symposium on the Foundations of Modern Physics*, eds. P. Lahti and P. Mittelstaedt (World Scientific, Singapore, 1985), 435.

19. D.C. Burnham and D.L. Weinberg, *Phys. Rev. Lett.* **25** (1970) 84.

20. G. Baym, *Lectures on Quantum Mechanics* (W.A. Benjamin, New York, 1969), especially p. 419.

21. Second item of Ref. 18.

22. P.W. Kwiat, W.A. Vareka, C.K. Hong, H. Nathel, and R.Y. Chiao, *Phys. Rev.* A **41** (1990) 2910.

23. P.A.M. Dirac, *The Principles of Quantum Mechanics* 4th ed. (Clarendon Press, Oxford, 1958), p.7.

24. C.K. Hong, Z.Y. Ou, and L. Mandel, *Phys. Rev. Lett.* **59** (1987) 2044; Z.Y. Ou and L. Mandel *Phys. Rev. Lett.* **61** (1988), 50; Z.Y. Ou and L. Mandel *Phys. Rev. Lett.* **61** (1988) 54; Y.H. Shih and C.O. Alley

Phys. Rev. Lett. **61** (1988) 2921; J.G. Rarity and P.R. Tapster, in *Photons and Quantum Fluctuations*, eds. E.R. Pike and H. Walther (Adam Hilger, Bristol, 1988); J.G. Rarity and P.R. Tapster *Jour. Opt. Soc. Am.* B **6** (1989) 1221; Z.Y. Ou and L. Mandel *Phys. Rev. Lett.* **62** (1989) 2941; Z.Y. Ou. L.J. Wang, X.Y. Zou, and L. Mandel, *Phys. Rev.* A **41** (1990) 566; J.G. Tapster and P.R.Tapster, in *Proceedings of the Sixth Rochester Conference on Coherence and Quantum Optics* (1990). These papers refer to many supporting theoretical analyses. In all these experiments the down-conversion radiation emerges from the crystal in two beams, except for the last two, which use four beams. The advantages of four-beam experiments have been argued by M.A. Horne, A. Shimony, and A. Zeilinger, *Phys. Rev. Lett.* **62** (1989) 2209, and in *Sixty-two Years of Uncertainty: Historical, Philosophical and Physical Inquiries into the Foundations of Quantum Mechanics*, ed. A. Miller (Plenum, New York, 1990). Recently a four-beam arrangement has been used by J.G. Rarity and P.R. Tapster to perform the first non-polarization test of a Bell Inequality (to be published).

25. J.D. Franson, *Phys. Rev. Lett.* **62** (1989) 2205.

26. Z.Y. Ou and L. Mandel, *Phys. Rev. Lett.* **61** (1988) 54.

27. W. Pauli, *Z. f. Phys.* **80** (1933) 573; A. Wightman, *Revs. Mod. Phys.* **34** (1962) 845 and references therein.

28. S. Friberg, C.K. Hong, and L. Mandel, *Phys. Rev. Lett.* **54** (1985) 2011.

29. Also for simplicity, we assume equal amplitudes and zero relative phases between the terms of Eq. (17). In the most general Franson-Kwiat et al. type experiment, one could envisage, for the *four separated beams*: different geometrical lengths, different phase plates, different filters, and different attenuators.

30. Personal communication from R. Chiao.

31. Z.Y. Ou, L.J. Wang, X.Y. Zou, and L. Mandel, *Phys. Rev.* A **41** (1990) 566.

32. Z.Y. Ou, X,Y, Zou, L.J. Wang, and L. Mandel, "Observation of Non-local Interference in Separated Photon Channels" (to be published).

A MEASUREMENT CIRCUIT TO TEST BELL'S INEQUALITIES IN A MACROSCOPIC SUPERCONDUCTING DEVICE

C. D. TESCHE
IBM Research Division, P. O. Box 218
Yorktown Heights, NY 10598

ABSTRACT

A measurement circuit for the flux state of an rf SQUID is described which has elements analogous to those used in a conventional EPR experiment. These elements include dc SQUID switches which act like polarizers, and dissipative dc SQUID magnetometers which act like particle detectors. These elements can be used in an experimental search for violations of macrorealism and non-invasive measurability in a macroscopic system.

1. Introduction

In a typical Einstein-Podolsky-Rosen (EPR) experiment,[1] a scattering event or decay is used to generate a pair of correlated particles. Each one of these particles is made to pass through a polarizer followed by a particle detector. Measurements on an ensemble of pairs of particles are used to determine a joint probability density for detection as a function of the polarizer settings. A set of joint probability densities for a range of polarizer settings can be used to determine if the correlations displayed by the probability densities are consistent with the predictions of quantum mechanics or with those of a theory based on some very general assumptions about the behavior of macroscopic systems.[2,3]

As an example of the latter, Bell has derived a set of inequalities for the joint probability densities of the polarizer settings which must be satisfied for any theory which assumes separability (no influence may travel faster than the speed of light) and realism (objects have well defined properties even if they are not measured).[2] However, if the decay is determined by a quantum mechanical process, it is possible to choose polarizer settings such that these inequalities are violated.

Recently, Leggett and Garg have suggested that a search for violations of a set of inequalities similar to Bell's inequalities might be performed in a macroscopic system using a single junction rf SQUID.[4] The macroscopic variable is the net magnetic flux linking the superconducting loop of the SQUID. In their analysis, the assumption of separability in the derivation of Bell's inequalities for the EPR experiment performed with photons is replaced by that of non-invasive measurability. This "common sense" assumption requires that the net flux through the SQUID, which is generated by an externally applied magnetic field and a "macroscopic" current flowing around the SQUID loop which is on the order of a few μA, could be determined in principle at arbitrary times t_i without affecting the subsequent dynamics of the device. Thus, a second measurement of the net flux at time $t_j > t_i$ could be made with the same result as if the system had not been disturbed.

In this gedanken experiment, the analog to the continuously variable polarization settings for the two particles in the EPR experiment are the times t_i and t_j at which the flux is determined. The analog to the decay or scattering process which correlates the polarization state of the two particles is the free development in time of the flux state of the rf SQUID. It is convenient to choose SQUID parameters such that it is overwhelmingly likely that the flux at time t_i is one of two values labeled by the quantity $Q_i = +1$ or -1. In this case, the analog to Bell's inequalities for the EPR experiment are[4]

$$K_{12} + K_{23} + K_{31} \geq -1 \tag{1a}$$

and

$$K_{12} + K_{23} + K_{34} - K_{14} \leq 2 \tag{1b}$$

where the correlation functions

$$K_{ij} = \langle Q_i Q_j \rangle . \tag{1c}$$

In order to determine if this gedanken experiment can be realized in the laboratory, we need to find realistic values for the rf SQUID and, most importantly, we need to develop a suitable measurement circuit which contains elements analogous to polarizers and particle detectors for the flux state of the SQUID.

2. The rf SQUID

The rf SQUID is a superconducting loop which is interrupted by a thin insulating layer (Josephson tunnel junction). The conduction electrons in the superconductor are paired, and thus encounter negligible dissipation within the body of the superconductor. The pairs tunnel through the insulating barrier. In general, there is some dissipation associated with this process, as well as a capacitance determined by the geometry of the junction.

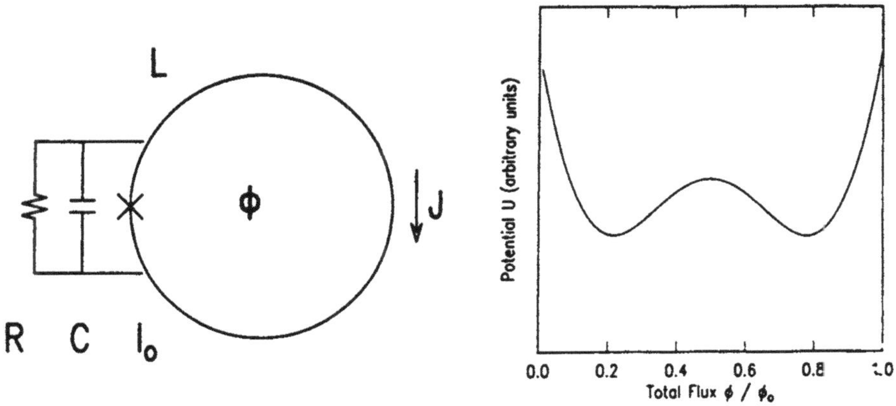

Figure 1. (a) Lumped circuit element model for the rf SQUID. (b) Potential energy as a function of the net flux for the rf SQUID.

A lumped circuit model for this device is shown in Figure 1 (a). A superconducting screening current J flows around the SQUID loop inductance L in response to an externally applied magnetic flux ϕ_X. For suitably chosen device parameters, the net flux obeys an equation of motion similar to that of a particle moving in a double well potential, with the junction capacitance C and conductance 1/R correspond to the particle mass and dissipation[5]

$$C\ddot{\phi} + (1/R)\dot{\phi} = -\,\partial U/\partial\phi\,. \tag{2}$$

The potential U is given as a function of the net flux ϕ and the externally applied flux ϕ_X as

$$U = (\phi - \phi_X)^2/2L - (I_o\phi_o/2\pi)\cos(2\pi\phi/\phi_o)\,. \tag{3}$$

At a flux bias of $\phi_0/2$, the screening current is equal in magnitude in both wells, but opposite in direction (Figure 1b). This is the optimal bias point for the observation of coherent tunneling effects.

The net flux may be considered to be a macroscopic variable in the sense that the condensate of superconducting pairs is described by the product of a large number of pair wave functions which, for a suitable choice of parameters, can be highly localized in one of the potential wells. A linear superposition of two such product wave functions localized on opposite sides of the barrier cannot be reduced into a wave function which is a simple product of pair wave functions distributed between the wells. Thus, quantum oscillations of the flux state of the SQUID between the potential wells is the result of the coherent tunneling of the entire superconducting condensate.

Although this effect has not yet been observed, thermal activation and incoherent tunneling between distinct flux states have been seen.[6,7] These experiments have been used to verify many of the features of detailed models which describe the rf SQUID. It is also possible to produce continuous variations in the pair wave function by changing some parameter such as the external bias flux. This effect is routinely observed in the operation of conventional rf and dc (double junction) SQUID magnetometers. Although useful information about SQUID device parameters can be obtained from these observations, the effect is not a manifestation of a macroscopic quantum phenomena.

3. How to Get an rf SQUID into a Coherent State

Although violations of Bell's inequalities can, in principle, be observed independent of the validity of any particular model, it is helpful to use the existing models to determine appropriate parameters for the rf SQUID. The most important results which are required are the dependence of the coherent oscillation frequency on the SQUID circuit parameters, the effect of dissipation on the decay of the coherent oscillations, and the sensitivity of the oscillations and decay to variations in the flux bias point. The coherent oscillation frequency Δ is exponential in the height ΔU of the tunnel barrier[8]

$$\Delta = 4\sqrt{3}\,\omega_0(S_0/2\pi\hbar)^{1/2}\exp(-S_0/\hbar) \tag{4}$$

where the action S_0 is

$$S_o/\hbar = (16/3)\Delta U/\hbar\omega_o \quad (5)$$

and ω_o is the frequency of small oscillations in the well.

The crossover temperature T^* for the onset of coherent oscillations is determined by the coherent oscillation frequency, the width $\Delta\phi$ of the tunneling barrier, and the junction resistance R[9,10]

$$T^* = \hbar\Delta/\alpha\pi k_B \quad (6a)$$

where

$$\alpha = (\Delta\phi)^2/2\pi\hbar R. \quad (6b)$$

The externally applied flux must be very well controlled. The required flux stability is a function of the oscillation frequency. Variations in bias flux about the bias point of $\phi_o/2$ which fall within the range of 10^{-7} - $10^{-8}\phi_o$ are tolerable for coherent oscillation frequencies on the order of a few MHz.

Reasonable values of oscillation frequency, crossover temperature and flux stability can be obtained for a suitably chosen set of SQUID parameters.[11]. An rf SQUID with critical current on the order of 2.54 μA, capacitance of 0.17 pF, and loop inductance of 150 pH is predicted to have a crossover temperature of 3.3×10^{-8} (R/Ω) K. These are experimentally realizable values provided the intrinsic resistance is in excess of a few MΩ. Recent measurements of the intrinsic quasiparticle resistance in high quality Josephson junctions have yielded values in this range.[12]

4. An "Ideal Negative Result" Measurement

It is essential that the joint probability densities for the flux state be determined by a measurement circuit which does not destroy the coherent oscillations prior to time t_2, where $t_2 > t_1$. Leggett and Garg have addressed this difficulty by introducing the idea of an experiment with an "ideal negative result". The measurement circuit is assumed to interact strongly with the rf SQUID at the first time t_1 only if the SQUID is in one of the two flux states (i.e. state Q = +1 corresponding to a clockwise current). This divides the resulting data into two groups. In the first group, the flux state at time t_1 was (+), with little subsequent change in the dynamics of the system. In the second group, the flux state at time t_1 is in the (-) state, with a significant change after t_1 of the dynamics of the SQUID.

The possible dramatic change in the development of the flux state of the SQUID is compensated for by using only data from the first group in the analysis. This requirement can, in principle, be satisfied even for a quantum mechanical system. This can most easily be seen by considering the analogy between the particular choices of the times t_i and the setting of the polarizers in the EPR experiment. The only particles which are used in the final analysis are those which have passed through the polarizers relatively unscathed, and are ultimately detected by the counters.

This constraint on the final data set is to be contrasted with the lack of constraints on the data set for a system which satisfies the assumption of "non-invasive measurability at the macroscopic level". In this case, it is assumed that the state of the SQUID could be determined for both flux states with arbitrary precision without affecting the subsequent dynamics. A system which satisfies both this assumption and the requirement of macroscopic realism would be incapable of violating the Bell's inequality relationships which have been derived for the rf SQUID.[4]

5. Can an "Ideal Negative Result" Measurement be Performed?

The feasibility of this experiment has been questioned on the grounds that it is impossible within the framework of quantum mechanics to perform a non-invasive measurement of magnetic flux for either flux state because any measurement of the flux state which is completed at time t_1, even if the result is negative, must necessarily destroy the coherent oscillation of the rf SQUID.[13] This is a valid criticism if one imagines that the measurement circuit either introduces so much dissipation into the rf SQUID environment at time t_1 that the coherence is effectively destroyed, or if the measurement circuit acts so as to project the rf SQUID into a well defined flux state at time t_1. However, neither one of these possibilities must necessarily occur at time t_1 in a well designed experiment.

It is useful to return to the conventional EPR experiments for inspiration on this point. In the EPR experiment, the two particles are passed through two independent polarizers and then either detected or not by a pair of counters. Whether an object is or is not a "good" polarizer is determined experimentally. In particular, although polarizers are capable of strongly affecting the states of the particles, a pair of "good" polarizers does not destroy the coherence between that fraction of the particles which pass through the polarizers and are subsequently counted in the detectors. Clearly, replacing one of these "good" polarizer with a highly dissipative detector which emits a single uncorrelated photon will result in a loss of coherence.

We need to invent the analog of "good" polarizers and separate, dissipative particle detectors for the rf SQUID. The "polarizers" may strongly affect the dynamics of only one of the flux states of the rf SQUID at time t_1. However, the "particle detectors" must not be fully effective at time t_1. This would be the equivalent of interposing a detector which emits a single photon before the second polarizer in the EPR experiment.

6. DC SQUID Magnetometers as "Particle Detectors"

The most sensitive detector of magnetic flux now available is the dc SQUID magnetometer.[14] The dc SQUID consists of a superconducting loop interrupted by two Josephson junctions (Figure 2 c). The SQUID is biased by some constant current I_o. The voltage across the parallel combination of the junctions is then a function of the applied magnetic flux. The linearity of the device in the small signal limit is achieved by the use of highly dissipative shunt resistances in parallel with the Josephson junctions. These shunts not only generate broadband noise, but also provide power gain when biased at I_o.

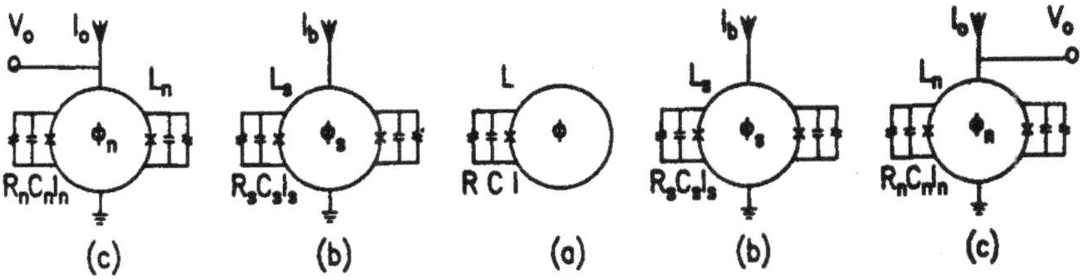

Figure 2. The superconducting measurement circuit for the non-invasive determination of the joint probability density for the rf SQUID. The state of the rf SQUID (a) is determined by the combined action of two hysteretic dc SQUID switches (b) and two dissipative dc SQUID magnetometers (c). The switches are activated during the time interval Δt at times t_1 and t_2. The output voltage is integrated over the time interval T_a.

Recent advances in device design and fabrication make it possible to operate well coupled dc SQUID magnetometers with noise properties determined by quantum effects at temperatures on the order of a few hundred mK. However, even in this limit, the shunt resistors connected across the junctions provide broadband noise which is inductively coupled back into the source of the signal flux (in this case, an rf SQUID). The effect of this broadband noise over a measurement interval which is long enough to permit the determination of the state of the rf SQUID above the broadband output voltage noise is that the coherent oscillations in the rf SQUID are effectively destroyed.[15] Thus, the conventional dc SQUID magnetometer is a good candidate for an ideal (in the sense that it is limited by "quantum noise" processes) "particle detector", but cannot be considered as a "polarizer".

7. Hysteretic DC SQUIDs as "Polarizers"

The destruction of the coherent oscillations in the rf SQUID by an inductively coupled dc SQUID can be avoided by removing the shunts from the dc SQUID. However, the unshunted dc SQUID is highly hysteretic, and can be operated only as a flux-activated switch. However, this is exactly the property which is needed to develop a "polarizer" for an rf SQUID. The dc SQUID switch (without additional shunt resistors), is operated in the following mode. The flux bias through the dc SQUID is set at close to $\phi_o/2$. In this case, there are two states for the circulating current available to the dc SQUID switch corresponding to a clockwise and counterclockwise screening current in the dc SQUID loop. Call these two states (L) and (R). These states are quite similar to those of the rf SQUID, except that the potential barrier between the two potential wells in the dc SQUID can be altered by changing an external bias current I_b. An external bias flux ϕ_s is chosen such that one of the wells is much deeper than the other. As a result, one of the states of the dc SQUID switch, (L), is metastable. The potential barrier is high, however, so tunneling out of the well is suppressed over the time scale of the experiment. In addition, the inductive coupling between the rf and dc SQUID is chosen such that the effect of any flux coupled into the dc SQUID by the circulating current of the rf SQUID is negligible. As a result, not only is the coherent oscillation frequency of the rf SQUID relatively unaltered, but the decay of the oscillations is not significantly increased by the presence of the dc SQUID switch.

The dc SQUID switch can be made more sensitive to the flux state of the rf SQUID at some time t_i by the application of a current bias pulse which suddenly depresses the potential barrier of the dc SQUID switch. As a result, if the rf SQUID is in the (+) state, the dc SQUID can be made to "switch" from the original state (L) into the state (R). However, if the rf SQUID is in the (-) state, the additional change in the bias flux of the dc SQUID only serves to increase the stability of the metastable state (L). Hence, no appreciable change in either the dc SQUID switch or its effect on the rf SQUID occurs.

We now have a good candidate for an rf SQUID "polarizer". First, most of the dissipation has been removed from the dc SQUID switch. Since there is no substantial coupling to dissipation in the combined system, the coherent behavior of the combined system (rf and dc SQUID) is preserved. Second, we can turn the "polarizer" on and off. This is roughly analogous to locating a conventional polarizer in some specific region of space. Third, we have a system which produces no significant action on the rf SQUID provided that it is in the (-) state. This is analogous to the action of a polarizer on a state which is transmitted unchanged. Forth, a dramatic back-action does occur if the rf SQUID is in the (+) state. In particular, a large transient in the dc SQUID screening current is generated which disrupts the flux bias point of the rf SQUID. It is interesting that this large transient illustrates that the rf SQUID has not been "prepared" into either the(+) or (-) state. Furthermore, the switching of the dc SQUID makes identification of this rf SQUID state unambiguous. Note that if the subsequent behavior of the rf SQUID in the (+) state is of interest, then the quantum system must be enlarged to include the dc SQUID switch. This is possible without too much complication in principle, since the combined system consists of a device with five Josephson junctions biased below threshold with very little intrinsic dissipation. However, as we have seen, understanding correlations in this data set is not needed in the final analysis.

8. The Complete Measurement Circuit

We now have both an rf SQUID "polarizer" (the dc SQUID switch activated at time t_i) and an rf SQUID "particle detector" (a dc SQUID magnetometer which is weakly coupled to the dc SQUID switch and extremely weakly coupled to the rf SQUID). The complete circuit consisting of a pair of "polarizers" and "particle detectors" (Figure 2). The settings of the "polarizers" are the times t_1 and t_2 at which the two switches are briefly activated. The dc SQUID magnetometers detects the states of the dc SQUID switches only after an integration time $T_a >> t_i$. The back-action of the dc SQUID magnetometers over this long interval is sufficient to effectively destroy the coherent oscillations in the rf SQUID, even though the dc SQUID magnetometer and rf SQUID are extremely weakly coupled. However, the back-action of the dc SQUID magnetometers over the much shorter time t_i has a negligible effect on the coherent oscillations of the rf SQUID. Thus, the "measurement" of the flux state of the dc SQUID is not completed first at time t_1, and then again at time t_2, but rather the flux state at both t_1 and t_2 are "measured" after time T_a. A qualitative description of the action of the dc SQUID magnetometers can be found in ref. 11.

9. Conclusions

A measurement circuit for the flux state of an rf SQUID has been described which has elements analogous to those used in a conventional EPR experiment. These elements include dc SQUID switches which act like polarizers, and dissipative dc SQUID magnetometers which act like particle detectors. In addition, experimentally feasible values for the parameters of an rf SQUID which is predicted to undergo coherent oscillations have been determined. These elements are essential for the development of an experimental search for violations of analogs of Bell's inequalities for a macroscopic system.

References

1. A. Einstein, B. Podolsky and N. Rosen, Phys. Rev. 47: 777 (1935).
2. J. S. Bell, Physics 1: 195 (1964).
3. J. F. Clauser, M. A. Horne, A. Shimony and R. A. Holt, Phys. Rev. Lett. 23: 880 (1969).
4. A. J. Leggett and Anupam Garg, Phys. Rev. Lett. 54: 857 (1985).
5. J. Kurkijarvi, Phys. Rev. B6: 832 (1972).
6. L.D. Jackel, W.W. Webb, J.E. Lukens, and S.S. Pei, Phys. Rev. B 9 115 (1972).
7. D.B. Schwartz, B. Sen, C.N. Archie and J.E. Lukens, Phys. Rev. Lett. 55 1547 (1985).
8. S. Chakravarty and A.J. Leggett, Phys. Rev. Lett. 52: 5 (1984).

9. S. Chakravarty and S. Kivelson, Phys. Rev. Lett. 50 1811 (1983).

10. H. Grabert and U. Weiss, Phys. Rev. Lett. 53: 1787 (1984).

11. C.D. Tesche, Phys. Rev. Lett. 64: 2358 (1990).

12. J.R. Kirtley, C.D. Tesche, W.J. Gallagher, A.W. Kleinsasser, R.W. Sandstrom, S.I. Raider, and M.P.A. Fisher, Phys. Rev. Lett. 61 2372 (1988).

13. Asher Peres, Phys. Rev. Lett. 61: 2019 (1988).

14. J.C.Jaklevic, J. Lamb, A.H.Silver, J.E. Mercereau, Phys. Rev. Lett. 12:159 (1964).

15. C. D. Tesche, in New Techniques and Ideas in Quantum Measurement Theory, ed. Daniel M. Greenberger (Annals of the New York Academy of Sciences, New York, 1986) Vol.480, p.36.

PRESENTATIONS AT THE POSTER SESSION

R.R. ALDRINGER, Gettysburg College, *Geometrical SO(4,1) Gauge Formulation of Relativistically Extended Objects and the Mass Spectrum*

T.B. ANDREWS, Brooklyn NY, *A Wave System Theory of Quantum Mechanics.*

E. ARTHURS and M.S. GOODMAN, Bell Communications Research, Morristown, NJ, *A Quantum Experiment Uncertainty Principle.*

Kayoko AWAYA, Akasuki Gakuen Junior College, *An Analysis of Theories of Measurement.*

Simon BERKOVICH and S. LEE, George Washington University, *Investigation of quantum mechanics features in a classical cellular automaton model.*

F. A. BLOOD, Jr., Rutgers University, Camden, NJ, *An underlying equation for quantum mechanics?*

George E. CARVER, Capital University, *Accelerated Vacuum State and Quantum Operators of a Charged Klein-Gordon Field.*

A. COMTET, Division de Physique Théorique, IPN, Orsay, and S. OUVRY, CERN, *The Virial Expansion and the 2-Dimensional Axial Anomaly.*

A. COMTET, Y. GEORGELIN, and S. OUVRY, Université Paris, *Statistical aspects of the anyon model.*

A. CORDERO, Queens College, City University of New York, *Properties After Kochen/Specker and Bell.*

K. EDER, R. GÄHLER, M. GRUBER, W. MAMPE, and A. ZEILINGER, Technical University of Vienna, Technical University of Munich, and ILL, *Interferometry With Very Slow Neutrons.*

Michael R. GALLIS and Gordon N. FLEMING, Pennsylvania State University, *Environment and spontaneous localization*

L.C. GARCIA de ANDRADE, University of Alabama in Huntsville, *Asymptotic Bevavior of Gravitational-Torsion Waves*

J.D. GILLASPY, K. DEILAMIAN, and D.E. KELLEHER, National Institute of Standards and Technology, *Precision Test of the Pauli Exclusion Principle and the Symmetrization Postulate of Quantum Mechanics*

Franz HASSELBACH and Mark NICKLAUS, Institut für Angewandte Physik, Tübingen, *Observation of the rotational phase shift of electron waves (Sagnac Effect)*

J.B. KENNEDY, Stanford University, *Controllable Quantum Signalling: Unification and Criticism of No-Signalling Proofs.*

J.B. KENNEDY, Stanford University, *Spacetime, EPR, and the Geometry of Gauge Theory.*

Donald H. KOBE, University of North Texas, *Invariance of the Berry Phase Under Unitary Transformations: Application of the Time-Dependent Generalized Harmonic Oscillator.*

Donald H. KOBE, University of North Texas, and J.Q. LIANG, Shanxi University, *Scattering from a Magnetic Flux Line Due to the Lorentz Force of the Return Flux.*

S. KUDAKA, S. MATSUMOTO, K. KAKAZU, University of the Ryukyus, *Generalized Coherent State Approach to Quantum Measurement.*

Egon MARX, National institute of Standards and Technology, *Electromagnetic Potentials in Relativistic Quantum Mechanics.*

Conrad MIZIUMSKI, Millersville University, *Possible Three-Measurement Spin Correlations Having Testable Predictive Significance.*

Qian NIU and Franco NORI, University of California, Santa Barbara, *Theory of superconducting wire networks and Josephson-junction arrays in magnetic fields.*

H. PENG, University of Alabama in Huntsville, *A Gravitational Analog of the Electric Aharonov-Bohm Effect.*

H. PENG, University of Alabama in Huntsville, *Proposal Experiment on Searching for Gravitomagnetic Vector Field.*

Simon W. SAUNDERS, Oxford University, *Locality in Gallean and Lorentz (Quantum) Theory.*

J. SCHMEIDMAYER and J. SUMMHAMMER, Atominstitut Österreichischer Universitäten, *Electron and Ion Point Sources for Quantum Experiments.*

D.M. SNYDER, Berkeley, California, *Is Quantum Mechanical Measurement Purely Physical?*

E.J. SQUIRES, University of Durham, *Wavefunction Localization and Lorentz Invariance.*

C.D. TESCHE, IBM, Yorktown Heights NY, *Invasive versus Non-Invasive Measurement of Magnetic Flux.*

A.B. VAN OOSTEN, University of Groningen, *Lorentz-Force Versus Aharonov-Bohm Effect.*

H. WEINFURTER and G. BADUREK, Hahn-Meitner Institut and Technical University Vienna, *Geometrical Phase Shifts for Non-Cyclic Evolutions.*

Bernard ZYGELMAN, Center for Astrophysics, Harvard University, *Non-Abelian Geometric Phase in Atoms and Molecules*

LIST OF PARTICIPANTS

David Z. Albert
245 East 19th Street
Apt 67
New York NY 10003

JSahin Aktas
Dept. of Physics and Astronomy
The University of South Carolina
Columbia SC 29208

Yakir Aharonov
Dept. of Physics and Astronomy
The University of South Carolina
Columbia, SC 29208

Michael S. Alexander
Dept of Chemistry
The University of South Carolina
Columbia SC 29208

Randolph R. Aldinger
Dept. of Physics
Gettysburg College
Gettysburg, PA 17325

C.O. Alley
Dept. of Physics and Astronomy
University of Maryland
College Park MD

M. Keramat Ali
Dept. of Physics
University of Lethbridge
Lethbridge, Alberta, CANADA
T1K3M4

Jeeva Anandan
Dept. of Physics
The University of South Carolina
Columbia, SC 29208

Thomas B. Andrews
3828 Atlantic Ave.
Brooklyn, NY 11224

Ping Ao
Dept. of Physics
University of Ill at U-C
1110 W. Green St.
Urbana, IL 1161801

Edward Arthurs
Belcore
445 South Street
Morristown, NJ 07960

C.K. Au
Dept. of Physics and Astronomy
The University of South Carolina
Columbia, SC 29208

Frank Avignone
Dept. of Physics and Astronomy
The University of South Carolina
Columbia, SC 29208

Kayoko Awaya
Akatsuki Gakuen Junior College
238 Shiroyama, Kayou-cho,
Yokkaichi-City
MIE, 512, JAPAN

Jose L. Balduz, Jr.
5835 Darlington Rd., #4
Pittsburgh, PA 15217

Leslie Ballentine
Physics Dept.
Simon Fraser University
Burraby, BC Canada

Lawrence Bennett
600 Clemson Rd.
Columbia, SC 29223

Stephan Berko
Brandeis University
Waltham, MA 02254

Simon Berkovich
Dept. of EE&CS
The George Washington University
Washington, DC 20052 USA

Michael Berry
H.H. Wills Physics Laboratory
Tyndall Ave.
Bristol B581TL
U.K.

L.C. Biedenharn
Department of Physics
Duke University
Durham, NC 27706

David Bohm
Physics Dept.
Birbeck College
Malet St.
London, England
WCE 7HX

Frank A. Blood
Dept. of Physics
Camden College of Arts and Sciences
Camden, NJ 08102

Ronald E. Byers
R.R. 3 Box 441
Pella, Ia. 50219

David Boulware
University of Washington
Seattle, WA 98195

Felix A. Buot
Naval Research Laboratory
Washington, DC 20375

Paul Camp
Physics
University of South Carolina
Columbia SC 29208

Russell C. Casella
Theory, Rm A106, Bldg. 235
National Inst. of Standards & Tech.
Gaithersburg, MD 20899

Lay Nam Chang
Physics Dept. Virginia Tech.
Blacksburg, VA 24061

Ngee-Pong Chang
City College
Convent Avenue at 138th St.
New York NY 10031

Vladmir Chaloupka
Physics Dept.
Univ. of Osslington
Seattle, WA

Joy Christian
Boston University
Boston, MA

Raymond Chiao
Univ. of California, Berkeley
Berkeley, CA 94720

Eric Chisolm
59 Fern Valley Rd.
Brandon, MS 39042

David E. Chyba
Dept. of Physics
Bryn Mawn College
Bryn Mawn, PA 19010

Theodore Claiborne
Doty Scientific
600 Clemson Rd.
Columbia, SC 29066

David E. Chyba
Dept. of Physics
Bryn Mawr College
Mryn Mawr, PA 19010

Michael Clingan
Doty Scientific
600 Clemson Rd.
Columbia, SC 29223

Juan I. Collar
Dept. of Physics
Univ. of South Carolina
Columbia, SC 29208

Alberto Cordero
Dept. of Philosophy
Queens College, CUNY
Flushing, NY 11367

Richard Creswick
Dept. of Physics and Astronomy
The University of South Carolina
Columbia SC 29208

Susan S. D'Amato
Physics Department
Furman University
Greenville, SC 29163

Timir Datta
Dept. of Physics and Astronomy
The University of South Carolina
Columbia SC 29208

Kavoos Deilamian
Physics /A167
N.I.S.T.
Gaithersburg, MD 20879

Luiz Carlos Garcia DeAndrade
Dept. of Physics
University of Alabama in Huntsville
Huntsville, AL 35899

Christopher Dewdney
Dept. of Applied Phys.
Portsmouth Polytechnic
Portsmouth, P0`1 207 ENGLAND

Stanley Deser
Brandeis University
Waltham, MA 02254

Allen Clark Dotson
St. Laurence Presbyterian College
Lauenburg, NC 28352

F. David Doty
600 Clemson Rd.
Columbia, SC 29223

Ronald D. Edge
Dept. of Physics and Astronomy
The University of South Carolina
Columbia SC 29208

Catherine Epp
Reed College Box #299
3203 SE Woodstock, Portland, OR
97202

Lawrence W. Fagg
Physics Dept.
Catholic University of America
Washington, DC 20064

Horacio A. Farach
Dept. of Physics
Univ. of South Carolina
Columbia, SC 29208

Larry Ford
Physics Dept.
Tufts Univ.
Medford, MA 02155

Paul Frampton
University of North Carolina
Chapel Hill, NC 27514

Christopher Fuchs
207 O Branson St.
Chapel Hill, NC 27514

Anupim Garg
Northwestern University
Evanston, IL 60208

Michael R. Gallis
104 Davey Lab, Box 177
University Park, PA 16302

Steve Giddings
Dept. Physics
Harvard University
Cambridge, MA 02138

John D. Gillaspy
NIST - Physics Bldg. (221), RM.
A167
Gaithersburg, MD 20899

Benjamin M. Gimarc
Department of Chemistry
The University of South Carolina
Columbia SC 29208

Matthew Goodman
Bellcore
445 South Street
Morristown, NJ 07960

Larry Gould
Physics Dept.
University of Hartford
West Hartford, CT 06117

Marek Grabowski
Dept. of Physics
Virginia Tech
Blacksburg, VA 24061

Daniel Greenberger
Physics
City College of the City
University of New York
New York, NY 10031

Krish Gounder
Dept. of Physics
Univ. of South Carolina
Columbia, SC 29202

James W. Green, III
Route 5, Box 720
Winnsboro, SC 29180-9516

Martin P. Greiter
Institute for Advanced Study
Princeton, NJ 08540

David Griffiths
Physics Dept.
Reed College
Portland, Oregon 97202

C. Richard Hagen
Univeristy of Rochester
Rochester, NY

Franz Hasselbach
Institute f. Argewardle Physik
Aüf der Morgersweilis
074 Tübingen
Federal Republic of Germany

Tim Havens
1300 W Palmetto Street
Florence SC 29501

ames Hartle
hysics Dept.
Jniv. of California
Santa Barbara, CA 93106

Vasslios Hassapis
Physics Dept.
Georgia Southern College
Statesboro, GA 30460

Thomas M. Helliwell
Physics Dept. Harvey Mudd College
Claremont, CA 91711

Vicki Homer
Dept. of Physics and Astronomy
The University of South Carolina
Columbia SC 29208

G.T. Hooft
nstitute for Theoretical Physics
P.O. Box 80-006
NL 3508 TA Ulrecht, The Netherlands

effrey Lynn Hopkins
610 Timberlane Dr. F-1
Columbia, SC 29205

Ralph Howard
Dept of Mathematics
The University of South Carolina
Columbia SC 29208

Neal Jahren
Dept. of Physics, 371 MWAH
University of Minnesota at Duluth
Duluth, MN 55817-2496

evin L. Jensen
Code 6877, STD
Naval Research Lab
Washington, DC 20375-5000

oseph E. Johnson
Dept. of Physics and Astronomy
The University of South Carolina
Columbia SC 29208

R. Michael Jones
Wave Propagation Lab
R/E/WP1
25 Broadway
Boulder, Colorado 80303

Dr. P.N. Kaloyerou
Sub-Faculty of Philosophy
0 Merton St.
Oxford University
Oxford, ENGLAND

J.B. Kennedy
Department of Philosophy
Stanford University
Standford CA 94305

L.A. Khalfin
Academy of Sciences
Leningrad, USSR

Y.S. Kim
Dept. of Physics
University of Maryland
College Park, MD 20742

Asif Khandker
Dept. of Physics
Millsaks College
Jackson, Mississippi 39210

Igor Klebanov
Physics Dept.
Jadwin Hall
Princeton University
Princeton, NJ 08544

James Knight
Dept. of Physics
The University of South Carolina
Columbia, SC 29208

Donald H. Kobe
Department of Physics
University of North Texas
Denton, Texas 76203 USA

Mohammed Kolahchi
Dept of Physics
University of Kentucky
Lexington KY 40506

David Kraines
Dept. of Math
Duke University
Durham University, NC 27706

Kuniharu Kubodera
Dept. of Physics and Astronomy
The University of South Carolina

Dr. Prasun K. Kundu
Department of Physics and Astronomy
Ohio University
Athens, OH 45701

Michael Lahana
7039 W. Fairview Drive
Littleton, Colorado 80123

Jean LeTourneux
Laboratoire de Physique Nucleaire
Universite' de Montre'al
Monte'al, Canada H3C3J7

Barbara Levi
Physics Today
335East 45 Street
New York City NY

John Linder
Physics Department
The College of Wooster
Wooster, OH 44691

Seth Lloyd
452-48 Caltech
Pasadena, CA 91125

Oscar A. Lopez
Dept. of Physics and Astronomy
The University of South Carolina
Columbia SC 29208

Richard B. MacKenzie
Laboratoire de Physique Nucleaire
Universite' de Montre'al
Montre'al, Cananda H3C3J7

Louis Marchildon
Physics Dept. Universite Du Quebec
Trois-Rivieres Que CANADA
G9A5H7

Egon Marx
Met. A117
NIST
Gaithersburg, MD 20899
Egon Marx

Alden Mead
Dept of Chemistry
University of Minn
Minneopolis MN 55455

Manuel Mesa
1415 Blossom St, UT-166
Columbis SC 29201

Marc C. Nicklaus
National Cancer Institute, NIH
Building 37, Room 5C28
Bethesda MD 20892

Conrad Miziumski
Millersville University
Millersville, PA 17551

Fred Myhrer
Dept. of Physics
The University of South Carolina
Columbia, SC 29208

Yuval Ne'eman
Univ. of Tel Aviv
Tel Aviv Israel

Dr. Marc C. Nichelaus
Institut fur Angewardle, Physik
Universitat Tubingen
Tubingen
Federal Republic of Germany

Borge Nodland
90 Suncoast Blvd.
Wesley Chapel, FL 33544

Franco Nori
Institute for Theoretical Physics
University of California
Santa Barbara, CA 93106

Y. Jack Ng
Dept. of Physics and Astronomy
University of North Carolina,
Chapel Hill, NC 27599

Waichi Ogura
Physics Dept.
Virginia Polytechnic Tnst.
Blacksburg, VA 24061

Professor Geoffrey I. Opat
School of Physics
The University of Melbourne
Parkville, Victoria 3052
Australia

Stephane Ouvry
IPN Division de Physique
Thierique Fr-giu06 Orsay
France

A. Jay Palmer
7041 Fairways Dr.
Longmount, CO 80501

Manu B. Paranjape
Laboratoire de Physique, Nucleaire
Universite' de Monte'al
Montre'al, Canada H3C3J7

Dr. Huei Peng
The University of Alabama in
Huntsville
Research Institute C-1B
Huntsville, AL 35899

Philip C. Peters
Dept. of Physics FM-15
Univ. of Washington
Seattle, WA 98195

H. Pilloff
Code 1112LO
ONR
800 N. Quincy Street
Arlington, VA 22217

Alex Pines
Dept. of Chemistry
University of California,
Berkeley, CA 94720

Charles P. Poole, Jr.
Dept. of Physics and Astronomy
The University of South Carolina
Columbia SC 29208
Sandu Popescu
Tel Aviv University
Tel Aviv, Israel

Glulio Pozzi
Dept. of Materials Science
University of Lecce
Italy

Lon N. Pringle
1078 Lanier Blvd.
Atlanta, GA 30306

Paavo Pylkkänen
Dept. of Philosophy
Unionink.40B
University of Helsinki
00170 Helsinki, Finland

Dr. John R. Ray
303 Kinard Lab
Physics and Astronomy Dept.
Clemson University
Clemson, SC 29634-1911

Amita Raval
328 Barron Street
West Columbia, SC 29169

Tricia M. Reeves
School or Natural Applied Science
USC Coastal Carolina College
P.O. Box 1954
Conway, SC 29526

B. Resnick
Tel Aviv University
Tel Aviv
Israel

Alberto Rimini
Dip. Di Fisica Nucleare E Teorica
Via Bassi 6
27100 Pavia
ITALY

Carl Rosenfeld
Dept. of Physics
The University of South Carolina
Columbia, SC 29208

John L. Safko
Dept. of Physics
The University of South Carolina
Columbia, SC 29208

Brian K. Saltzer
329 Bayshore Blvd. Apt. 319
Tampa, FL 33606

Tiina Seppala
Koskelante 52 F34
00610 Helsinki Finland

Jörg Schmiedmayer
Atominst. ÖST. Universitaten
Schöttelstrasse 115 A-1020 Wien
Austria

Dr. S.W. Saunders
Wolfson College
Oxford University
Oxford OX26UD, UK

Dr. O.F. Schuette
Physics Dept.
USC
Columbia, SC 29208

Abner Shimony
Boston University
Boston, MA

Yanhua Shim
Dept. of Physics
University of Maryland
Baltimore, MD 21228

Chopin Soo
Physics Dept.
Virginia Tech.
Blacksgurg, VA 24061

Nick C. Steph
Dept. of Phyiscs-Franklin College
Franklin, IN 46131

Ady Stern
Physics Department
Tel Aviv University
Tel Aviv, Israel 69978

Allen Stern
Dept. of Physics and Astronomy
Univ. of Alabama,
Tuscaloosa, AL 35487

E.J. Squires
Dept. of Mathematical Sciences
University, Durham City, DH1 1TX
ENGLAND

Duvry Stephane
IPN Division de Physique
Theorique Orsay FR-91606
FRANCE

Leo Stodolsky
Max-Planck Institute for Physics
Munich
Federal Republic of Germany

George Sudarshan
University of Texas
Austin, TX 78712

Daniel E. Sudarsky
3118 Quarry Dr.
Lafayette, IN 47905

Lenny Susskind
Dept of Physics
Standford University
Standford CA 94305-4060

Sidney Teitler
Naval Research Lab
Washington, DC 20375-5000

C.D. Tesche
IBM Research Cntr.
P.O. Box 218
Yorktown Heights, NY 10598

Akira Tonomura
Advanced Research Lab,
Hitachi Ltd.,
Kokubunji, Tokyo 185, Japan

Roger D. Truesdale
2500 Burney Drive
Columbia, SC 29205

Zoltan Trocsanyi
Physics Dept. Virginia Tech.
Blacksburg, VA 24061

Chia H. Tze
Physics Dept.
Virginia Tech.
Blacksburg VA 24001

Y. Umino
Dept. of Physics and Astronomy
The University of South Carolina

R.A. Uritam
Physics Dept.
Boston College
Chestnut Hill, MA 02167

Tim Usher
Physics Department
The University of South Carolina
Columbia, SC 29208

Lev Vaidman
Dept. of Physics and Astronomy
The University of South Carolina
Columbia, SC 29208

S.R. Valluri
c/o Dept. of Philosophy
Univ. of Western Ontario
London, ON N6A 3K7, CANADA

A.B. van Oosten
16 Nyenborgh
9747 AG Groningen
The Netherlands

Erik Verlinde
100 Einstein Drive
Princeton, NJ 08540

Dr. Udaykumar D. Vyas
Dept of Math/Computer Sciences
Winston-Salem State University
Wiston-Salem, NC 27110

Angel Wang
Dept of Physics
The University of South Carolina
Columbia SC 29208

Jin-shan Wang
Doty Scientific Inc.
600 Clemson Rd., Columbia, SC
29223

Richard A. Webb
IBM
T.J. Watson Research Center
P.O. Box 218 Yorktown Heights,
N.Y. 10598

Frederick G. Werner
Physics Dept.
Xavier University
Box 124, Winter Road
Silver Lake, New Hampshire 03875

Mark Whitlow
600 Clemson Rd.
Columbia, SC 29223

Frank Wilczek
Institute for Adv. Study
Princeton University
Princeton, NJ 08540

Tony Zee
Institute for Theortical Physics
Univ. of California at Santa Barbara
Santa Barbara CA 91306

Dr. Bernard Zygelman
Harvard College Observatory
60 Garden St. Cambridge, MA 02138

www.ingramcontent.com/pod-product-compliance
Lightning Source LLC
Chambersburg PA
CBHW041304210926
51851CB00002B/6
9789810202569